SUSTAINING GLOBAL SURVEILLANCE AND RESPONSE TO EMERGING ZOONOTIC DISEASES

Gerald T. Keusch, Marguerite Pappaioanou, Mila C. González,
Kimberly A. Scott, and Peggy Tsai, *Editors*

Committee on Achieving Sustainable Global Capacity for Surveillance
and Response to Emerging Diseases of Zoonotic Origin

Board on Global Health
Institute of Medicine

Board on Agriculture and Natural Resources
Division on Earth and Life Studies

INSTITUTE OF MEDICINE *AND*
NATIONAL RESEARCH COUNCIL
OF THE NATIONAL ACADEMIES

THE NATIONAL ACADEMIES PRESS
Washington, D.C.
www.nap.edu

THE NATIONAL ACADEMIES PRESS 500 Fifth Street, NW Washington, DC 20001

NOTICE: The project that is the subject of this report was approved by the Governing Board of the National Research Council, whose members are drawn from the councils of the National Academy of Sciences, the National Academy of Engineering, and the Institute of Medicine. The members of the committee responsible for the report were chosen for their special competences and with regard for appropriate balance.

This study was supported by the U.S. Agency for International Development Award No. GHN-G-00-07-00001-00. Any opinions, findings, conclusions, or recommendations in this document are those of the authors and do not necessarily reflect the views of the organizations or agencies that provided support for the project. Mention of trade names, commercial products, or organizations does not constitute their endorsement by the sponsoring agency.

Library of Congress Cataloging-in-Publication Data

Institute of Medicine and National Research Council (U.S.). Committee on Achieving Sustainable Global Capacity for Surveillance and Response to Emerging Diseases of Zoonotic Origin.

Sustaining global surveillance and response to emerging zoonotic diseases / editors, Gerald T. Keusch ... [et al.] ; Committee on Achieving Sustainable Global Capacity for Surveillance and Response to Emerging Diseases of Zoonotic Origin, Board on Global Health, Institute of Medicine, Board on Agriculture and Natural Resources, Division on Earth and Life Studies.

p. ; cm.

Includes bibliographical references.

ISBN 978-0-309-13734-8 (pbk.)

1. Zoonoses. 2. Public health surveillance. 3. Global health. I. Keusch, Gerald. II. Title.

[DNLM: 1. Communicable Diseases, Emerging—prevention & control. 2. Zoonoses—epidemiology. 3. Biosurveillance—methods. 4. Disease Outbreaks—prevention & control. WA 110 I585s 2009]

RA639.I57 2009

362.196'959--dc22

2009044034

Additional copies of this report are available from the National Academies Press, 500 Fifth Street, NW, Lockbox 285, Washington, DC 20055; (800) 624-6242 or (202) 334-3313 (in the Washington metropolitan area); Internet, http://www.nap.edu.

For more information about the Institute of Medicine, visit the IOM homepage at: **www.iom.edu**.

Printed in the United States of America

Front cover, from top: Angus cattle on pasture. Photo by Scott Bauer, courtesy of USDA. Laboratory technician with diagnostic materials at the Washington Animal Disease Diagnostic Laboratory. Photo by Charlie Powell. Designation of H1N1 isolate digitally inserted by Photoshop. Researcher administers a new medication for bird flu to a young chicken. Photo by Steve Snowden, courtesy of iStockphoto. Laboratory scientist analyzes data at the Washington Animal Disease Diagnostic Laboratory. Photo by Henry Moore.
Back cover, from top: Fruit bat surveillance. Photo courtesy of Wildlife Trust. A young male with a puppy on Independence Day in India. Photo by Jay Graham, courtesy of Photoshare. Deer runs through a suburban neighborhood. Photo by Lillis Photography, courtesy of iStockphoto. A girl carries two lambs in rural Bolivia. Photo by Enriqueta Valdez-Curiel, courtesy of Photoshare.

Suggested citation: IOM (Institute of Medicine) and NRC (National Research Council). 2009. *Sustaining global surveillance and response to emerging zoonotic diseases*. Washington, DC: The National Academies Press.

"Knowing is not enough; we must apply.
Willing is not enough; we must do."

—Goethe

INSTITUTE OF MEDICINE

OF THE NATIONAL ACADEMIES

Advising the Nation. Improving Health.

THE NATIONAL ACADEMIES

Advisers to the Nation on Science, Engineering, and Medicine

The **National Academy of Sciences** is a private, nonprofit, self-perpetuating society of distinguished scholars engaged in scientific and engineering research, dedicated to the furtherance of science and technology and to their use for the general welfare. Upon the authority of the charter granted to it by the Congress in 1863, the Academy has a mandate that requires it to advise the federal government on scientific and technical matters. Dr. Ralph J. Cicerone is president of the National Academy of Sciences.

The **National Academy of Engineering** was established in 1964, under the charter of the National Academy of Sciences, as a parallel organization of outstanding engineers. It is autonomous in its administration and in the selection of its members, sharing with the National Academy of Sciences the responsibility for advising the federal government. The National Academy of Engineering also sponsors engineering programs aimed at meeting national needs, encourages education and research, and recognizes the superior achievements of engineers. Dr. Charles M. Vest is president of the National Academy of Engineering.

The **Institute of Medicine** was established in 1970 by the National Academy of Sciences to secure the services of eminent members of appropriate professions in the examination of policy matters pertaining to the health of the public. The Institute acts under the responsibility given to the National Academy of Sciences by its congressional charter to be an adviser to the federal government and, upon its own initiative, to identify issues of medical care, research, and education. Dr. Harvey V. Fineberg is president of the Institute of Medicine.

The **National Research Council** was organized by the National Academy of Sciences in 1916 to associate the broad community of science and technology with the Academy's purposes of furthering knowledge and advising the federal government. Functioning in accordance with general policies determined by the Academy, the Council has become the principal operating agency of both the National Academy of Sciences and the National Academy of Engineering in providing services to the government, the public, and the scientific and engineering communities. The Council is administered jointly by both Academies and the Institute of Medicine. Dr. Ralph J. Cicerone and Dr. Charles M. Vest are chair and vice chair, respectively, of the National Research Council.

www.national-academies.org

COMMITTEE ON ACHIEVING SUSTAINABLE GLOBAL CAPACITY FOR SURVEILLANCE AND RESPONSE TO EMERGING DISEASES OF ZOONOTIC ORIGIN

GERALD T. KEUSCH (*Co-Chair*), Boston University, MA
MARGUERITE PAPPAIOANOU (*Co-Chair*), Association of American Veterinary Medical Colleges, Washington, DC
CORRIE BROWN, University of Georgia, Athens
JOHN S. BROWNSTEIN, Harvard Medical School, Boston, MA
PETER DASZAK, Wildlife Trust, New York
CORNELIS de HAAN, The World Bank (retired), Washington, DC
CHRISTL A. DONNELLY, Imperial College London, United Kingdom
DAVID P. FIDLER, Indiana University, Bloomington
KENNETH H. HILL, Harvard University, Cambridge, MA
ANN MARIE KIMBALL, University of Washington, Seattle
RAMANAN LAXMINARAYAN, Resources for the Future, Washington, DC
TERRY F. McELWAIN, Washington State University, Pullman
MARK NICHTER, University of Arizona, Tucson
MO SALMAN, Colorado State University, Fort Collins
OYEWALE TOMORI, Redeemer's University, Ogun State, Nigeria
KEVIN D. WALKER, Michigan State University, East Lansing
MARK WOOLHOUSE, University of Edinburgh, United Kingdom

Study Staff

KIMBERLY A. SCOTT, Study Director
PEGGY TSAI, Program Officer
MILA C. GONZÁLEZ, Research Associate
SARAH JANE BROWN, Senior Program Assistant
JULIE WILTSHIRE, Financial Officer
PATRICK W. KELLEY, Director, Board on Global Health
ROBIN A. SCHOEN, Director, Board on Agriculture and Natural Resources

BOARD ON GLOBAL HEALTH[1]

[1]IOM boards do not review or approve individual reports and are not asked to endorse conclusions and recommendations. The responsibility for the content of the report rests with the authoring committee and the institution.

BOARD ON AGRICULTURE AND NATURAL RESOURCES

Acknowledgments

This report has been reviewed in draft form by persons chosen for their diverse perspectives and technical expertise in accordance with procedures approved by the National Research Council's Report Review Committee. The purpose of the independent review is to provide candid and critical comments that will assist the institution in making its published report as sound as possible and to ensure that the report meets institutional standards of objectivity, evidence, and responsiveness to the study charge. The review comments and draft manuscript remain confidential to protect the integrity of the deliberative process. We wish to thank the following for their review of this report:

Sir George Alleyne, Pan American Health Organization
Scott Barrett, Paul H. Nitze School of Advanced International Studies, Johns Hopkins University
Ron Brookmeyer, Johns Hopkins Bloomberg School of Public Health
Donald S. Burke, Graduate School of Public Health, University of Pittsburgh
Seth Foldy, Division of Public Health, State of Wisconsin
Lawrence O. Gostin, Georgetown University
David Harlan, Global Animal Health and Food Safety, Cargill, Inc.
James M. Hughes, School of Medicine and Rollins School of Public Health, Emory University
Anni McLeod, Food and Agriculture Organization of the United Nations
Melinda Moore, RAND Corporation

Mark E. White, Division of Global Preparedness and Program Coordination, Centers for Disease Control and Prevention
Tilahun Yilma, International Laboratory of Molecular Biology for Tropical Disease Agents, University of California, Davis

Although the reviewers listed above have provided constructive comments and suggestions, they were not asked to endorse the conclusions or recommendations, nor did they see the final draft of the report before its release. The review of this report was overseen by **David Challoner**, Vice President for Health Affairs, Emeritus, University of Florida and **James Fox**, Massachusetts Institute of Technology. Appointed by the National Research Council and Institute of Medicine, they were responsible for making certain that an independent examination of this report was carried out in accordance with institutional procedures and that all review comments were carefully considered. Responsibility for the final content of this report rests entirely with the author committee and the institutions.

Preface

In April 2009, as the committee was preparing to respond to reviewer input and finalize this report, a multi-country outbreak of a new influenza A(H1N1) virus was being reported. First detected as a cluster of cases of severe respiratory illness with multiple deaths in Mexico, a unique influenza A virus was isolated that was originally reported as having genes of swine, avian, and human origin and therefore it was immediately referred to as "swine flu." Influenza A(H1N1) virus has since spread to 74 countries and, as of June 11, 2009, the World Health Organization declared it the first pandemic in more than 40 years. Although the virus is now circulating in humans, the presumed link with swine led to public confusion on how the virus was being spread, consequently leading to pork industry losses of approximately $28 million dollars per week and the banned importation of pigs and pork products by at least 15 countries. The specifics of when and how this virus emerged, in what populations, how long its circulation has gone undetected, and the identity of the source of exposure remain the focus of ongoing investigations. While it is not possible to fully analyze the progression and impact of events with the benefit of time and hindsight before completing the work on this report, this outbreak serves to illustrate many of the issues discussed in this report.

The committee's consensus report traces the need and existing capacity for global, sustained, integrated zoonotic disease surveillance and response capacity; discusses the current gaps, challenges, and inadequacies with existing systems; and suggests new approaches to more effectively achieve the requirements of an "ideal" system. Looking forward with the benefit of past experience, including what we know about the current influenza A(H1N1)

2009 outbreak, we see a future of continued zoonotic disease agent emergences, perhaps at an even more rapid rate given the sheer increases in human and animal populations, their encroachment on each other's habitat, continuing changes in climate, the intensification and consolidation of agriculture, and the rapid movement of increasingly more people and goods around the world. With the prominence of these drivers of emergence, when a new zoonotic pathogen that is also readily transmitted from person to person is detected first in humans, it will be extremely difficult to achieve containment, even when everything that can be done is done efficiently and effectively. Thus, looking for ways to prevent emergence and to detect these pathogens at the first point possible in animal populations deserves serious consideration.

The questions that ultimately must be asked in dissecting the influenza A(H1N1) 2009 outbreak are: what surveillance systems could have identified the problem more quickly, whether those systems could have triggered a global response to limit its spread and/or impact in a more timely way, and what lessons can be drawn from the experience and extrapolated to other potential emergent disease agents—some of which are unknown at the present time. Although the time from the detection of a cluster of severe pneumonia cases in Mexico to the identification of the cause as influenza A(H1N1) 2009 and global awareness and a patchwork global response was shorter than that experienced in previous outbreaks, we believe the urgency will only grow to create an even more effective system for sustained, integrated, early human and animal disease detection that is immediately followed by and intimately linked to a timely and appropriately targeted response. Achieving such a system is not easy: If it were, it would have been accomplished decades ago. But given the inevitability of disease emergence occurring again and again, the solution requires strong leadership and commitment to ensure that multiple disciplines from different sectors will work closely together to address the myriad complex and sophisticated challenges they will pose.

For this reason, the committee believes it is high time for national and international public health leadership, as recommended in this report, to address how global and effectively integrated zoonotic disease surveillance can be achieved. The recently announced USAID Predict and Respond initiatives are a good start, but more will be required from actors of all levels to address a global concern. Little comfort can be taken in the fact that SARS turned out to be readily controlled by simple barrier and sanitary measures, that highly pathogenic avian influenza (HPAI) H5N1 influenza virus has yet to acquire the necessary attributes for efficient human-to-human transmission, and that influenza A(H1N1) 2009 does not, at this time, seem to be both readily transmitted and highly virulent in humans. Each of these agents may still evolve to become the highly pathogenic pandemic strains of the future, or others may arise that are far more challenging to address.

We thank the committee for serving as individuals and away from their institutional affiliations and obligations, and for giving so willingly and collegially of their time and effort. On behalf of the committee, we would like to thank the Institute of Medicine and National Research Council staff who worked tirelessly throughout the development of this report, providing support and assistance in organization, planning, execution, writing, and more. This report would not have been possible without their participation.

Gerald T. Keusch, *Co-Chair*
Marguerite Pappaioanou, *Co-Chair*
Committee on Achieving Sustainable Global Capacity for Surveillance and Response to Emerging Diseases of Zoonotic Origin

Contents

List of Tables, Figures, and Boxes

TABLES

FIGURES

BOXES

Acronyms and Abbreviations

AHI Facility	Avian and Human Influenza Facility
APHIS	Animal and Plant Health Inspection Service of the U.S. Department of Agriculture
BSE	bovine spongiform encephalopathy
BSL	biosafety level
CAFO	concentrated animal feeding operation
CDC	U.S. Centers for Disease Control and Prevention
CITES	Convention on International Trade in Endangered Species of Wild Fauna and Flora
Codex	WHO/FAO Codex Alimentarius Commission
DHS	U.S. Department of Homeland Security
DoD	U.S. Department of Defense
DoD-GEIS	U.S. Department of Defense-Global Emerging Infections Surveillance and Response System
DoI	U.S. Department of the Interior
DoS	U.S. Department of State
ENSO	El Niño-Southern Oscillation
EU	European Union
FAO	Food and Agriculture Organization of the United Nations

FDA	U.S. Food and Drug Administration
FELTP	Field Epidemiology and Laboratory Training Program
FETP	Field Epidemiology Training Program
FMD	foot-and-mouth disease
GAINS	Global Avian Influenza Network for Surveillance
GATT	General Agreement on Tariffs and Trade of the WTO
GDP	gross domestic product
GLEWS	Global Early Warning System
GOARN	Global Outbreak Alert and Response Network
GPAI	Global Program for Avian Influenza
GPHIN	Global Public Health Intelligence Network
HHS	U.S. Department of Health and Human Services
HLSC	House of Lords Select Committee
HPAI H5N1	highly pathogenic avian influenza H5N1
IDSR	Integrated Disease Surveillance and Response
IFI	International Finance Institution
IGO	intergovernmental organization
IHR 1969	International Health Regulations 1969
IHR 2005	International Health Regulations 2005
INCLEN	International Clinical Epidemiology Network
IOM	Institute of Medicine
IPCC	Intergovernmental Panel on Climate Change
IT	information technology
KEMRI	Kenya Medical Research Institute
NAHLN	U.S. National Animal Health Laboratory Network
NGO	nongovernmental organization
NRC	National Research Council
OIE	Office International des Epizooties, also World Organization for Animal Health
PCR	polymerase chain reaction
PEPFAR	President's Emergency Plan for AIDS Relief
PHS	Public Health Service Act
ProMED	Program for Monitoring Emerging Diseases
PVS	Performance of Veterinary Services tool
RVF	Rift Valley fever

SARS	severe acute respiratory syndrome
SPS Agreement	Agreement on the Application of Sanitary and Phytosanitary Measures
SSAFE	Safe Supply of Affordable Food Everywhere
TBT Agreement	Agreement on Technical Barriers to Trade
UN	United Nations
UNSIC	United Nations System Influenza Coordinator
USAID	U.S. Agency for International Development
USDA	U.S. Department of Agriculture
USFWS	U.S. Fish and Wildlife Service
USGS	U.S. Geological Survey
vCJD	variant Creutzfeldt-Jakob disease
WAHIS	World Animal Health Information System
WHO	World Health Organization
WHO-AFRO	World Health Organization Regional Office for Africa
WNV	West Nile virus
WTO	World Trade Organization

Summary

Infectious disease surveillance systems play an important role in safeguarding human and animal health. By systematically collecting data on the occurrence of infectious diseases in humans and animals, investigators can identify new outbreaks, track the spread of disease, and provide an early warning to human and animal health officials nationally and internationally for follow-up and response. Unfortunately, for several reasons, the disease surveillance systems operating around the world are not very effective or timely in alerting officials to newly emerging zoonotic diseases—diseases transmitted between humans and animals.

Emerging zoonoses are a growing concern given multiple factors. First, zoonoses are often novel diseases that society is medically unprepared to treat, as was the case with HIV/AIDS and variant Creutzfeldt Jakob disease (vCJD), better known as mad cow disease. Second, zoonoses are unpredictable and have variable impacts on human and animal health. For example, different strains of influenza A virus—such as highly pathogenic avian influenza (HPAI) H5N1 and pandemic H1N1 2009—have different host ranges and cause illnesses of different degrees of severity. Third, zoonotic diseases outbreaks are increasing in number: At least 65 percent of recent major disease outbreaks have zoonotic origins. Fourth, because of increasing international trade, travel, and movement of animals, zoonotic diseases can emerge anywhere and spread rapidly around the globe, as demonstrated by the recent outbreak of severe acute respiratory syndrome (SARS) and the ongoing 2009 influenza pandemic. Fifth, the spread of zoonotic diseases can take a major economic toll on many disparate industries, including those in the agricultural, manufacturing, travel, and hospitality sectors, and can threaten the peace and economic stability of communities both directly and indirectly connected to disease outbreaks. The economic cost of HPAI H5N1 between 2003 and 2006 was estimated to equal nearly 2 percent of the regional gross domestic product of China, Taiwan, Hong Kong, and Singapore.

In response to concern about the global spread of zoonotic diseases, the U.S. Agency for International Development (USAID) approached the Institute of Medicine and the National Research Council for advice on how to achieve more sustainable global capacity for surveillance and response to emerging zoonotic diseases. A study committee was formed to review global responses to zoonotic diseases over the past several decades and to examine the current state of global zoonotic disease surveillance systems in light of the underlying causes of disease emergence and spread. The committee was asked to examine how an investment in global disease surveillance should be considered relative to funding emergency (critical situation) responses, and to make recommendations for improving coordination between different surveillance systems, different governments, and different international organizations.

After its review,[1] the committee found that, in the United States and elsewhere, traditional systems of infectious disease surveillance in humans operate separately from those for animals. This separation impedes communication between human and animal health officials on zoonotic disease occurrences that can threaten human health. For example, during the 1999 West Nile virus outbreak in the United States, a veterinarian tried to notify human health authorities about the possible connection between bird die-offs in a New York zoo and human outbreaks of febrile illness occurring in the same area. However, human health officials did not act upon the alert to investigate the potential threat to humans in a timely manner. Another problem is the mismatch of surveillance capabilities in locations where diseases are most likely to emerge. The industrialized world has the most robust surveillance systems for both human and animal health; however, most recent zoonotic diseases have emerged in the developing world, where surveillance systems are weaker.

Disease surveillance is essential to ensure that information is passed on to authorities to implement an efficient, early response, averting the need for a large emergency response after the disease has spread. A previous assessment estimated that an investment of $800 million per year is needed for global disease surveillance and early response capabilities; however, the

[1]The committee's review was based on its data-gathering sessions, survey data, expertise of committee members, and *Achieving Sustainable Global Capacity for Surveillance and Response to Emerging Diseases of Zoonotic Origin: Workshop Summary* (IOM and NRC, 2008).

economic losses from emerging, highly contagious zoonotic diseases have reached more than $200 billion over the past decade. Therefore, a global zoonotic disease surveillance system to reduce the emergence of zoonotic diseases in humans and to help detect other livestock diseases early could help to prevent the staggering economic losses associated with zoonotic disease outbreaks. It was beyond the committee's charge to comprehensively assess how best to implement appropriate evidence-based responses to an emerging zoonotic disease in human and animal populations; therefore, significant further review and study of integrated emergency response systems is needed.

Detecting and responding to zoonotic diseases is challenging, the committee found, because the underlying drivers of zoonotic disease emergence and spread result from an evolving complex of biological, genetic, ecological, political, economic, and social factors. One catalyst for disease emergence is the increasing demand for meat in developing countries where there are also many challenges in proper animal production management. In those countries, human populations and urban centers are expanding, with housing and agriculture competing with existing wildlife habitat. The movement of goods and people across borders—such as trade in food animals and exotic pets, international travel, and the movement of refugees into compromised living conditions—has increased the risk of disease spread. Climate change models suggest that wildlife migration patterns could change and that precipitation increases could lead to an expansion of insect- and water-borne diseases. The convergence of these diverse and nuanced drivers can create zoonotic disease "hotspots."

However, effective surveillance systems rely on local and national participants' ability and willingness to accurately report disease outbreaks, and their capability to implement local and national responses. Early identification of zoonotic disease emergence is essential to rapidly contain outbreaks, yet many local and national authorities lack the human and technical capability, capacity, and supporting financial resources to do so. Tensions increase when reporting can lead to international health and economic consequences, such as trade sanctions, travel warnings, animal culling, and declining public confidence in products, as was the case with pork products during the influenza A(H1N1) 2009 outbreak. Local and national incentives for reporting disease outbreaks help alleviate an individual or a country's fears about bearing such consequences alone and can diminish the temptation to conceal or withhold information.

The drivers of zoonotic disease emergence and the measures to prevent their emergence and spread are global in nature. The issues are important to the international community and cannot be addressed by individual countries acting alone. Confronting the threat of zoonotic disease emergence benefits governments and people of all states, thus the committee concluded that a global zoonotic disease surveillance system[2] is a global public good.[3] While disease surveillance and response are the responsibility of every nation, a system providing sustainable global coverage will only be possible with the efforts of nearly all nations and will require active national and international collaboration with relevant private and public stakeholders.

The committee concluded that because the U.S. government is among the world leaders in disease surveillance and has a considerable stake in preventing the emergence and limiting the spread of zoonotic diseases, it should lead efforts to coordinate a globally integrated and sustainable zoonotic disease surveillance system. However, improving global zoonotic disease surveillance cannot be achieved without the proactive engagement of the World Health Organization (WHO) and the World Organization for Animal Health (Office International des Epizooties, OIE), the global standard-setting bodies for human and animal health, respectively. It is imperative for the Food and Agriculture Organization of the United Nations (FAO), the World Trade Organization, and private industry to be involved because of their roles in global food safety and security through trade agreements among their member countries, and because of their roles in implementing disease surveillance to meet respective goals and missions.

RECOMMENDATIONS

Achieving an effective zoonotic disease surveillance system that is global, sustainable in funding and capacity, and integrated across disciplines and sectors will require technical, economic, and political improvements (see Table S-1). Recommendations assigned as high priority are foundational for a global, integrated, zoonotic disease surveillance and response system. The remaining recommendations are considered priority, although not listed in rank order. While resources and leadership sufficient for carrying out these recommendations may result in different implementation timetables, each of the 12 recommendations is essential to achieve and sustain a successful global system.

[2]The committee defines "zoonotic disease surveillance" as the ongoing systematic and timely collection, analysis, interpretation, and dissemination of information about the occurrence, distribution, and determinants of diseases transmitted between humans and animals. Zoonotic disease surveillance reaches its full potential when it is used to plan, implement, and evaluate responses to reduce infectious disease morbidity and mortality through a functionally integrated human and animal health system.

[3]The International Task Force on Global Public Goods defines "global public goods" as "issues that are broadly conceived as important to the international community, that for the most part cannot or will not be adequately addressed by individual countries acting alone and that are defined through a broad international consensus or a legitimate process of decision-making" (2006, p. 13).

TABLE S-1 Recommendations by Priority and Category

	Technical	Economic	Political
	Strengthen Surveillance and Response Capacity	*Financing and Incentives for Surveillance and Response*	*Governance of Global Efforts to Improve Surveillance and Response Capabilities*
High priority	Establish surveillance and response strategies (*Recommendation 1-1*)	Establish sustainable funding strategies (*Recommendation 2-1*)	Create a coordinating body for global zoonotic disease surveillance and response (*Recommendation 3-1*)
Priority	Improve use of information technology to support surveillance and response activities (*Recommendation 1-2*)	Create an audit and rating framework for surveillance and response systems (*Recommendation 2-2*)	Deepen the engagement of stakeholders (*Recommendation 3-2*)
	Strengthen the laboratory network to support surveillance and response activities (*Recommendation 1-3*)	Strengthen incentives for country and local reporting (*Recommendation 2-3*)	Revise OIE governance strategies (*Recommendation 3-3*)
	Build human resources capacity to support surveillance and response efforts (*Recommendation 1-4*)		Mitigate disease threats from wildlife and trade (*Recommendation 3-4*)
	Establish a zoonotic disease drivers panel (*Recommendation 1-5*)		

NOTE: OIE = World Organization for Animal Health.

High-Priority Recommendations

The committee examined several infectious disease surveillance systems already in operation to identify some effective systems, uncover gaps in efforts, and examine improvements to existing systems to achieve the desired global disease surveillance system. Table 4-2 in the report presents a summary of current system gaps and challenges.

Technical: Strengthen Surveillance and Response Capacity

The committee found that the United States and Europe are greatly overrepresented in reports of emerging disease outbreaks, which is certainly related to disease surveillance and laboratory capacity. However, irrespective of resource availability, the committee was unable to identify a single example of a well-functioning, integrated zoonotic disease surveillance system across human and animal health sectors. The committee found large gaps in existing disease surveillance networks, including coverage across species and across geographic space. Of concern is that the cause for 90 percent of human infectious disease cases could not be identified, even in developed countries.

Recommendation 1-1: The U.S. Departments of Health and Human Services, Agriculture, Homeland Security, and the Interior should collaborate with one another and with the private sector and nongovernmental organizations to achieve an integrated surveillance and response system for emerging zoonotic diseases in the United States. In addition, these government agencies, including the U.S. Department of State and USAID, should collaborate with WHO, FAO, and OIE to spearhead efforts to achieve a more effective global surveillance and response system, learning from and informing the experiences of other nations.

Given finite resources and the complexity of the challenge, an integrated zoonotic disease surveillance and response system can succeed only if the U.S. government and its partners, informed by best practices documented to date, develop strategic approaches and strengthen the needed capacities at both the national and global levels. Such strategic approaches would include

(a) Work with researchers to develop science-based criteria to determine the magnitude and distribution of disease drivers.

(b) Immediately strengthen surveillance in human populations at high-risk for zoonotic diseases (for example, livestock and poultry workers) in countries where disease surveillance in animal populations is weak.

(c) Develop and strengthen surveillance systems in animal populations so that outbreaks are detected early in animal populations rather than discovered later through secondary human outbreaks.

(d) Synchronize and share surveillance information from both human and animal populations in an integrated system, in as close to real time as is possible.

(e) Engage science-based nongovernmental organizations as valuable partners that provide the wide geographic reach and field-expertise needed for more comprehensive surveillance and response activities.

Economic: Financing and Incentives for Surveillance and Response

Funding needs will be significant to develop and sustain a global disease surveillance system for emerging and reemerging zoonotic diseases. Existing international aid architecture is fragmented and donor funding is unpredictable, especially during a global economic crisis. The committee concluded that the long-term infrastructure for disease surveillance and response has been underfunded in part due to the historical practice of time limited donor funding for specific diseases.

Recommendation 2-1: USAID—in partnership with international finance institutions and other bilateral assistance agencies—should lead an effort to generate sustainable financial resources to adequately support the development, implementation, and operation of integrated zoonotic disease surveillance and response systems. An in-depth study of the nature and scope of a funding mechanism should be commissioned by these agencies, and the study should specifically consider a tax on traded meat and meat products as a potential source of revenue.

Given the benefits the international community derives from early detection of a potential health or economic (trade) risk, countries with greater resources need to show leadership by supporting low-income countries and international organizations. Whatever the source for sustainable financing, it should be tied to activities that can increase the risk of zoonotic disease emergence and spread, such as trade. The proposed levy on traded meat and meat products places the burden on the wealthier importing countries. Access to funding could be dependent on the recipient country's commitment to and development of national surveillance capabilities.

Political: Governance of Global Efforts to Improve Surveillance and Response Capabilities

Recent concerns about a potential highly virulent human influenza pandemic have resulted in coordinated international action to help countries improve their ability to detect disease outbreaks. In 2006, the UN appointed a System Influenza Coordinator (UNSIC), which has been a key factor in the development of strong partnerships among technical agencies such as WHO, FAO, OIE, and other bilateral and multilateral partners, including the World Bank. UNSIC provides a useful model for the governance of a global zoonotic disease surveillance system.

Recommendation 3-1: USAID, in cooperation with the UN and other stakeholders from human and animal health sectors, should promote the establishment of a coordinating body to ensure progress toward development and implementation of harmonized, long-term strategies for integrated surveillance and response for zoonotic diseases.

A streamlined architecture for global health governance on zoonoses would benefit from structured coordination of critical intergovernmental bodies. Establishing a permanent zoonotic disease coordinating body with the authority and means to bring together technical agencies, including WHO, FAO, and OIE, will ensure that all relevant stakeholders are consulted and involved. The mechanism could also draw attention to problems and challenges faced in implementation of the International Health Regulations (IHR) 2005, OIE agreements, and OIE/FAO strategies, and identify additional funding streams for zoonotic disease control.

Priority Recommendations

Technical: Strengthen Surveillance and Response Capacity

Improve Use of Information Technology Information technology is essential for early disease detection, monitoring, and surveillance by enabling real-time collection and sharing of detailed information about outbreaks. Technological breakthroughs have led to new ways to collect and transmit epidemiological, clinical, demographical, and other information in the field. These include the use of handheld computers, cell phones, remote sensing, and web-based data streams, which are used to capture and disseminate information from even the most remote and resource-challenged countries.

> **Recommendation 1-2: With the support of USAID, international organizations (such as WHO, FAO, OIE, and the World Bank) and public- and private-sector partners should assist nations in developing, adapting for local conditions, and implementing information and communication technologies for integrated zoonotic disease surveillance. Effective use of such technologies facilitates acquisition, integration, management, analysis, and visualization of data sources across human and animal health sectors and empowers information sharing across local, national, and international levels. To establish, sustain, and maintain this technologically sophisticated system, both leadership and investment are critically needed.**

Technology development should focus on bidirectional information sharing with specific attention to data aggregation technology, open source development, transparency, privacy, and standards to facilitate improved communication within and between human and animal health sectors and across borders. Leadership and investment is needed within each country and will require partnership with key nongovernmental actors such as private philanthropies, industry partners, and nongovernmental organizations.

Strengthen Laboratory Network Identifying the cause of emerging outbreaks is a vital part of any disease surveillance system. Existing reference laboratories lack broad capabilities in disease diagnosis because they often have only agent specific expertise and may lack a specific mandate for zoonotic disease surveillance. The committee found that no resource exists to provide data on existing global zoonotic disease diagnostic laboratory capability and capacity for both human and animal health sectors. Moreover, no model is available for a workable global laboratory network infrastructure for integrated zoonotic disease diagnosis and reporting. What is clear is the overall geographic mismatch between reference laboratory and collaborating center locations and hotspot regions (Figure S-1).

> **Recommendation 1-3: USAID should promote and initially fund the establishment of an international laboratory working group charged with designing a global laboratory network plan for zoonotic disease surveillance. The working group's objective would be to design a laboratory network that supports more efficient, effective, reliable, and timely diagnosis, reporting, information sharing, disease response capacity, and integration of human and animal health components. In addition, a long-term coordinating body for zoonotic diseases, perhaps modeled after the UN System Influenza Coordinator's office (see Recommendation 3-1), should implement the global laboratory network plan, manage it, and assess its performance in consultation with the international laboratory working group.**

Local and advanced reference technical laboratory capacity needs to be organized into national, regional, and global networks. An international working group—with representation from national human and animal health laboratories from the public, private, and military sectors, international nongovernmental organizations, professional associations, and wildlife health—should be tasked to strategically outline steps to assess, plan, and fund the needed global network laboratory capacity. Implementation of the plan can be modeled on the U.S. Integrated Consortium of Laboratory Networks.

Build Human Resources Capacity To produce and retain a skilled multidisciplinary workforce capable of conducting integrated surveillance and response, new and existing personnel need to be trained in field-based, integrated emerging zoonotic disease surveillance and response.

> **Recommendation 1-4: Given the need for increased human capacity to plan, conduct, and evaluate integrated zoonotic disease surveillance and response, U.S. government agencies should take the lead in developing**

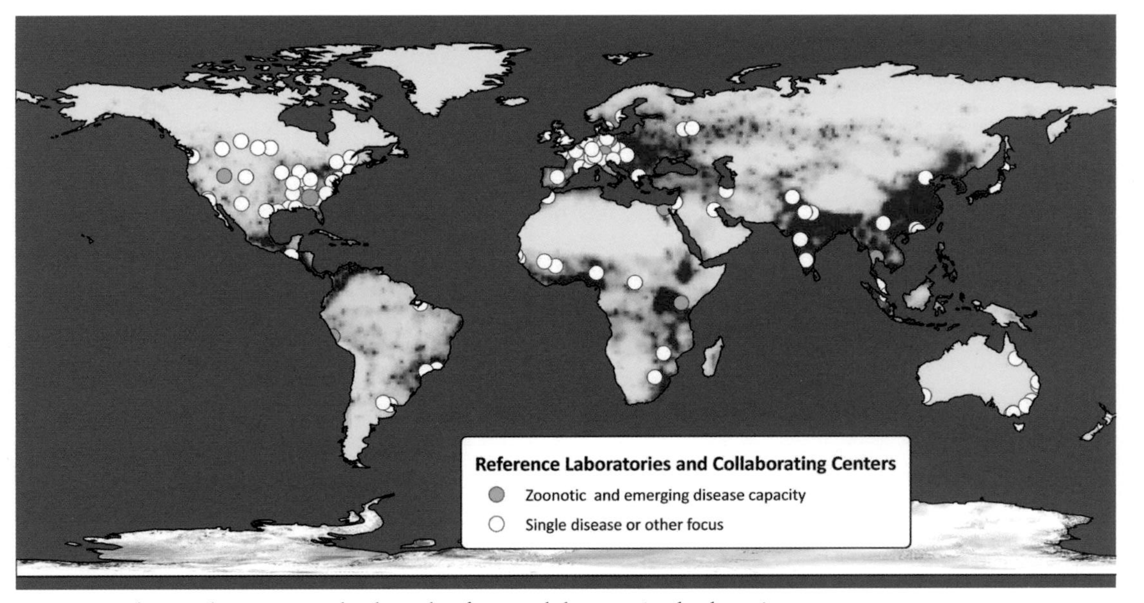

FIGURE S-1 Zoonotic disease hotspots and selected reference laboratories by location.
NOTE: The white dots signify the location of identified World Health Organization, Food and Agriculture Organization of the United Nations, World Organization for Animal Health, and U.S. Department of Defense reference laboratories and collaborating centers, many of which have a single disease or other focus mandate. Green dots are laboratories that have a broader function in zoonotic and emerging diseases. Locations shaded in red and orange represent hotspot regions. The map does not include university-based research and other laboratories working in the area of emerging disease detection and characterization. Numerous other private-sector and national laboratories may be able to provide laboratory support capability (e.g., those of the Institute Pasteur and Meriux Alliance), but were not included on this map.
SOURCE: Hotspot location data derived from Jones et al., 2008. Reference laboratory data received from committee's communication with Stephane de La Rocque, Tracy DuVernoy, Cassel Nutter, Alejandro Thiermann, and Chris Thorns (2008).

new interdisciplinary educational and training programs that integrate human and animal health and allied fields. Existing national and regional training programs in field epidemiology, clinical, and laboratory diagnosis supported by the U.S. Departments of Health and Human Services, Agriculture, and the Interior should be improved to include a better balance of human and animal health concerns, incorporate contributions from laboratory and social science professionals, and connect with one another where appropriate.

The National Institutes of Health's Fogarty International Center—in partnership with the Centers for Disease Control and Prevention (CDC), U.S. Department of Agriculture (USDA) Agricultural Research Service, USDA Animal and Plant Health Inspection Services, USDA National Institute of Food and Agriculture (the former Cooperative State Research, Education, and Extension Service), and U.S. Geological Survey—should be funded to provide leadership and partner with educational institutions and relevant ministries to develop these programs. The new curricula and training programs need to include human and animal health professionals, paraprofessionals, and community and public health professionals for maximal opportunities to improve interdisciplinary communication.

Establish a Zoonotic Disease Drivers Panel The drivers of zoonotic disease are individually and collectively complex, and the measures for controlling them are transnational in nature. Although some of these drivers are understood in isolation or in simpler, temporal interactions with each other (e.g., food-insecure people resorting to hunting wild animals for bushmeat, which in turn exposes them to HIV), the complex ways in which they change and interact over time are not well understood. This is a serious and noticeable gap in current global zoonotic disease surveillance and response efforts.

Recommendation 1-5: The U.S. Department of State, in collaboration with WHO, FAO, OIE, and other international partners, should impanel a multidisciplinary group of technical experts to regularly review state-of-the-science information on the underlying drivers of zoonotic disease emergence and propose policy and governance strategies to modify and curb practices that contribute to zoonotic disease emergence and spread.

The zoonotic disease drivers panel would regularly review scientific information to inform national and global policymakers of strategic actions to mitigate consequences of driver interaction that can lead to disease emergence. The group should be composed of the recommended coordinating body for zoonotic diseases and international representatives with demonstrated technical expertise to examine the broad set of drivers. It could be modeled after the Intergovernmental Panel on Climate Change. The Science and Technology Advisor to the President and the Department of State's Science and Technology Advisor to the Secretary could co-lead the effort and bring the results of the panel's findings to the attention of important stakeholders and diplomatic forums, including the UN, Group of Eight (G8), Group of Twenty (G20), and regional intergovernmental organizations.

Economic: Financing and Incentives for Surveillance and Response

Create an Audit and Rating Framework Countries participate in assessments of national human and animal health systems under the IHR 2005 and OIE programs, respectively. At present, there is no independent mechanism to review progress towards achieving integrated surveillance and response system capabilities, increasing the likelihood of uneven or incomplete progress.

> **Recommendation 2-2: USAID should convene a technical working group to design and implement, by the end of 2012, an independent mechanism to audit and rate national surveillance system capacities for detecting and responding to emerging zoonotic disease outbreaks in humans and animals.**

The technical working group needs representation from WHO, FAO, OIE, academia, nongovernmental organizations, national governments, and private-sector partners. The 2012 deadline coincides with the target date for full implementation of IHR 2005. Assessing both country risk and reliability of reporting disease outbreak can help stakeholders identify barriers to improve national and global capabilities. National surveillance capacity information should be made publicly available by each country and such information should be subject to independent audit and verification by the audit framework. Because information on national risk is a public good, resources to support this activity should be sourced through the global funding mechanism described in Recommendation 2-1. This audit and rating framework would be housed within an independent global technical consortium.

Strengthen Incentives for Country and Local Reporting An important lesson from disease outbreaks such as HPAI H5N1 is that the ability of the global human and animal health systems to respond is only as good as the ability and willingness of local and national systems to detect and report outbreaks. Bilateral aid agencies and international organizations have not yet paid enough attention to reducing the tendencies of countries to conceal outbreaks. Such measures would include designing economic incentives for reporting outbreaks, providing adequate compensation to cover economic impacts of response, and assuring that implemented control measures are based on scientific evidence.

Recommendation 2-3: To reduce incentives to conceal outbreaks and mitigate the negative social and economic repercussions of early disease reporting (e.g., stigma of disease, food safety concerns, culling, and trade and travel disruptions), financial incentives at the following levels are needed through partnerships among bilateral aid agencies, the international community, and national governments:

(a) Country level: USAID—in partnership with international finance institutions and other bilateral assistance agencies—should implement economic incentives to encourage middle- and low-income countries to report human, animal, and zoonotic disease outbreaks.

(b) Local level: National governments, with added support from the international community, should identify and provide the resources needed for financial incentives to promote early disease reporting and to engage in effective responses at the local level.

The international community can also minimize the unnecessary cost of sanctions at both levels by using existing regulatory mechanisms, like zoning and compartmentalization, where appropriate. International community application and acceptance of these initiatives allow for continued trade of safe products from countries or zones that have reported a disease. In addition to funding for upgrading surveillance capacity, guaranteed assistance with outbreak containment needs emphasis, including the availability of diagnostic kits and vaccines for humans or animals. Without such support, countries have fewer incentives to report disease outbreaks, regardless of international legal obligations.

National governments need to make explicit plans to increase incentives by allocating financial resources for adequate reparation to those who stand to lose from reporting, while decreasing disincentives by reviewing and reducing the unwarranted use of outbreak control measures such as travel restrictions, quarantines, and culling.

Political: Governance of Global Efforts to Improve Surveillance and Response Capabilities

Deepen Engagement of Stakeholders The complexity of achieving sustainable, integrated national and global surveillance and response systems for zoonotic diseases requires deliberate and intensified efforts to engage and connect all relevant stakeholders at each governance level—local, national, and global. Moreover, high stakes for trade or industry groups—as illustrated by the detection of bovine spongiform encephalopathy (BSE) in three cows in the United States between 2003 and 2006, causing great economic harm to that industry with a total loss of $11 billion—necessitate their involvement as well.

Recommendation 3-2: In its work on zoonotic disease surveillance and response, USAID—in collaboration with WHO, FAO, and OIE—should convene representatives from industry, the public sector, academia, nongovernmental organizations (NGOs), as well as smallholder farmers and community representatives to determine how best to build trust and communication pathways among these communities in order to achieve the efficient bi-directional flow of both formal and informal information needed to support effective, evidence-based decisionmaking and coordinated actions.

The public desires higher levels of health and less risk of disease; governments have a political interest in the trade-off between improving the levels of sanitary health on behalf of citizens and the freedom of international commerce; and industry has an economic interest in the trade-off between quality and yield. Despite these often mutually beneficial interests, different sectors can still be resistant to working together. To overcome such barriers, it is critical to engage relevant stakeholders from all levels to help build transparency and trust.

Revise OIE Governance Strategies The committee analyzed similarities and differences in the governance strategies and legal obligations embedded within WHO's IHR 2005 and OIE's approaches, rules, and resolutions. Although they have more similarities than some comparative analyses have recognized, the committee concluded that the OIE rules lack important provisions found in IHR 2005 that should be operative to promote animal health.

Recommendation 3-3: To protect animal health and international trade, and to contribute significantly to the reduction of human and animal health impacts from zoonotic diseases, OIE members states should take the necessary steps to:

(a) Adhere to Resolution 17 (adopted on May 28, 2009), which reminds OIE member states of their obligation to make available to OIE all information on relevant animal diseases, including those that are of zoonotic potential.

(b) Create legally binding obligations for OIE members to develop and

maintain minimum core surveillance and response capabilities for animal health risks, including zoonotic diseases.

(c) Authorize OIE to publicly disseminate information received from nongovernmental sources, in the event OIE member states fail to confirm or deny such information in a timely manner, or when denials of such information run counter to persuasive evidence that OIE has obtained from other sources.

(d) Empower the OIE Director-General to declare animal health emergencies of international concern with respect to emerging or re-emerging zoonotic diseases that constitute a serious animal or public health risk to other countries and issue recommendations about how countries should address such emergencies.

Adopting these four outlined principles will strengthen OIE's ability to ensure that its member nations have the minimal capacity for effective surveillance and response to animal diseases, enabling them to control animal diseases before they decimate animal populations and impact human health. These four recommendations provide a stronger foundation for coordinating and collaborating among human and animal health organizations, ministries, and experts.

Mitigate Disease Threats from Wildlife and Trade The legal and illegal trade in wildlife and wildlife products is an often ignored conduit for zoonotic pathogens, and it is apparent that the ability to monitor and control this trade is limited. There is also a noted lack of coordination, even within the United States, for disease detection in livestock and animal product imports and in wildlife.

Recommendation 3-4: To mitigate and decrease the threat of zoonotic diseases emerging from wildlife, U.S. government entities and their international partners, especially OIE, should proactively take the following initiatives:

(a) Conduct a comprehensive review of federal and state laws on trade in wildlife as a prelude to optimizing the policy and regulatory options to identify gaps and weaknesses in such laws, and to enact new legislation, regulations, or administrative rule changes to strengthen the government's ability to protect human and animal health from diseases carried by wildlife traded through foreign or interstate commerce.

(b) Incorporate efforts and initiatives that support actions to prevent, prepare for, protect against, and respond to threats to human and animal health into current and new international negotiations and cooperative processes that address drivers of zoonotic diseases

(e.g., exotic pet trade, food safety and security, environmental degradation, and climate change).

(c) **Pursue negotiations for a new international agreement on trade in wildlife species that improves international collaboration on reducing the threat that such trade presents to human and animal health. The objectives of the negotiations and the agreement would be to make wildlife-related zoonotic disease prevention and control a higher priority in the international management and control of legal and illicit trade in wildlife species, the production and distribution of food and animals, and environmental protection.**

(d) **Incorporate wildlife diseases and zoonoses into the OIE World Animal Health Information System and integrate reporting on wildlife diseases and zoonoses in the Global Early Warning System. OIE should also expand the role and capability of its Working Group on Wildlife Diseases in order to more effectively meet the growing zoonotic threat that wildlife diseases represent.**

U.S. government entities including the Departments of Agriculture, Commerce, Health and Human Services, Homeland Security, and the Interior should take the lead for these recommendations. Other relevant entities include the U.S. Postal Service and the U.S. Trade and Development Agency. To overcome the current fragmentation of responsibility in the United States, a first step would be establishing an inter-agency working group to recommend a collaborative strategy for improved oversight and action. Internationally, OIE should adopt a broader view of its remit by forming an *ad hoc committee* to assess the most significant disease risks in the international trade in wildlife, including those of potential impact to human, livestock, and environmental health.

CONCLUSION

Minimizing morbidity and mortality in human and animal populations and protecting national and global security, international trade, and individual livelihoods through a sustainable and integrated zoonotic disease surveillance system is both a global public good and in the self-interest of all nations. Steadfast global dedication of attention and resources from multiple collaborating sectors is needed to achieve such a system, and it will also require unprecedented collaboration across all levels, sectors, and professional disciplines. Implementing all of the committee's recommendations would also strengthen the global implementation of IHR 2005, WHO's legal mechanism for improving disease surveillance and response capacities for its member countries. The committee's recommendations reflect elements and resources needed to strengthen global efforts to improve zoonotic disease surveillance and response.

1

Introduction

"The confluences of human and animal health, along with wildlife, create new opportunities for pathogens to emerge and reemerge."

—Animal Health at the Crossroads: Preventing Detecting, and Diagnosing Animal Diseases (National Research Council, 2005a)

Zoonotic[1] pathogens have caused the majority of the emerging infectious disease events in the past six decades (see Figure 1-1) (Woolhouse and Gowtage-Sequeria, 2005; Jones et al., 2008). These diseases have the potential to cause significant morbidity and mortality in humans and animals, with resulting implications for international trade, travel, economies, and national security. Global interconnectivity has increased opportunities for disease emergence and rapid disease transmission, and the various linkages in the global economy also enable systemic social, political, and economic consequences (World Economic Forum, 2006). Public awareness and concern have grown dramatically as the potential for a global pandemic of influenza was heightened by the emergence of highly pathogenic avian influenza (HPAI) H5N1 infections and with the arrival of pandemic H1N1 in 2009. There is a need and possible momentum for new country-led initiatives and international collaborations aimed at managing this global threat (Murphy, 2008).

CHARGE TO THE COMMITTEE

Statement of Task

The Committee on Achieving Sustainable Global Capacity for Surveillance and Response to Emerging Diseases of Zoonotic Origin was convened

[1] A zoonotic disease or infection is transmissible between animals and humans. Zoonoses may be bacterial, viral, or parasitic, and may involve unconventional agents (IOM, 2003; WHO, 2008).

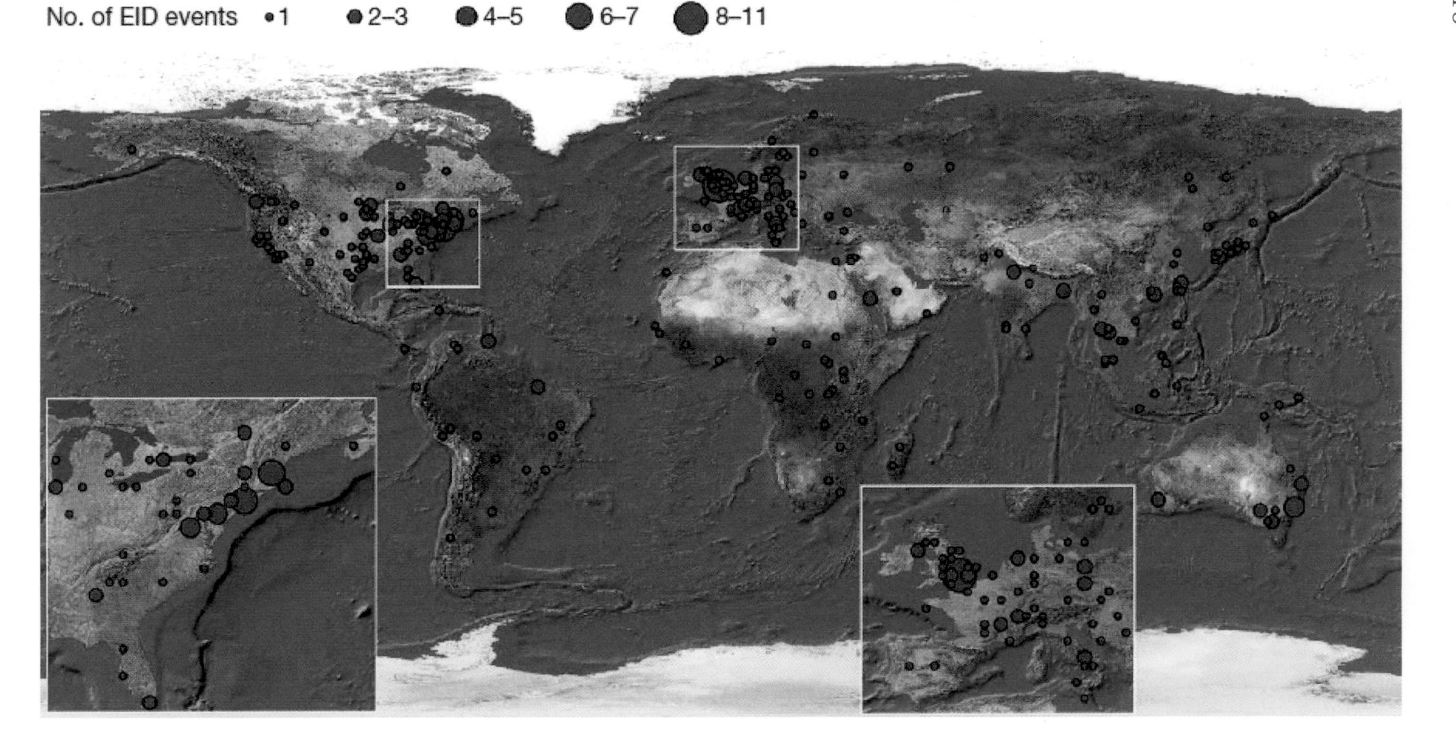

FIGURE 1-1 Emerging infectious disease events detected from 1940 to 2004. The map is derived for disease events caused by all pathogen types. Circles represent one-degree grid cells, and the area of the circle is proportional to the number of events in the cell.
SOURCE: Jones et al. (2008). Reprinted with permission from Macmillan Publishers LTD: *Nature*.

by the Institute of Medicine (IOM) and the National Research Council (NRC) at the request of the U.S. Agency for International Development to examine the needs and challenges associated with building sustainable global disease surveillance and response for zoonotic diseases. This included a review of the diseases that have emerged in the past several decades and the drivers associated with their emergence and reemergence; a review of the current state of existing global disease surveillance systems for zoonotic disease; and an examination of policy and regulatory options to mitigate or decrease the threat of zoonotic diseases globally. The committee was also asked to recommend ways to strengthen and improve coordination of the human and animal health systems and the mechanisms that govern them to achieve sustainable and timely disease surveillance worldwide that could improve the prevention of and response to these disease threats (see Box 1-1 for the Statement of Task).

Limitations on the Scope

Security threats can be caused by the intentional introduction of microbes for deliberate disease emergence. While the committee recognizes the dual-purpose nature of zoonotic pathogens and its potential for biosecurity concerns, this report is instead focused on nondeliberate disease emergence and events.

In addition, the report predominantly addresses surveillance concerns rather than focusing on response measures. The committee understood the importance of acting on surveillance information to prevent and control emerging zoonotic disease outbreaks. However, given the serious gaps and challenges that currently preclude early detection and reporting and the limitations of the committee's charge, the committee primarily focused its efforts to address these surveillance gaps and challenges. Significant additional review, discussion, and consideration would be needed at a future time to comprehensively assess how best to implement appropriate evidence-based responses following the detection of an emerging zoonotic disease in human and animal populations.

The Committee's Approach to Its Task

Several publications from the IOM and the NRC have examined the topics of infectious diseases and microbial threats to health and security (IOM, 1992, 2003), and the challenges and resources needed to strengthen animal health infrastructure, including the training of veterinarians (NRC, 2004, 2005a,b). This report builds on perspectives outlined in the report *Animal Health at the Crossroads: Preventing, Detecting, and Diagnosing Animal Diseases* (NRC, 2005a).

BOX 1-1
Statement of Task

The charge to the committee was to provide consensus advice on the challenge of achieving sustainable global capacity for disease surveillance and response to emerging diseases of zoonotic origin. Specifically, the committee was to address the following issues:

1. Review the emergence and spread over the past several decades of a diverse range of agents of zoonotic origin.
2. Summarize what is known about the causes underlying this growing phenomenon, trends in these factors, and the implications for long-term domestic and international development and security.
3. Assess the evolving nature, extent, and risks of animal and human interactions, focusing specifically on recent infectious disease events of international significance, such as highly pathogenic avian influenza H5NI.
4. Review the historic human and animal health responses to emergent zoonotic diseases along with lessons learned that may be applicable to future threats.
5. Review the current state of and gaps in global systems for disease surveillance of zoonotic infections in human and animal populations.
6. Develop conclusions on the appropriate balance between emergency response to threats and establishing sustainable global disease surveillance capacity for early detection, mitigation, and characterization of known, changing, and unknown threats.
7. Identify and prioritize for the international context recommendations to strengthen and improve coordination of the human and animal health systems to achieve a sustainable and integrated institutional capacity for timely disease surveillance that could improve prevention of and response to zoonotic diseases across both realms.
8. Explore options—including policy and regulatory options, such as international agreements—to mitigate and decrease the threat of emerging zoonotic diseases worldwide, and to improve coordination between governments and other relevant international organizations.

The Committee on Achieving Sustainable Global Capacity for Surveillance and Response to Emerging Diseases of Zoonotic Origin met over 10 months. A 2-day workshop was held in conjunction with the first committee meeting in June 2008 in Washington, DC. At the data-gathering workshop, invited speakers and experts discussed aspects of building capacity for disease surveillance and response to emerging zoonotic diseases. Speakers and participants included representatives from international organizations, U.S. government agencies, and researchers and academicians from the Americas, Asia, and Africa. A summary of the workshop proceedings, *Achieving Sustainable Global Capacity for Surveillance and Response to Emerging Diseases of Zoonotic Origin: Workshop Summary*, was published in December 2008 (IOM and NRC, 2008).[2] The committee collected more information through four additional committee meetings, two teleconference meetings with invited experts, and multiple conference calls and electronic communications.

The committee defined several crucial terms for the purpose of this report, and the definitions are found in Appendix A. The committee considers public health to include both human and animal health. When human health officials, clinicians, researchers, or policymakers are referenced in discussions, the reader should also assume the committee intends to include their equivalents in the animal health realm, although such interactions are not yet routine. The committee refers to integrated systems to convey the importance of connecting and engaging both human and animal sectors in addressing the problem of emerging zoonotic infectious diseases.

INTERNATIONAL CONTEXT FOR ZOONOTIC DISEASE SURVEILLANCE AND RESPONSE

An important development in the past decade, driven by the emergence of HPAI H5N1 and severe acute respiratory syndrome (SARS), has been a transformation in how governments, international governmental organizations, and nongovernmental actors think about emerging zoonotic disease surveillance and response capacities. Human and animal health threats—and their intersections—have risen in public concern to become subjects of foreign policy and diplomacy. In this rise to political prominence, the committee recognized conceptual innovations in the way stakeholders think about disease surveillance and response capacities, and why they are important. Through foreign policy and diplomacy, governments attempt to achieve four objectives:

1. To protect the nation's security;
2. To advance the nation's economic well-being and power;
3. To foster development in countries and regions important to the nation's security and economic interests; and
4. To protect human dignity through humanitarianism and human rights (Fidler, 2008).

Although past governance efforts against human and animal health threats have touched on some of these functions, they have never been systematic or conducted in ways that really mattered in the "high politics" of national or international politics. That may explain why the international regimes for human and animal health developed as devices to reduce the economic

[2] Available online through the National Academies Press at www.nap.edu.

burden of outbreaks, even though the World Health Organization eventually linked human health to human rights by advocating "Health for All" as a right under the Alma Ata declaration (WHO, 1978). Some key international governmental organizations relevant to the discussion on global surveillance and response of zoonotic diseases are described in Box 1-2.

The formal legal obligations that countries have to report emerging human and animal infectious disease events are only one part of the international institutional frameworks that guide the behavior of actors at the global level; also important are the set of informal norms, rules, and expectations they share. Because the economic, political, military, or even moral power relationships between nations are commonly asymmetric, it is essential to have international governance structures in place to limit the impact of the hierarchy of power among the participating nations, particularly if global public goods[3]—that is shared objectives for the good of all—are ever to receive support over more narrow national interests. International "institutions," including the "persistent and connected sets of rules (formal or informal), that prescribe behavioral roles, constrain activity, and shape expectations" (Keohane, 1984; Ostrom, 2005) can play this role by guiding the interactions of actors towards the achievement of shared objectives. These institutions are distinct from the actors involved, which may be states, government agencies, organizations, corporations, foundations, or even individuals.

While the institutions and the actors can be stable for long periods of time, some events can so perturb the institutional framework that it becomes necessary to find and negotiate a new set of rules and roles. Emerging zoonotic infectious diseases represent such a redefining event with respect to tourism, travel, and trade of food and animal products across national borders. Moreover, emerging zoonotic infectious diseases are not currently predictable, and so the "global institutions" that will govern the interactions between sovereign states and non-state actors (firms, nongovernmental organizations [NGOs], individuals) will need to have flexibility built in and be able to evolve to allow the involved actors to effectively meet the challenges of governance as they arise. The committee believes that it is important to distinguish between the institutions, in the context described, and the actors that must participate in building and supporting a global surveillance and response system to address emerging zoonotic infectious diseases. For example, as extensively discussed later in

[3]The International Task Force on Global Public Goods defines "global public goods" as "issues that are broadly conceived as important to the international community, that for the most part cannot or will not be adequately addressed by individual countries acting alone and that are defined through a broad international consensus or a legitimate process of decision-making" (2006, p. 13).

BOX 1-2
International Institutions and Actors

WHO: The World Health Organization (WHO), created by the United Nations (UN) in 1948, is the directing and coordinating authority for health within the UN system. It is responsible for providing leadership on global health matters, shaping the health research agenda, setting norms and standards, articulating evidence-based policy options, providing technical support to countries, and monitoring and assessing health trends. The World Health Assembly is the supreme decisionmaking body for WHO and is attended by delegations from all 193 member states. The Secretariat of WHO is staffed by some 8,000 health and other experts and support staff on fixed-term appointments, working at headquarters in Geneva, Switzerland, in the six regional offices, and in countries. WHO is headed by the Director-General, who is appointed by the Health Assembly on the nomination of the Executive Board. WHO collaborates with more than 800 institutions in 90 countries to carry out its programs and activities (www.who.int).

FAO: The Food and Agriculture Organization of the United Nations (FAO), created in 1945, has the mission of raising levels of nutrition, improving agricultural productivity, bettering the lives of rural populations, and contributing to the growth of the world economy. The organization, headquartered in Rome, Italy, is directed by the Director-General, elected by the Conference. FAO employs more than 3,600 staff members (1,600 professional and 2,000 general service staff) and maintains 5 regional offices, 9 subregional offices, 5 liaison offices, and 74 fully fledged country offices—excluding those hosted in regional and subregional offices (www.fao.org/about/mission-gov/en/).

OIE: The Office International des Epizooties (OIE, also known as the World Organization for Animal Health) is responsible for improving animal health worldwide. It was created in 1924 by the ratification of an agreement by member states of the League of Nations, and it is recognized as a reference organization by the World Trade Organization. As of June 2009, OIE had a total of 174 member states. The daily operations are managed by the Director-General, elected by the OIE International Committee, from the Paris, France, headquarters. The organization has approximately 40 health experts and support staff. OIE maintains permanent relations with 36 other international and regional organizations and has regional and subregional offices on every continent (www.oie.int).

WTO: The World Trade Organization (WTO) is an international organization established in 1995 with the primary purpose to open trade for the benefit of all. It provides a forum for negotiating agreements aimed at reducing obstacles to international trade and ensuring a level playing field for all, thus contributing to economic growth and development. It also provides a legal and institutional framework for the implementation and monitoring of 16 different multilateral agreements (to which all WTO members are parties) and two different plurilateral agreements (to which only some WTO members are parties), as well as for settling disputes arising from their interpretation and application. Decisionmaking is generally by consensus of the entire membership (currently 153 members, of which 117 are developing countries or separate customs territories). The organization is led by the Director-General. The Secretariat is in Geneva, Switzerland, with 700 staff members (www.wto.org).

this report, the revised International Health Regulations 2005 (IHR 2005) have been ratified by 194 nations and thus represent a legal requirement for compliance under the IHR protocol. However, the underlying institutions that will guide behavior as new challenges arise are less clear and less well understood. Without some debate and agreement on a basic set of rules and expectations, implementation of IHR 2005 may lag, and a truly effective global governance arrangement will remain elusive.

ORGANIZATION OF THE REPORT

The report presents the committee's findings, conclusions, and recommendations on achieving a sustainable global zoonotic disease surveillance and response system. Chapters 2 and 3 provide background context for exploring the magnitude of the challenges and threats posed by zoonotic diseases to human and animal health, macro- and microeconomies, global trade, and the sociocultural-political impacts and interactions for disease prevention and mitigation. Chapter 4 analyzes the current global capacity for zoonotic disease surveillance and response, while Chapter 5 examines the incentives and protections for improving disease reporting at various levels. Financing challenges for sustaining global disease surveillance are discussed in Chapter 6. Chapter 7 describes the governance mechanisms, processes, and innovations the committee deems critical to strengthening disease surveillance and response capabilities for human and animal health. Finally, Chapter 8 provides recommendations for sustaining global surveillance and response to zoonotic diseases and also examines some possible challenges that will need to be overcome in effectively implementing and strengthening efforts to protect human and animal health.

REFERENCES

Fidler, D. P. 2008. *Pathways for Global Health Diplomacy: Perspectives on Health in Foreign Policy* (WHO Globalization, Trade and Health Working Paper Series, June).

International Task Force on Global Public Goods. 2006. *Meeting global challenges: International cooperation in the national interest.* Final Report. Stockholm, Sweden: International Task Force on Global Public Goods.

IOM (Institute of Medicine). 1992. *Emerging infections: Microbial threats to health in the United States*, edited by J. Lederberg, R. E. Shope, and S. C. Oaks, Jr. Washington, DC: National Academy Press.

IOM. 2003. *Microbial threats to health: Emergence, detection, and response*, edited by M. S. Smolinski, M. A. Hamburg, and J. Lederberg. Washington, DC: The National Academies Press.

IOM and NRC (Institute of Medicine and National Research Council). 2008. *Achieving sustainable global capacity for surveillance and response to emerging disease of zoonotic origin: Workshop summary.* Washington, DC: The National Academies Press.

Jones, K. E., N. G. Patel, M. A. Levy, A. Storeygard, D. Balk, J. L. Gittleman, and P. Daszak. 2008. Global trends in emerging infectious diseases. *Nature* 451(7181):990–993.

Keohane, R. O. 1984. *After hegemony: Cooperation and discord in the world political economy.* Princeton, NJ: Princeton University Press.

Murphy, F. A. 2008. Emerging zoonoses: The challenge for public health and biodefense. *Prev Vet Med* 86(3–4):216–223.

NRC (National Research Council). 2004. *The national need and priorities for veterinarians in biomedical research.* Washington, DC: The National Academies Press.

NRC. 2005a. *Animal health at the crossroads: Preventing, detecting, and diagnosing animal diseases.* Washington, DC: The National Academies Press.

NRC. 2005b. *Critical needs for research in veterinary science.* Washington, DC: The National Academies Press.

Ostrom, E. 2005. *Understanding institutional diversity.* Princeton, NJ: Princeton University Press.

WHO (World Health Organization). 1978. Declaration of Alma Ata. International conference on primary health care, Alma-Ata, USSR, September 6–12. http://www.who.int/hpr/NPH/docs/declaration_almaata.pdf (accessed July 15, 2009).

WHO. 2008. *Zoonoses.* http://www.who.int/topics/zoonoses/en/ (accessed November 11, 2008).

Woolhouse, M. E., and S. Gowtage-Sequeria. 2005. Host range and emerging and reemerging pathogens. *Emerg Infect Dis* 11(12):1842–1847.

World Economic Forum. 2006. *Global risks 2006.* Geneva, Switzerland: World Economic Forum.

2

Making the Case for Zoonotic Disease Surveillance

"The difficulty of uncertainty is that we are dealing with things that are likely to emerge at some time and that need attention. We have to persuade decision-makers to invest in surveillance systems and other actions to deal with these uncertainties in a flexible and responsive way without being able to tell them, with an absolute precision, when they are going to emerge and what their economic or social cost might be."

—*Dr. David Nabarro*
Senior United Nations System Coordinator
for Avian and Human Influenza
Special Interview with the Committee
(September 11, 2008)

Recent emerging zoonotic diseases have had significant impacts in industrialized countries, despite well-developed health systems and sanitary infrastructures (Vorou et al., 2007; Jones et al., 2008; Murphy, 2008), and their impacts have been even more devastating for middle-income and developing countries. When emerging diseases become endemic, they not only continue to cause morbidity and mortality in human and animal populations, but also represent a threat of future epidemics if conditions for explosive transmission are reestablished. Emerging infectious disease trends suggest that the frequency of such disease events that are zoonotic in nature will not lessen in the future (McMichael, 2004; Woolhouse and Gaunt, 2007; Jones et al., 2008). If anything, with increasing human and animal populations and changing environments, the trends are more consistent with continual increases in the pace of emergence; however, it is simply unknown where or when they will occur (King, 2004; Morens et al., 2004).

Disease surveillance represents the eyes and the ears of the global public health effort, systematically generating information that informs actions to contain, control, and mitigate the consequences in at-risk humans and animals. Detecting diseases early through surveillance and implementing early response measures can reduce the scope, magnitude, and cost of emergency response measures downstream. To better predict and prevent zoonotic disease outbreaks, scientific approaches are needed to gather and understand information about the nature of disease appearance and spread, and to understand genetic-, population-, social-, and ecological-level characteristics that enable zoonotic pathogens to jump species and spread easily to humans. National and international support is also critical in addressing this global issue.

SOCIOECONOMIC FACTORS AFFECTING ZOONOTIC DISEASE EMERGENCE

Humans and animals can serve as pathogen reservoirs and vectors, and pathogens that may have resided in one part of the world can be carried or spread across long distances to become established in another part of the world. Technological advances now allow humans, animals, animal products, and their disease vectors to circumnavigate the globe in the span of 24 hours. Distance is no longer a barrier to disease. For example, in the first half of 2003, the United States saw concurrent importation of two zoonotic agents never before seen in the country—severe acute respiratory syndrome (SARS) and human monkeypox—as well as the establishment of new geographical niches for West Nile virus (WNV), an agent new to the United States and now endemic across the country. That same year, the United States also dealt with its first diagnosed case of bovine spongiform encephalopathy (BSE) despite more than 10 years of broad preventive efforts by the government and industry. In 2008, international tourist arrivals reached 924 million (UNWTO, 2009), a number that is estimated to grow annually by 5 percent over the next 20 years (FAO et al., 2008).

Globalization and Trade

Today, more goods, people, technology, and financial resources flow between countries than ever before, making countries less self-reliant and more dependent on each other. The level of economic interdependence among countries has increased dramatically on a global scale, especially in the past decade, as illustrated in Figure 2-1. In 2008, total global trade stood at $32.5 trillion, almost equally divided between imports and exports (WTO, 2009). In 2008, the total value of food imported into the United States was $75 billion or about 7.5 percent of total imports (Collins, 2007), and more than 25,000 shipments of food regulated by the U.S. Food and Drug Administration[1] arrived daily in the United States from more than 100 countries (Koonse, 2008).

In particular, the international trade of live animals and animal products

[1]The U.S. Food and Drug Administration inspects and monitors the safety of all foods, domestic and imported, except for meat, poultry, and egg products, which are regulated by the U.S. Department of Agriculture.

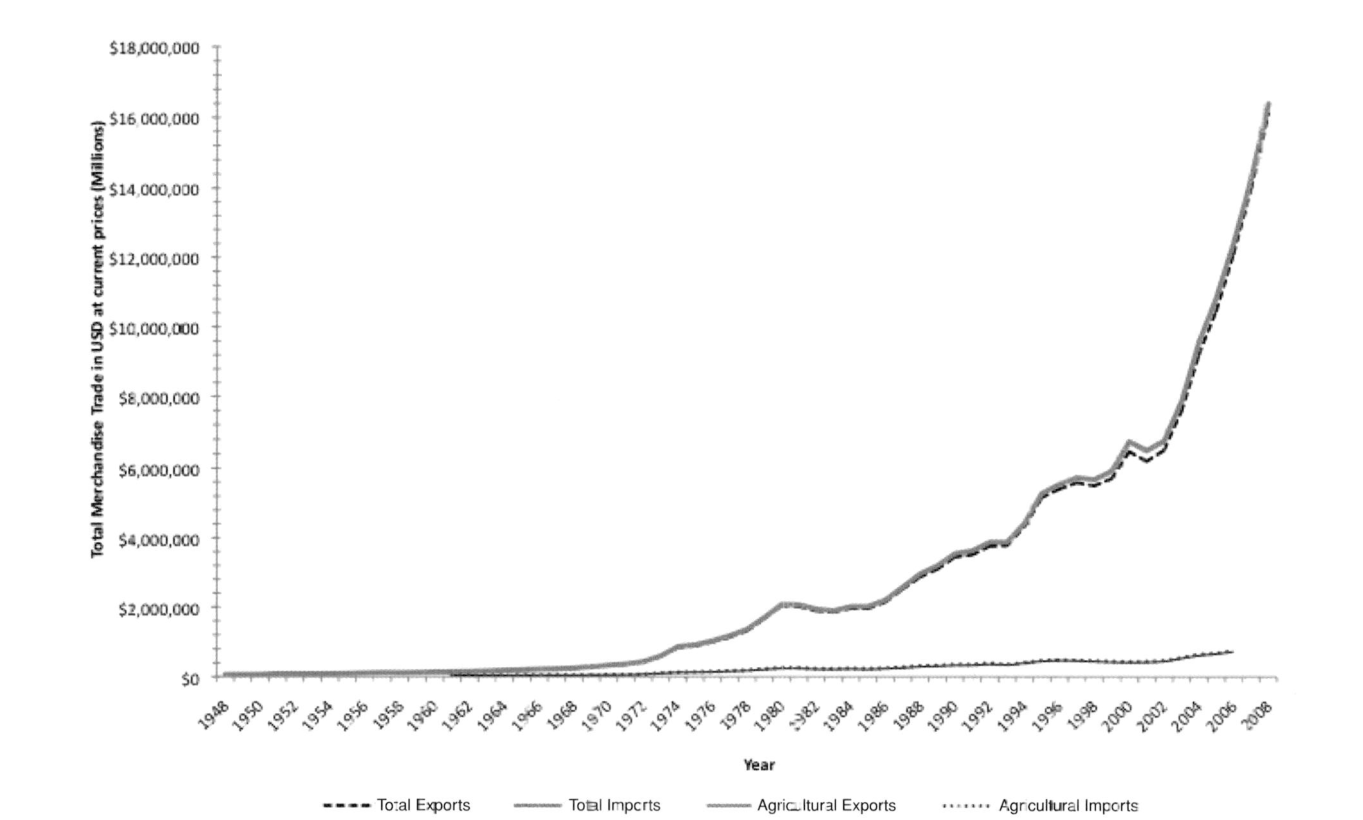

FIGURE 2-1 Total trade versus total agricultural trade. Agricultural trade data only from 1961–2006. In 2008, total trade imports were valued at $16.1 trillion USD and total trade exports were valued at $16.4 trillion USD.
SOURCES: FAO (2009), WTO (2009).

has sharply increased over the past decade (Figure 2-2). Increased trade brings increased movement of animals and animal products, thereby increasing the potential for disease emergence from zoonotic pathogens.

The global food production system is highly competitive and increasingly mobile. With attractive export markets, it often pays for exporting countries to establish the necessary veterinary infrastructure to meet the sanitary requirements of the importing country, as shown by countries such as Thailand for poultry and Brazil for beef. However, even competitive market economies do not necessarily reward additional investments in animal health infrastructure or encourage disease surveillance to track changing risk factors that might signal the potential emergence of a new disease. This failure to build veterinary capacity is even more relevant in countries where the food-animal production sectors primarily serve the local economy. Only with time and adverse experience are some countries and companies now grappling with disease threats across their production and distribution supply chains, including the possibility of full-fledged disease outbreaks.

Evolving Animal Agriculture and Trade

To remain economically viable in highly competitive environments and to produce affordable animal protein for the growing global population, there is continued pressure to seek out economies of size and scale, including expanding or establishing operations in those parts of the world offering favorable cost structures. Thus, the geographic distance between where animals are produced and where ultimate consumption occurs continues to expand. North America currently supplies one quarter of global meat exports (FAO, 2006). Asia has approached the Americas in volume of poultry production in a little more than a decade (see Figure 2-3). Brazil is now the largest single country for poultry and beef exports, and its diversified export market enables the movement of products to more than 150 countries (FAO, 2009).

Starting with more developed agricultural economies, such as the United States, but then spreading to other countries, much of the agricultural products that flow into international trade originate from increasingly capital intensive enterprises and well-coordinated supply chains. On the supply side, improvements in technology, infrastructure, and animal health have all contributed to this growth. Along with improvements in other areas such as genetics, nutrition, and management, the growing recognition of animal and herd health programs has enabled expansion and growth of large-scale animal agriculture. Large-scale production with animal crowding and unsanitary conditions in some settings has contributed to the use of antibiotics to fight disease, with secondary effects on selection for antibiotic-resistant microorganisms and environmental contamination.

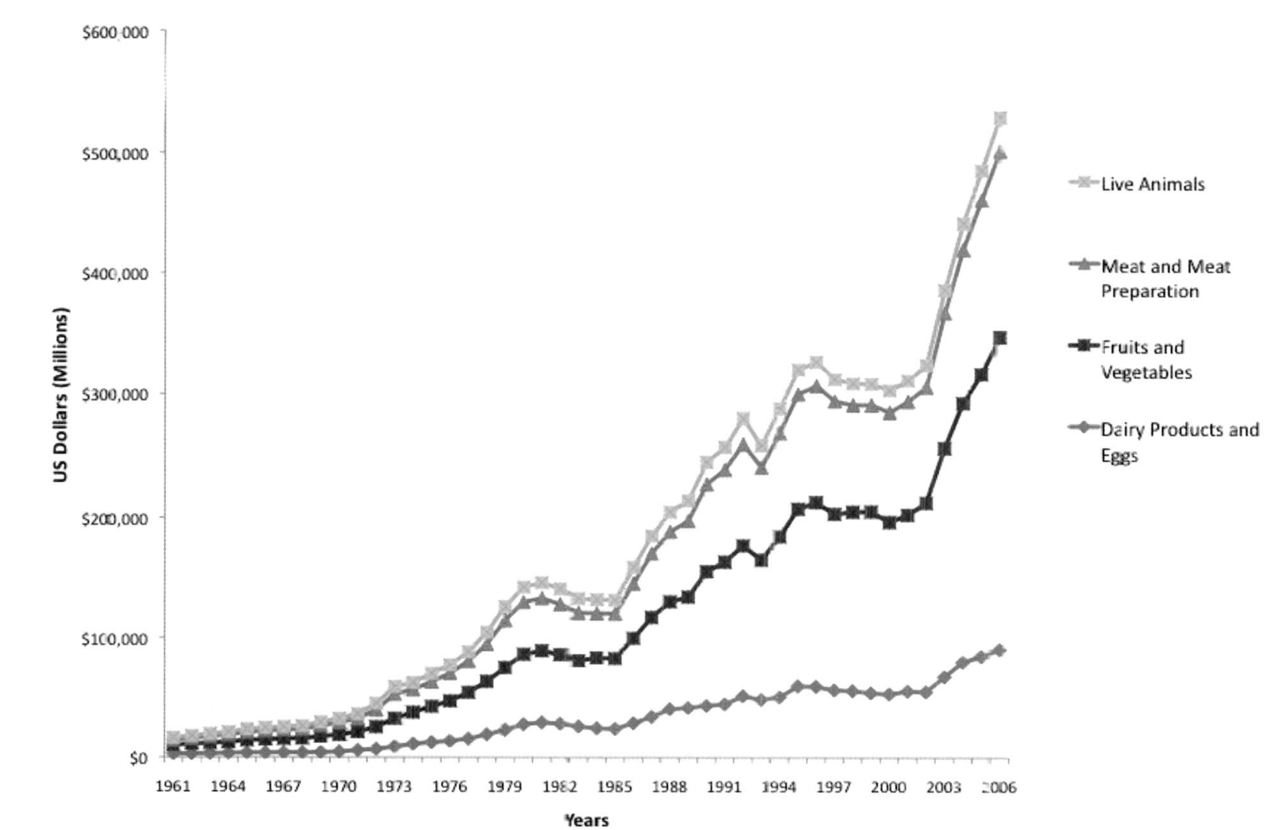

FIGURE 2-2 International agricultural trade by commodity type, 1961–2006.
SOURCE: FAO (2009).

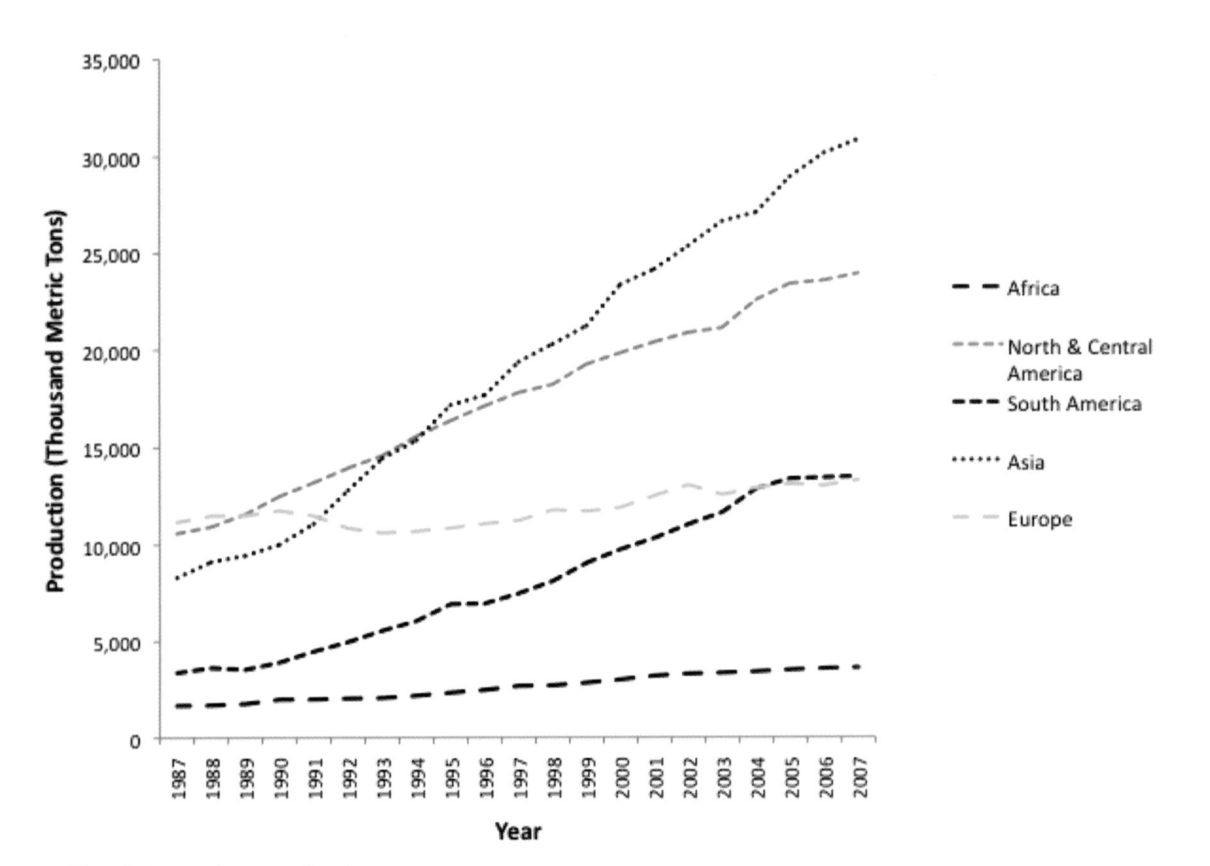

FIGURE 2-3 Trends in poultry production.
SOURCE: FAO (2009).

For countries such as the United States, the recognition that herds free of selected diseases could be translated into broader social and economic benefits has led to the support and implementation of national disease eradication campaigns. Freedom from brucellosis and tuberculosis not only contributes to the improvement of human and animal health, but has also lowered production costs, thereby establishing an international marketing advantage over countries that are not elevating their level of sanitary health. The public investment in animal health infrastructure includes the capacity to carry out disease surveillance, diagnosis, and treatment, and helps to facilitate export growth by enabling the movement of disease-free animals and related products into new markets and countries. To ensure that such improvements are not jeopardized or compromised, imports of susceptible animals or products are restricted from those countries that have not eliminated disease or achieved comparable levels of sanitary health. This allows certain exporting countries to further grow production capacity for domestic and international markets, largely through the adoption of standards formulated through the World Organization for Animal Health (OIE).[2]

The higher level of sanitary infrastructure has provided benefits to both producers and consumers. Producers benefit through factors such as decreased costs of production (e.g., the extra cost of raising healthier animals is compensated by survival, weight gain, and increased market price), real or perceived increases in product quality, and the ability to meet consumer demand. Consumers benefit from the reduced risk of exposure to zoonotic pathogens.

In many parts of the world, the public investment in national animal health infrastructure has not been commensurate with agricultural development. South and Central America provide more than one-fourth of the world's agricultural exports (WTO, 2008), yet only 5 percent or so of national government outlays go into agriculture support. Moreover, only 5 to 10 percent of that finds its way into animal and plant health programs, and that is for a limited array of existing pathogens and pests (Pomareda, 2001). In sub-Saharan Africa, where food-animal production contributes about 30 percent of the agricultural gross domestic product (GDP) and is a part of the livelihood of about 150 million people, public expenditure on food-animal production research and development is less than 10 percent of the total public agricultural research expenditure (World Bank, 2008a).[3] In addition, private-sector expenditure for agricultural research is low,

---

[2]The Office International des Epizooties (OIE) is also known as the World Organization for Animal Health. OIE formulates standards related to animal health through committees consisting of representatives from member countries that are later adopted in its general assembly. OIE is recognized as a technical reference organization on animal health by the World Trade Organization.

[3]Adapted from agricultural expenditure data in the 2008 World Development Report.

although philanthropic organizations have more recently emerged to support crop and livestock research.

Emerging Market Economies

In 2000, emerging market economies accounted for 56 percent of the global middle class. By 2030, that figure is expected to reach 93 percent; China and India alone will account for two-thirds of this expansion. Rising incomes and growing demand can increase total trade while altering existing and/or creating new trade flows, resulting in new or changing risk factors. For instance, rapidly growing economies fuel an increase in individual wealth, which also increases the demand for meat. In 2007, the average Chinese consumer ate 50 kg of meat, which is more than twice the amount consumed in 1985 (The end of cheap food, 2007). In 2008, an estimated 21 billion food animals were produced for a global population of 6.5 billion people (FAO et al., 2008).

Market dynamics also led to more live animal auctions where animals are brought together and then shipped across great distances and traditional "wet markets" where local farmers market their live animals to local consumers. These trends contribute to an increase in animal densities and closer contact between humans and animals, with a considerably greater risk of dispersing pathogens. International trade can transcend geographical barriers that in the past may have naturally slowed the spread of disease. The global market economy can also amplify disease effects through market instability as characterized by price volatility, shifts in consumption patterns, and variability in supplies.

International Wildlife Trade

Globalization has also impacted the movement of live, wild animals. From 2000 to 2004, more than 1 billion live animals were legally imported into the United States from 163 countries (Jenkins et al., 2007; Marano et al., 2007). In 2007 alone, the U.S. Fish and Wildlife Service processed 188,000 wildlife shipments worth more than $2.8 billion, and recorded more than 200 million legally imported live wildlife (CRS, 2008a; Einsweiler, 2008). These animals and animal products were imported for zoo exhibitions, scientific research,[4] food and products, and increasingly for the growing commercial pet trade, including many exotic animals (Marano,

[4]The U.S. Centers for Disease Control and Prevention (CDC) prohibited the importation of most monkeys as companion animals in 1975, but some imported for research are now being sold in the pet trade. CDC and other enforcement agencies do not track where animals go after quarantine (Ebrahim and Solomon, 2006).

2008). Most of these animals are not required under U.S. law to be screened for zoonotic diseases before or after entering the country (Marano et al., 2007). The effect of this, compounded by the lack of coordination among U.S. government agencies involved in regulating different aspects of wildlife imports, are important reasons for the failure to prevent the introduction of new pathogens into the country (Stephenson, 2003). Some exotic animals and wildlife that are banned from import are able to enter through the illegal wildlife trade.[5] These are likely to include less healthy, more risky animals that pose a greater threat to human health and security (CRS, 2008a; U.S. House of Representatives, 2008). Even so, most of the zoonotic diseases reported to be caused by wildlife trade involved imports of legal wildlife (see Appendix B on monkeypox).

The European Union (EU) is the top global importer of wildlife and wildlife products by value at €2.5 billion in 2005 (Engler and Parry-Jones, 2007), and it is concerned that increasing demands for wildlife importation is a driver of illegal and unsustainable trade. EU member states have concluded that a major barrier to wildlife trade law enforcement and implementation is their lack of a coordinated strategic approach to monitor compliance (Theile et al., 2004; Engler and Parry-Jones, 2007). A review of the socioeconomic factors that drive the wildlife trade in Southeast Asia, which is both a consumer of wildlife products and a key supplier, revealed the inadequacy of policies and interventions aimed at decreasing the illegal and unsustainable trade of wildlife (World Bank, 2008b). Although poor populations in this region are often involved in wildlife trade, they do not necessarily drive this trade; therefore interventions for poverty reduction are not likely to reduce wildlife exports. Instead, many experts consider that the increased disposable income in consumer countries is the major contributor of demand for Southeast Asian wildlife, parallel to the increased access to these markets (World Bank, 2008b). These observations only serve to highlight the complexity of market forces. On the supply side, the illegal logging industry and the bushmeat trade has facilitated the extraction of certain wildlife species and threatened local wildlife populations (Chomel et al., 2007). Refugee camps set up in response to humanitarian crises, such as northwestern Tanzania, have led to serious forest degradation and have provided people with a greater proximity to wildlife habitats to hunt bushmeat, resulting in a decline of wildlife populations (Jambiya et al., 2007). The lack of a single international mechanism that captures data on wildlife trade represents a serious shortcoming of current national and international policies aimed at preventing illegal and unsustainable international wildlife trade (Gerson et al., 2008).

---

[5] The illegal wildlife trade is difficult to quantify, although some estimates range from $5 billion to more than $20 billion annually (CRS, 2008a).

The Need for Disease Surveillance in Food Animals

Improved prevention and disease control efforts in food-animal health has led to multiple benefits for human and animal populations, including reduced human morbidity and mortality, enhanced food security, improved market access for products, economic gains, and savings on potential outbreak costs (Caspari et al., 2007). Many countries have strengthened their border controls and quarantine procedures, but the advances and benefits in improving animal health through actions such as disease eradication, prevention, and education have not been uniform across all countries. However, as education has advanced and become more available, surveillance and prevention efforts have also advanced and become specialized in areas such as vaccines and diagnostics. Although significant investments are needed to build infrastructure and institutional and regulatory capacity, necessary investments have not yet been made to implement food-animal disease surveillance, diagnosis, and treatment.

Countries such as the United States and Australia have made available significant financial and technical resources for international disease eradication or control campaigns, especially in the past 5 years for the control of highly pathogenic avian influenza (HPAI) H5N1 in Southeast Asia. In 2006, the U.S. Agency for International Development (USAID) provided $161.5 million for disease surveillance and pandemic preparedness for avian influenza (CRS, 2008b). In 2009, USAID will award $260 million over 5 years for the Predict and Respond initiatives aimed at four regions of the world prone to zoonotic disease emergence (Grants.gov, 2009a,b). From 2003–2006, Australia's Agency for International Development committed $152 million to combat avian influenza and other emerging and reemerging zoonotic diseases (AusAID, 2009). The EU has supported major animal disease eradication campaigns in Asia and Africa: Specifically in Africa, the EU partnered with the Organization of African Unity in 1999, providing an overall budget of €72 million for 7 years for the Pan African Programme for the Control of Epizootics (PACE) (OAU-IBAR, 2009). PACE targeted establishing and strengthening sustainable animal disease surveillance in sub-Saharan Africa.

HEALTH AND ECONOMIC IMPACTS OF ZOONOTIC DISEASES

Human Health

Human mortality resulting from emerging zoonotic diseases has been relatively low compared to other leading causes of death from infectious diseases, with the exception of the 1918 influenza pandemic and HIV/AIDS, a zoonosis that now transmits readily among humans. Between 2003 and

2009, there were 421 confirmed human cases of avian influenza A(H5N1), and as of April 23, 2009, 257 deaths were reported to the World Health Organization (WHO) (Figure 2-4). In contrast, between November 2002 and July 2003, 8,096 individuals were diagnosed with SARS, which resulted in 774 deaths (WHO, 2004). As shown in Table 2-1, none of the recent major emerging diseases has led to large fatality numbers. The number of people infected or number of fatal cases, however, are not the only concerns. Impacts on trade and movement of people, economic stability, and panic and societal disintegration based on perception of danger can be seriously disruptive to the global order.

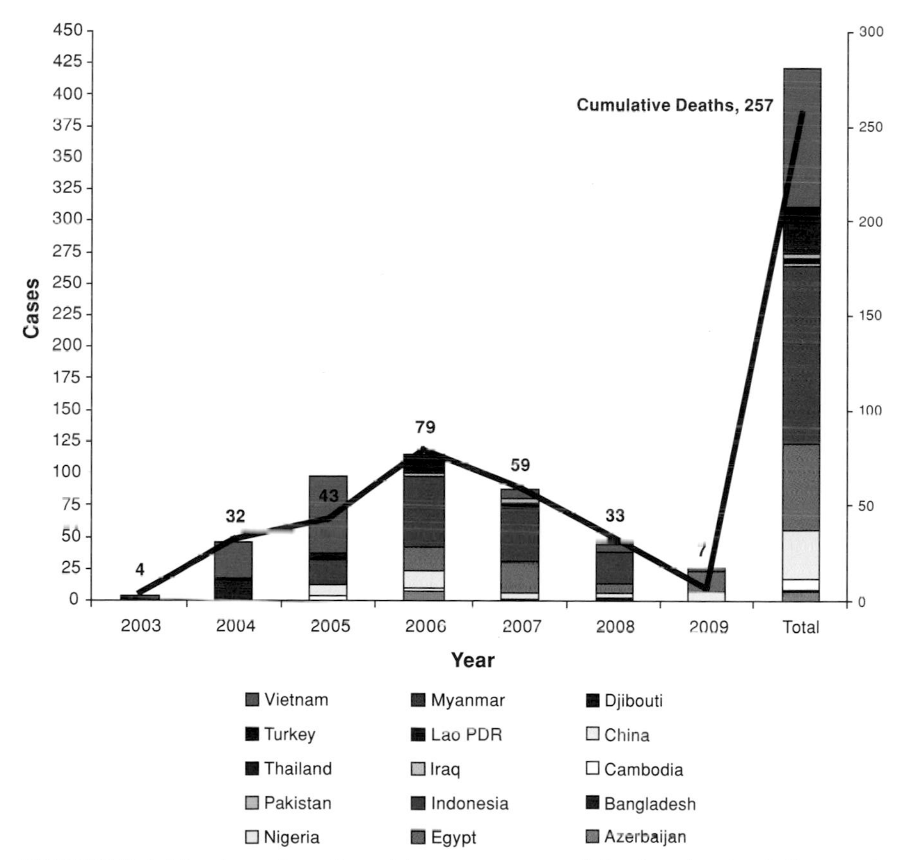

FIGURE 2-4 Number of confirmed human cases and deaths of avian influenza A (H5N1) reported to the World Health Organization by country and year. Confirmed cases (left axis) and cumulative deaths reported (right axis) as of April 23, 2009. SOURCE: WHO (2009).

TABLE 2-1 Selected Examples of Recent Zoonotic Outbreaks of International Significance

				Human Health Impacts			
Country	Disease	Period	Host	Cases	Fatalities	Animal Losses	Economic Impact
Malaysia	Nipah virus encephalitis	September 1998–April 1999	Swine, dogs, fruit bats	265[a]	105[a]	Swine: 1.1 million[b]	$617 million[b]
United Kingdom	Bovine spongiform encephalopathy (BSE)	1986–2009	Cattle	168 cases of vCJD[c]	164 cases of vCJD[c]	Cattle: 214,305[d] OTM: more than 8 million[e]	1986–1996: Direct costs $936 million per year; 1997–2000: $858 million per year[f]
United States	BSE	2003–2007	Cattle	3 cases of vCJD[g]	3 cases of vCJD[g]	Cattle: 3	$11 billion[h]
China, Taiwan, Hong Kong, and Singapore	Severe acute respiratory syndrome (SARS)	November 1, 2002, to July 31, 2003	Civets	7,667[i]	718[i]	N/A	$13 billion or 0.5–1.1 percent of GDP[j]; East Asian economies: 2 percent of regional GDP or US$200 billion[k]
Canada	SARS	November 1, 2002, to July 31, 2003	Civets	251[i]	43[i]	N/A	0.15 percent of GDP or C$1.5 billion[l]
Asia	Highly pathogenic avian influenza (HPAI)	January 24, 2004, to January 7, 2009	Poultry, wild fowl, and mammals	337[m]	222[m]	Birds: Estimated 250 million	Asia $10 billion (December 2003 to February 2006)[n]
Africa	HPAI	November 30, 2005, to January 7, 2009	Poultry, wild fowl, and mammals	56[m]	26[m]	N/A	N/A

Europe	HPAI	October 21, 2005, to January 7, 2009	Poultry, wild fowl, and mammals	0	0	N/A	N/A
Worldwide	Severe HPAI pandemic (estimate)	N/A	Poultry, wild fowl, and mammals	N/A	1–71 million[o]	N/A	Up to $3 trillion[o]
United States	West Nile virus fever	January 1999–December 2008	Birds and mosquitoes (vector)	28,975[p]	1,124[p]	Birds: >317 species have been infected; Horses: 23,755 cases	$400 million (1999–2007); Louisiana: $20.1 million (June 2002–February 2003)[q]
India	Plague	August–October 1994	Rodents	693[r]	52[r]	N/A	$600 million–$2 billion[s]
Kenya, Somalia, and Tanzania	Rift Valley fever	November 30, 2006, to May 3, 2007	Sheep, cattle, goats, water buffalo; mosquitoes (vector)	1,062[t]	315[t]	N/A	N/A

NOTES: N/A = not available; total = total cases worldwide; vCJD = variant Creutzfeldt-Jakob disease.

[a]FAO (2002); WHO (2007a).

[b]Estimate includes $35 million in compensation for the pigs destroyed; $136 million for the control program from the Department of Veterinary Services; $105 million in lost tax revenue from swine industry; $97 million for the 1.1 million pigs destroyed; $120 million due to loss of pig export trade; and $124 million for loss by pig farmers during the outbreak period (FAO and APHCA, 2002).

[c]Andrews (2009).

[d]Cattle slaughtered as a result of passive disease surveillance in Great Britain as of July 10, 2009 (Defra, 2009a).

[e]The Over Thirty Month (OTM) Rule bans meat from cattle aged over 30 months, which are more likely to have developed a significant amount of BSE agent in any tissue, from being sold for human consumption (Defra, 2009b).

[f]OECD and WHO (2003).

[g]The three cases of vCJD were acquired abroad (CDC, 2009).

continued

TABLE 2-1 Continued

[h]Losses in U.S. exports resulting from BSE-related restrictions (USITC, 2008).
[i]WHO (2004).
[j]Brahmbhatt and Dutta (2008).
[k]World Bank (2005).
[l]Darby (2003).
[m]WHO (2009).
[n]Based on survey of the United Nations Economic and Social Commission for Asia and the Pacific (Elci, 2006).
[o]McKibbin and Sidorenko (2006); Burns et al. (2008).
[p]CDC (2008).
[q]Zohrabian et al. (2004).
[r]CDC (1994).
[s]World Resources Institute, United Nations Environment Programme, United Nations Development Programme, and World Bank (1996); Cash and Narasimhan (2000); Gubler (2001).
[t]WHO (2007b).

BOX 2-1
Examples of the Underestimated Burden of Zoonotic Diseases

Rhodesiense sleeping sickness: According to this study, the actual mortality from sleeping sickness during an epidemic in southeast Uganda was approximately 12 times higher than reported. The authors considered that many sleeping sickness cases were likely to have been misdiagnosed as malaria in poorly resourced rural clinics and so were not properly treated; all such patients would have died (Odiit et al., 2005).

Rabies: These studies estimated that the actual incidence of human rabies in Tanzania was 10 times higher than reported through passive disease surveillance. Worldwide, the number of rabies deaths annually was estimated to be 32 times higher than the number reported to the World Health Organization (Fèvre et al., 2005; Knobel et al., 2005).

Leishmaniasis: The study reported that the actual incidence of visceral leishmaniasis in Bihar, India, was estimated to be 8 times higher than reported by passive disease surveillance (Singh et al., 2006).

Experience from past events and future projections based on contemporary events warn that low mortality is not a given for all disease events. The 1918 pandemic influenza virus killed tens of millions of people in a short time period, with estimates from 20 million to more than 50 million. Projections on the potential human losses from HPAI H5N1, should it attain a similar virulence as the 1918 virus, indicate that a severe pandemic of H5N1 virus could kill as many as 1 in 40 infected individuals or some 71 million (Barry, 2005; McKibbin and Sidorenko, 2006). Approximately 1 million individuals could die under a mild scenario (modeled after the Hong Kong influenza of 1968–1969), and 14 million under a moderate scenario (based on the characteristics of the 1957 Asian influenza) (McKibbin and Sidorenko, 2006). Looking at the same data, others suggest that as many as 180–260 million could die in a worst-case scenario (Osterholm, 2005). Furthermore, zoonoses can impose a significant human and animal health burden locally and, in many cases, that burden is underestimated (see Box 2-1).

Economic Impact

The economic impact of disease outbreaks depends on several critical factors, including public understanding and response, type of disease, and market scope. Measuring the economic impact of emerging zoonotic infections is complex because there are so many sources of losses and disproportionate impacts on different sectors and geographic regions (Kimball and

Davis, 2006). Table 2-1 provides estimated economic impacts associated with outbreaks for selected zoonotic diseases.

Emerging zoonotic diseases can cause economic losses as a result of morbidity and mortality among food animals, losses related to public interventions, and market losses at household, national, and global levels. Food-animal morbidity and mortality losses can be the result of the disease itself, or result from preventive actions such as culling of diseased, suspected, or at-risk animals. As of January 2009, 61 countries reported outbreaks of HPAI H5N1 in poultry, of which slightly more than half were developing countries. More than 250 million birds have died or been culled since the onset of the disease; however, this accounts for less than 1 percent of the 52 billion birds slaughtered annually. However, in Vietnam, which has implemented probably the most severe culling policy against HPAI H5N1, 50 million or 12 percent of the total annual poultry stock died or was culled, heavily impacting household and national economies.

Economic losses related to public interventions can be the result of efforts to prevent and eventually contain and eradicate the disease. Those efforts include quarantine and disease surveillance systems, hospital and medical services, and the cost and compensation for culling or eventual other losses experienced by the private sector. This can also include losses from unproductive "downtime" forced on affected poultry farms and measures to reduce human morbidity and mortality. During the SARS outbreak, 866 employees of the U.S. Centers for Disease Control and Prevention participated in the human and animal health response, totaling 46,214 person-days at a cost of well over $20 million in salary alone. This included deployments to 10 foreign countries and 19 domestic ports of entry (Marano, 2008). In the course of the 1994 outbreak of plague in India, trade and travel restrictions were imposed internally and externally, which led to economic impacts that shocked the region's stock markets with losses of nearly $2 billion (Price-Smith, 1998; Cash and Narasimhan, 2000; Gubler, 2001). That 1994 plague outbreak in India is described in more detail in Chapter 5. Similar travel and economic disruptions were seen with SARS: Figure 2-5 shows tourist arrivals in China and Thailand and compares the immediate impact of SARS with the 2004 Pacific Ocean Tsunami.

Losses through the market can result from changes in consumption patterns and trade, which directly affect prices and can last long beyond the period of risk. The spread of HPAI H5N1 caused international chicken prices to fluctuate in major poultry markets in Europe, Africa, and the Middle East (FAO, 2006). The EU's total ban of beef and cattle exports from the United Kingdom (UK) in March 1996 due to BSE (see Box 2-2) resulted in the loss of trade estimated at £700 million per year (DTZ Pieda Consulting, 1998; van Zwanenberg and Millstone, 2002; Kimball and Taneda, 2004).

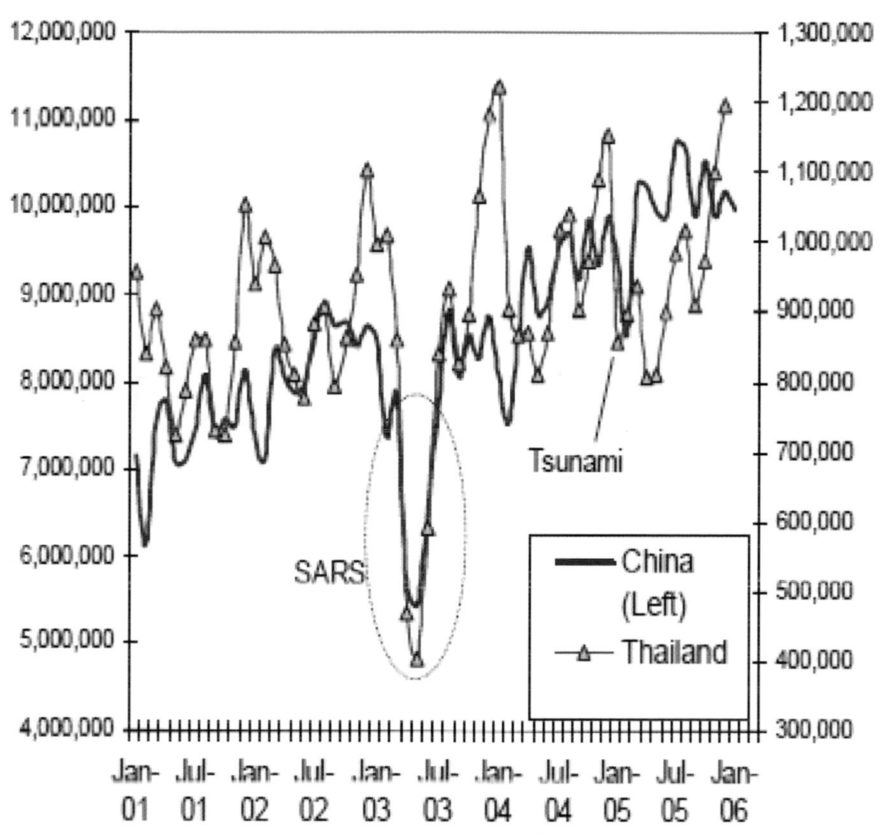

FIGURE 2-5 Tourist arrivals in China (left axis) and Thailand (right axis) between 2001–2006.
SOURCE: Brahmbhatt (2006). Reproduced with permission from the World Bank.

The combined economic impact of these losses indicates that outbreaks and epidemics of zoonotic diseases can cause short- and long-term economic consequences due to significant disruption of economic activities (Hanna and Huang, 2004). Detailed breakdowns of economic losses as described above are generally not available, but as shown in Table 2-1, total losses from emerging zoonotic diseases over the past two decades exceed $200 billion. Economic losses would be even higher if one had reached a severe pandemic scenario, which would amount to as much as 4.8 percent of global GDP (Burns et al., 2008). The serious economic effects of pandemic A(H1N1) 2009 have yet to be realized presuming there is a major global winter outbreak in the northern hemisphere. As shown in Figure 2-6, about

BOX 2-2
The Economic Impact of Bovine Spongiform Encephalopathy Outbreaks in the United Kingdom, the United States, and Canada

In 1986, the United Kingdom (UK) had a major outbreak of a novel disease in cattle, bovine spongiform encephalopathy (BSE) (Wells et al., 1987). By 1990, British scientists suggested a possible link between BSE and Creutzfeldt-Jakob disease (CJD); and thus, the UK government set up a new disease surveillance unit with the mandate to identify any change in the pattern of this disease that might be attributable to the emergence of BSE in humans. The existence of a novel variant of CJD (vCJD) was first reported in 1996. A series of experimental studies subsequently confirmed BSE transmissibility from animals to humans (The BSE Inquiry, 2000). The years it took for scientists to gather the necessary evidence to establish this linkage, however, delayed the introduction of measures to protect human and animal health.

The costs associated with the BSE outbreaks in UK cattle from 1986 to 1996 were reviewed by the BSE Inquiry, a committee created to investigate the response of the government to this animal disease. Based on this review, the public sector and ultimately the taxpayers bore the brunt of the economic consequences of BSE. Public expenditures due to BSE increased in the areas of biomedical research, compensation payments, and operational overheads incurred by different government agencies. From 1986 to 1996, the total expenditure on BSE-related research was £61 million, while other government expenditures, including compensation schemes and running costs, amounted to approximately £227 million (The BSE Inquiry, 2000). The private sector also suffered, particularly the production side of the beef industry and businesses (The BSE Inquiry, 2000). Before the European Commission introduced a ban of UK beef and cattle exports on March 27, 1996, the economic impact suffered by the beef- and cattle-related industries were relatively minor. The Inquiry concluded that the BSE-related costs suffered by farmers and businesses accelerated the decline of the industry's overall growth. The introduction of the 1996 ban resulted in the collapse of the industry that same year due to the loss of major export markets and related markets.

The United States and Canada suffered immense economic losses after BSE-infected animals were detected in 2003. In the United States, the value of U.S. beef exports dropped from $3.1 billion in 2004 to $2.5 billion in 2007 after the detection of a BSE-infected cow in December 2003. Net revenues declined by $1.5–2.7 billion per annum over the same period, resulting in a total loss to the sector of $11 billion USD (USITC, 2008). In Canada, the subsequent ban of Canadian beef and cattle imports by the United States and many other countries following the detection of a BSE-infected cow in May 2003 resulted in a drop in the value of beef and cattle exports of more than $1 billion in 2003, while domestic cattle prices fell 50 percent (FAO, 2006).

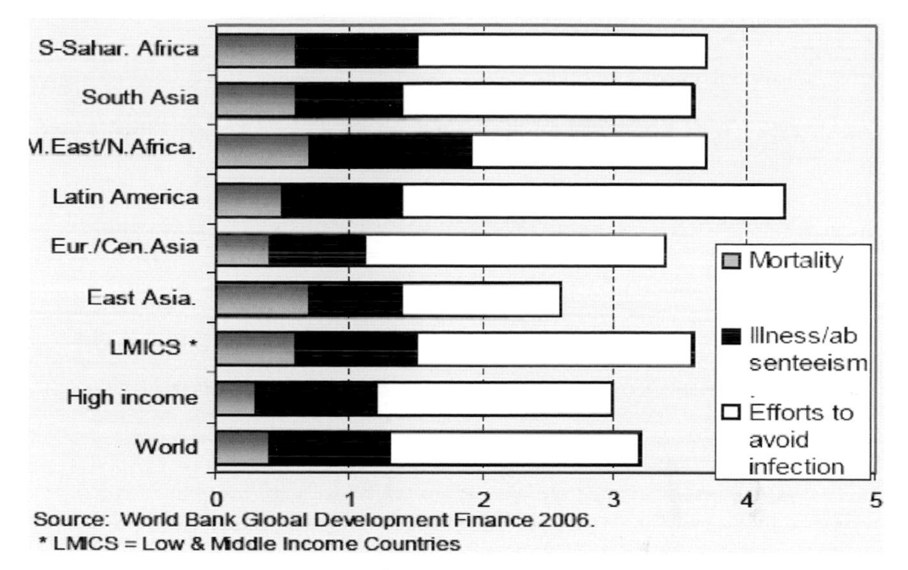

FIGURE 2-6 Economic impact of a potential human influenza pandemic by percentage of GDP (x-axis).
SOURCE: Brahmbhatt (2006). Reproduced with permission from the World Bank.

60 percent of the economic losses would be from efforts to avoid infection (e.g., minimizing face-to-face interactions). Although the economic impact estimates in the case of an influenza pandemic show a high mortality in humans, the largest impact might arise from the uncoordinated efforts of people to avoid infection and the economic losses resulting from the reduction in the size and productivity of the world labor force due to illness and death (Brahmbhatt, 2006).

Equity Impacts

In many of the least developed countries, both culling and the high mortality of birds have had a major impact on the livelihoods of poultry-dependent households. The poorest strata of rural households in developing countries derive a higher portion of their income from food-animal production than higher income households (de Haan et al., 2001). The importance of food-animal production for the poor is even more pronounced in poultry. In South Asian countries, more than 90 percent of flocks and 50–65 percent of birds are kept under an extensive "backyard" system. Village household surveys in Vietnam showed that income from the poultry sector was important for 99 percent of the poor households; losses because of death or

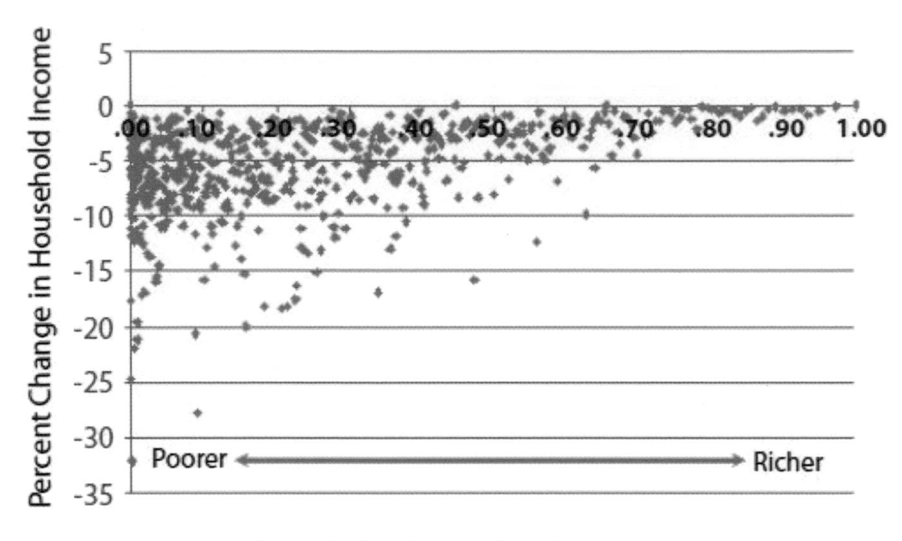

FIGURE 2-7 Household income and expenditure effects of a backyard poultry ban (percentage change in annual income).
SOURCE: Otte et al. (2006). Reproduced with permission from FAO.

culling of their flocks amounted to an average of $69 per household. A ban on poultry would cause losses of up to 30 percent of the income for the poorest households (see Figure 2-7). In Egypt, the poorest quintile of the population, with a monthly income of $35, earned 52 percent of their income from poultry, but suffered on average a loss of $22 from HPAI H5N1. Losses from emerging zoonotic diseases therefore disproportionately affect the poor (Otte et al., 2006).

DISEASE SURVEILLANCE TO MITIGATE EMERGENCY RESPONSE MEASURES AND COSTS

In a functionally integrated disease surveillance system for human and animal health, there are various opportunities for preventing, detecting, and responding to zoonotic disease emergence and transmission. Through early detection, a timely and effective response to zoonotic diseases in animal populations can prevent or minimize the likelihood of transmission to human populations (see Figure 2-8). After detecting a zoonotic disease event in either human or animal populations, surveillance data would inform human and animal health decisionmakers so they can plan, implement, and evaluate responses to reduce morbidity and mortality from zoonotic infections. Without capacity or willingness to activate an emergency response, surveillance merely occurs in a vacuum. Effective prevention and control of emerging zoonotic diseases require both disease surveillance and emergency response capabilities that include disseminating and communicating actionable disease surveillance information to officials who have the authority, motivation, and capability to implement a response. The relationship between disease surveillance and emergency response is typically in the size and efficacy of the two efforts: The more effective and timely the disease surveillance, the more likely it is to avert a relatively large emergency response. Large and effective surveillance programs will detect the first sign of a problem, then, if the actionable information is supplied to the proper authorities, a relatively small and targeted emergency response may effectively curtail spread and mitigate the threat. On the other hand, small and inadequate surveillance programs are likely to miss many new disease events, so by the time the disease is recognized, a much larger emergency response is necessary.

Surveillance information on zoonotic diseases in humans and animals, however, is highly variable under different scenarios, making the response to these zoonotic threats also variable. Box 2-3 and Appendix B provide some examples of the imbalance in the surveillance-response dynamic. It is also important to recognize that the threshold of detection will vary with the capacity of the laboratory. For instance, a newly emerged agent may be readily identifiable through basic technology widely available, such as bacterial culture of *Escherichia coli* O157:H7. A slightly more sophisticated laboratory, with the capability of embryonated egg inoculation, may be able to identify a new strain of highly pathogenic avian influenza. Identification of a disease entity such as BSE, which requires advanced technology such as Western blotting or immunohistochemistry, will be beyond the capacity of most laboratories, even if surveillance for other more easily detectable agents is extensive.

Using current approaches, the cost of emergency response is usually several times greater than the cost of disease surveillance. The more widespread the disease is before detection and implementation of response, the larger the cost of the control measures. Moreover, the case of HPAI H5N1 in Vietnam underscores the importance of continuous surveillance of this virus to prevent subsequent waves of outbreaks (see Appendix B). As discussed in more detail in Chapter 6, the investment in a well-functioning global disease surveillance system and in early response capability is roughly estimated to amount to about $800 million per year, whereas the economic losses from emerging, highly contagious zoonotic diseases have reached more than $200 billion over the last decade.

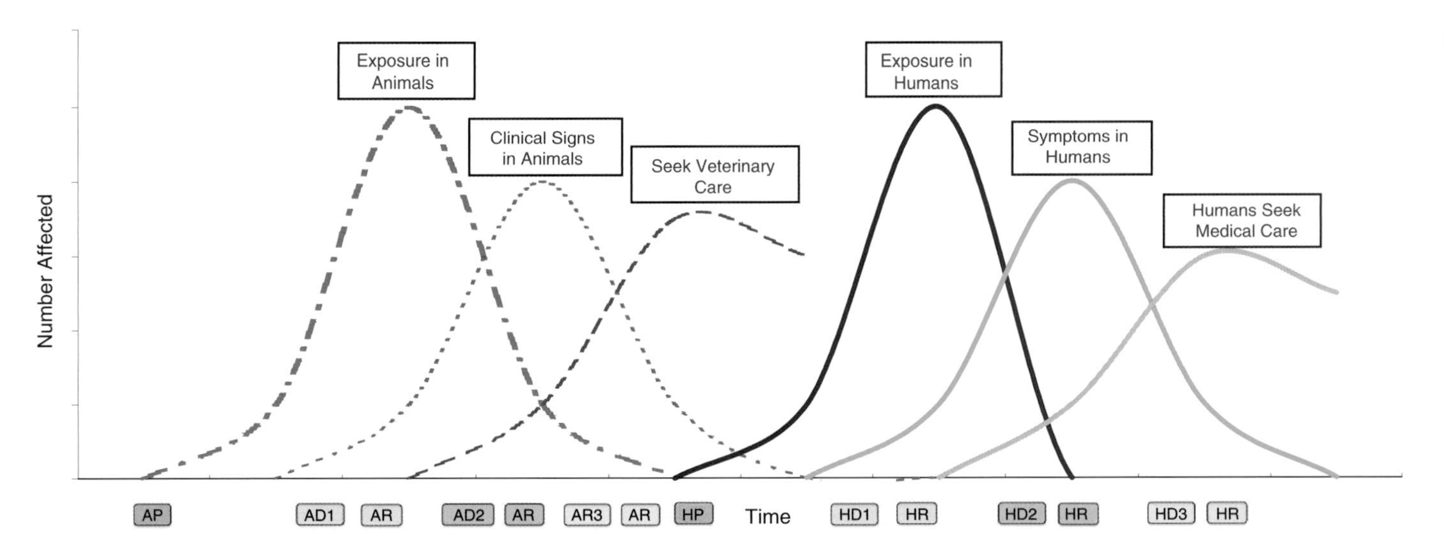

FIGURE 2-8 Opportunities to prevent, detect, and respond to the emergence and transmission of zoonotic diseases.
NOTES: The graph is a stylized representation of a zoonotic disease outbreak and is not meant to represent any specific infectious agent. The timeline along the X-axis indicates when during the outbreak various opportunities to intervene in animal and human populations occur, and is expanded to facilitate illustration.
AP = Preventing the emergence of zoonotic diseases in animal populations. Examples would include vaccination programs (rabies, avian influenza, other); effective live animal market sanitation and other management policies to prevent mixing of species; sanitation including manure and rodent/pest management on farms; biosecurity measures; testing and treatment programs of companion and stray animals for zoonotic diseases (e.g., *Echinococcus granulosus*); Hazards Analysis of Critical Control Points procedures in food systems; and regular preventive visits to veterinarians for companion animals.
AD = Detection of zoonotic diseases in animals:

- AD1, detection before clinical signs occur would result from testing animal specimens at diagnostic laboratories in ongoing serosurveillance programs, or in testing food animals at slaughter.

- AD2, detection in animal populations after clinical signs occur at the local level, by farmers, food-animal production workers, household flock and companion animal owners, and the public (e.g., wild animal die-offs).
- AD3, detection in animal populations after veterinary care is sought would occur with veterinarians and or animal health workers diagnosing and reporting disease.

AR = Response to control disease transmission in animal populations. This would include rapid outbreak investigation to determine etiologic agent, risk factors for spread, and points for control. Other response measures would include vaccination of animal populations in the face of an outbreak, testing and slaughter of food-animal populations, mass depopulation programs, strengthening biosecurity programs, and treating and curing zoonotic disease infection in household animals.

HD = Detection of zoonotic diseases in human populations:

- HD1, detection in humans before symptoms occur. This would include detection from screening programs, serosurveillance, etc.
- HD2/HD3, detection in humans after symptoms occur and before or at the time medical care is sought. Detection would occur by patient report, findings by village healthcare workers, at local health clinics, at district hospitals, at tertiary hospitals, by healthcare providers, and through laboratory testing.

HP = Preventing the transmission of zoonotic disease agents to human populations. This would include effective use of hand washing, use of personal protective equipment in abattoir workers, animal disease control personnel, and other people in regular close contact with animal populations; use of universal precautions and personal protection equipment by healthcare workers; vaccination programs; safe food preparation at the household level; restrictions on the importation, movement, and ownership of exotic pets; and regular prevention check-ups of humans in close contact with animals and animal populations.

HR = Response to control zoonotic disease in human populations. This would include outbreak investigation to identify etiologic agent, risk factors, points for control, treatment at diagnosis, instituting isolation and quarantine measures, targeted disease surveillance and vaccination programs, mass vaccination programs, targeted antimicrobial administration, and mass antimicrobial administration.

BOX 2-3
Selected Examples of the Balance and Imbalance Between Disease Surveillance and Emergency Response for Past Outbreaks

- **Limited surveillance not detecting a new disease, and once detection occurred, linkage with control is slow, so that emergency response is futile because it is so widespread:** HIV emerged in central Africa in the 1970s. Because of inadequate disease surveillance, authorities did not realize this was an emerging problem. The lack of recognition, combined with the long incubation period, allowed this disease to spread globally, so that it soon became the foremost infectious disease in many parts of the world. Then once recognized and associated with marginalized populations (homosexuals and drug abusers), effective control measures were slow to develop. Had early recognition occurred and been combined with effective controls, there could have been an effective global emergency response that might have prevented the majority of human morbidity and mortality.
- **Example of surveillance detecting a new disease locally, but without actionable information shared regionally and globally so that when the global spread of disease occurs, a global emergency response is necessary and very costly:** Severe acute respiratory syndrome (SARS) emerged in China, was not diagnosed until it moved to Hong Kong, and affected visitors from multiple continents. The issue for SARS was the lack of actionable information at early stages of the outbreak. Disease surveillance at the local level may have been effective, but the information did not reach the level required to implement a timely global emergency response. By the time it was recognized globally as a serious emerging health threat, emergency responses on several continents had to be activated.
- **Example of surveillance detecting a disease, but then no follow-through with appropriate emergency response, so the disease continues to spread:** In 2004, disease surveillance for highly pathogenic avian influenza (HPAI) H5N1 in Southeast Asia highlighted the presence of the HPAI H5N1 strain in chickens and its association with human mortalities. There were two problems here. First, disease surveillance detected the disease in humans and poultry, but only after the

Therefore, an effective global disease surveillance system can be expected to reduce the emergence of zoonotic diseases in humans and provide early detection of zoonotic diseases in livestock, thereby reducing billions in economic losses. In most emerging zoonoses, if the disease had been recognized much earlier (as would happen with well-functioning disease surveillance systems), effective emergency responses, if any, would have been smaller and cost effective. However, global disease surveillance systems have not been adequate to detect disease in timely fashion and limit impact, so more often than not massive and expensive emergency responses have been required.

disease had been observed in another region of the world. An emergency response was instituted that was weighted more toward surveillance and control in human populations rather than in poultry populations, thus allowing for continued spread and circulation in poultry. Second, because of the lack of integrated human, poultry, and wildlife expertise, considerable time was needed to identify disease transmission mechanisms. In the meantime, the virus continued to circulate, and eventually spread across Asia into Europe and Africa. Several countries improved their disease surveillance system after the first outbreak. For example, after two waves of HPAI H5N1, Thailand mounted an impressive disease surveillance system based on human and animal health village volunteers and about 1,000 joint (Ministries of Health and Agriculture) District Surveillance and Rapid Response teams, which has probably kept the third wave of outbreaks much more localized. Vietnam, after an initial delay in the reporting of the disease, also developed a community-based animal healthcare worker system for early alert, which has proven to be effective.

- **Example of good initial surveillance finding a disease, but delayed understanding of the disease epidemiology, then emergency response mounted is effective:** A new disease caused by Nipah virus surfaced in Malaysia in 1999. In this case, disease surveillance highlighted the presence of a neurological disease in pig farmers. The disease was initially misdiagnosed as Japanese encephalitis. After some delay, the true causative agent, Nipah virus, was identified and linked to infected swine, leading to the culling of 1.2 million pigs. It took longer to identify the fruit bat reservoir and the presence of fruit trees on the pig farms as a predisposing factor, and major economic losses could have been prevented. Another example is human monkeypox in prairie dogs in the United States in 2003. Detection of the zoonotic hazard was quickly followed by emergency responses to contain the threat. Both of these examples are from countries with advanced economic and healthcare systems, so both disease surveillance and emergency response were effective.

NOTE: For further details on surveillance and response of select zoonotic disease outbreaks, see Appendix B.

The reality is that procuring funding for large, expensive emergency response measures is easier than funding continual disease surveillance for detecting future and unknown diseases. This is unfortunate because a well-designed emerging zoonotic disease surveillance system is what will ultimately result in less human morbidity and mortality and fewer adverse economic impacts globally. It is widely recognized that emergency response is essential. Yet it is penny-wise and pound-foolish to continually invest in large emergency responses without investing in effective disease surveillance systems that would lead to smaller, less costly control efforts.

UNDERSTANDING ZOONOTIC DISEASE AGENTS AND TRENDS TO PREDICT ZOONOTIC DISEASE EMERGENCE

To accurately predict and detect when and where zoonotic pathogens might emerge, it is important to understand the biological pathways affecting their emergence. Data gathered from disease surveillance systems are crucial, enabling scientists to predict how and when pathogens may emerge and the extent of their spread and impact. This information allows decision-makers to more confidently allocate resources to prevent outbreaks from occurring. If a zoonotic disease outbreak should arise, such data become even more critical for informing effective control and response measures.

The Biology of Pathogen Emergence

Of approximately 1,400 species of human pathogens that are now recognized, more than 800 (nearly 60 percent) are known to be zoonotic (Woolhouse and Gaunt, 2007). Moreover, many nonzoonotic pathogens are known or believed to have origins in nonhuman animals (Table 2-2). Some of these have only recently emerged (e.g., HIV/AIDS, pandemic strains of

TABLE 2-2 Examples of Human Pathogens with Evolutionary Origins in Nonhuman Hosts

Disease	Pathogen	Original Host
AIDS	Human immunodeficiency virus-1	Chimpanzees
AIDS	Human immunodeficiency virus-2	Sooty mangabeys
SARS	SARS coronavirus	Bats/palm civets
Malaria	*Plasmodium falciparum*	Probably birds
Malaria	*Plasmodium vivax*	Asian macaques
Sleeping sickness	*Trypanosoma brucei* subspp.	Wild ruminants
Diphtheria	*Corynebacterium diphtheriae*	Probably domestic herbivores
Hepatitis	Hepatitis B virus	Apes
Viral lymphoma	Human T-lymphotropic virus-1	Primates (possibly Asian macaque)
(Unknown)	Human T-lymphotropic virus-2	Bonobos
Respiratory infection	Human coronavirus OC43	Bovine
Influenza	Influenza A virus	Wildfowl
Measles	Measles virus	Sheep/goats
Mumps	Mumps virus	Mammals (possibly pigs)
Smallpox	Variola virus	Ruminants (possibly camels)
Typhus	*Rickettsia prowazeckii*	Rodents
Plague	*Yersinia pestis*	Rodents
Dengue fever	Dengue fever virus	Old World primates
Yellow fever	Yellow fever virus	African primates

SOURCE: Adapted from Wolfe et al. (2007).

influenza A), and others have origins going back thousands or millions of years (e.g., plague, malaria).

Zoonotic disease emergence from nonhuman animals may be viewed as a series of steps from primarily animal diseases, such as rabies that occasionally are transmitted to humans, all the way to diseases originating in animals, such as HIV-1 that jumped species to humans and successfully transmitted from human to human without further involvement of the original animal host. There are five stages in this "pathogen pyramid" wherein the barriers to pathogens progressing from one stage to the next are both biological (functional constraints, often at the molecular level, to infection and transmission) and ecological (restricted opportunities to infect humans or transmit between humans) (Wolfe et al., 2007; see Figure 2-5 in IOM and NRC, 2008). Overcoming these barriers may involve evolution of the pathogen, although increased opportunities to infect or transmit between humans can arise purely from changes in human behavior or demography (e.g., intensification of food-animal production and increased trade of exotic species—see Chapter 3) or from changes in pathogen ecology (e.g., altered distribution of the reservoir host or vector). The example of HIV-1 suggests that a pathogen can rapidly progress through the stages of the pyramid over time scales of decades. High variability in virus genomes might generate high functional diversity, producing human-infective variants on a regular basis, some of which successfully "take off" in human populations (Woolhouse and Antia, 2008).

Pathogen Discovery

Analysis of emerging diseases from 1940 to the present demonstrates that the rate of emergence "events" rose significantly over this period (Jones et al., 2008) after correcting for trends in disease surveillance effort. The discovery of new human pathogen species continues at a rate of 3–4 species per year (see Appendix C). The discovery of new human pathogens has three components: (1) recognition of pathogens that have existed in humans for a long time, but have just been detected (e.g., hepatitis C); (2) pathogens that have existed for a long time, but have only recently had the opportunity to infect humans (e.g., Baboon cytomegalovirus); (3) newly evolved human pathogens that did not previously exist (e.g., pandemic A(H1N1) 2009 virus as a relatively recent example in humans; canine parvovirus as an animal example). Pathogens of all three kinds continue to be discovered.

The majority of recent discoveries of new human pathogens are viruses (see Figure 2-9 and Appendix C) (Woolhouse et al., 2008). The discovery rate of human non-virus pathogens is much slower and mainly involves rickettsia and microsporidia. There is every reason to expect current trends

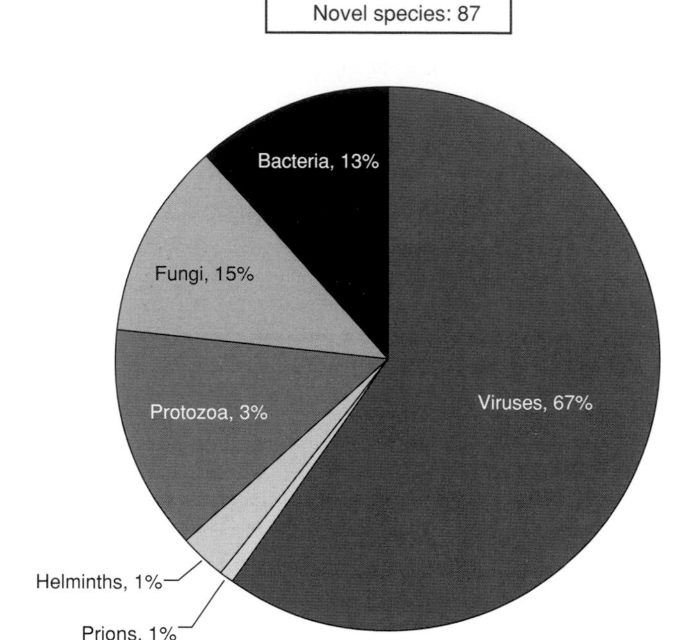

FIGURE 2-9 Patterns of pathogen discovery. Percentage of novel pathogen species by type.
SOURCE: Adapted from data by Woolhouse and Gaunt (2007).

of virus discovery to continue in the immediate future (Woolhouse et al., 2008). Although the rate of virus discovery has historically been remarkably consistent since the advent of tissue culture, the introduction of new technologies such as polymerase chain reaction and the advent of high-throughput sequencing has led to a substantial increase in the global capacity to identify novel pathogens. That, coupled with a great deal of interest in pathogen discovery, makes it possible that the rate of discovery, particularly of viruses, will accelerate as new efforts are made through surveillance programs.

The majority of newly discovered human pathogens are either zoonotic or have recent origins in nonhuman reservoirs. Most are associated with other mammalian hosts, a few with birds, and only rarely with other classes of vertebrates. The mammalian taxa most commonly associated with new zoonoses are ungulates, carnivores, and rodents. These patterns are similar to the known zoonoses; in other words, we share our new pathogens with the same kinds of reservoir with which we have always shared our pathogens (Woolhouse and Gowtage-Sequeria, 2005).

Many recent high-profile emerging zoonoses have spilled over from wildlife hosts to humans. The rate of emergence of these wildlife-origin zoonotic diseases also appears to have increased significantly over the past six decades, and pathogens of wildlife origin represent the majority of emerging pathogens in the 1990s (Jones et al., 2008). Animal susceptibility studies performed in laboratories worldwide in collaboration with WHO, the Food and Agriculture Organization of the United Nations (FAO), and OIE quickly identified a novel coronavirus as the etiological agent that caused the 2003 SARS outbreak. Moreover, these studies revealed that a variety of wild and domestic animals were harboring this agent (WHO, 2003).

Data Limitations and Information Gaps

The committee identified several issues in terms of the data limitations. First, monitoring is subject to massive ascertainment biases. There are vast differences in the efforts invested in different places and at different times, leading to important gaps in information whether at the level of species discovery, emerging disease "events," or disease outbreaks in humans. Adjusting for this bias is difficult. One-third of emerging disease events are reported from the United States, 10 times as many compared to China, India, Brazil, and other hotspot countries (see Figure 2-10), and that seems unlikely to represent the frequency of emerging events in these countries. Second, monitoring is ad hoc, not systematic, and is partly driven by responses to the most recent events (e.g., clusters of discoveries in eastern Australia; spate of discovery of coronaviruses following the SARS outbreak) and partly by availability of detection and identification technologies. Third, determining the number of pathogens that have not yet been identified or detected in mammalian and other reservoir hosts is difficult. The inventory of species pathogenic to humans is incomplete but still growing (Woolhouse and Gaunt, 2007). The inventory of species pathogenic to major domestic food-animal species, plus cats and dogs, is also incomplete. In addition, there is very limited knowledge of the pathogens for the vast majority of other mammal species, let alone birds or other vertebrates (Dobson and Foufopoulos, 2001). Fourth, questions on the frequency with which humans are exposed to animal pathogens (so-called "chatter") and what fraction of those exposures are capable of crossing the species barrier to cause human infection remain unanswered. Fifth, the determination of what constitutes a species barrier and what characteristics allow pathogens to overcome it (e.g., pathogen evolution, immunosuppression, new transmission routes) are important issues that need to be addressed. And sixth, whether a human infection that resulted from exposure to an animal pathogen can be transmitted (directly or via an indirect route) to another

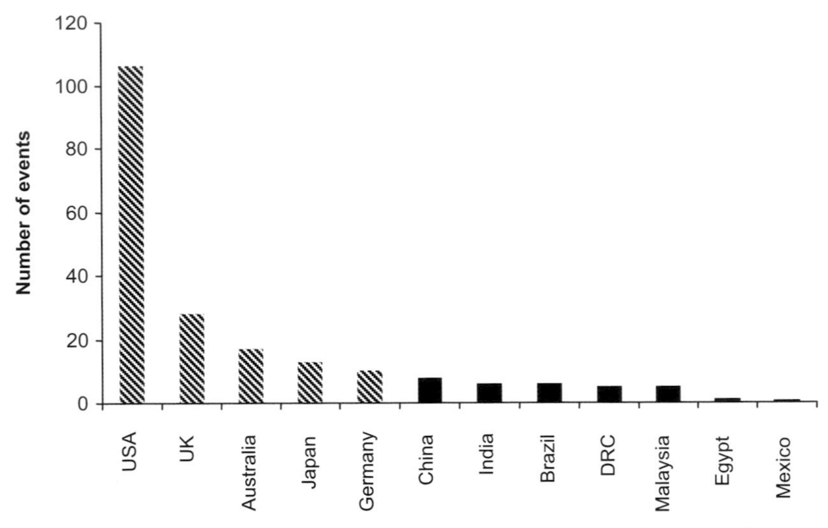

FIGURE 2-10 Patterns of reporting of emerging disease "events": five countries reporting the highest number of "events" (left) and selected others (right).
SOURCE: Woolhouse (2008a). Reproduced with permission from Macmillan Publishers LTD: *Nature*.

human is simply not known for hundreds of pathogen species (Taylor et al., 2001).

Domestic Animals and Wildlife Surveillance to Predict Zoonotic Disease Emergence

Given the desire to more effectively predict where the next zoonotic disease will emerge, there are many gaps in knowledge of potential emerging pathogens amidst the evidence of continuing events, underscoring the need for active disease surveillance in animal reservoirs for known zoonoses including domestic animals and also wildlife wherever possible. Improved disease surveillance is particularly important where the protection of human health depends wholly or partly on measures taken to prevent disease emergence or control disease in the reservoir (e.g., BSE, rabies, African sleeping sickness) and where the risk of outbreaks in humans is largely determined by the epidemiology of infection in the reservoir (e.g., Nipah virus, WNV, hantaviruses, plague). In addition, human resources and field capacity need to be developed to be able to conduct surveillance for zoonotic pathogens in animal reservoirs that often can be difficult to reach. Improved human resources and field capacity will greatly improve capacity to detect novel and emerging zoonoses.

Statistical Analysis, Modeling, and Predicting Future Trends

Once a zoonotic pathogen has emerged and been identified, surveillance data are critical for enabling researchers to predict the extent and magnitude of the outbreak. Statistical methods are needed to make reliable inferences and hypothesis testing from epidemiological findings and approaches (Jewell, 2003). The use of such analyses and disease models can better inform decisionmakers on how to effectively respond to disease outbreaks early on.

Statistical analysis and dynamical modeling have a long history of providing insights into the importance of infectious diseases and their transmission dynamics, beginning when Daniel Bernoulli modeled smallpox transmission in 1760 (Bernoulli, 1766). With dramatic increases in both computational power and detailed data on human and animal diseases in recent years, statistical analyses and quantitative studies have been undertaken in the wide range of issues related to zoonoses. These analyses and modeling utilize data from a variety of sources, including those from surveys (e.g., Easterbrook et al., 2007), from routine sentinel disease surveillance, and from detailed experiments with randomized treatments to identify and characterize key features of the epidemiological system. An example is a study of the use of antibiotics in food-animals to reduce bacterial illnesses in animals, thereby reducing subsequent human illness, with an associated risk of selecting for antibiotic-resistant bacteria, which could make food associated human infections harder to treat (Singer et al., 2007). Synthesis of statistical and mathematical methods has allowed transmission models to be based on robustly estimated parameter values. However, most modeling studies have been limited in scope to one host and one pathogen, even though most pathogens have multiple hosts (Woolhouse et al., 2001). A good example of multihost modeling is the rabies study in the Serengeti ecosystem of Tanzania (Lembo et al., 2008).

Uses of Statistical Analysis

Key statistical principles include those of quantitative hypothesis testing, parameter estimation (with corresponding measures of parameter uncertainty), and model fitting/criticism. Specific statistical methods have been developed to allow the integrated analysis of data sources that vary in source, type (e.g., combining retrospective studies of known outbreaks and disease surveillance of key disease events) (Burkom, 2003), and quality (rigor and relevance) (Turner et al., 2009). The analysis of all the relevant evidence relating to a particular disease can, however, lead to highly complex probability models. In such cases, particular care must be paid to model criticism and the detection of inconsistent or conflicting evidence

(Presanis et al., 2008), taking into consideration the assumptions of the model(s) underpinning the analyses.

A limitation of traditional statistical modeling and analysis (e.g., regression, survival analysis, and analysis of contingency tables) is that the insights are typically limited to comparisons and quantification of association rather than giving insights into the often complex mechanisms underlying the observed epidemiological patterns. Good models are difficult because epidemic diseases and especially emerging epidemic diseases are multisystem, dynamic, nonlinear, stochastic processes. Models for causal inference were developed to overcome some of these limitations (Holland, 1986). There are several examples from recent emerging infectious disease investigations (including Hendra virus, Nipah virus, coral diseases, and avian influenza) where techniques designed to infer causation—including epidemiological causal criteria, strong inference, causal diagrams, model selection, and triangulation—were successfully applied (Plowright et al., 2008).

Uses of Dynamical Modeling

Dynamical models of disease transmission are those developed to represent underlying epidemiological (and sometimes demographical) processes. Four main aims of such modeling have been identified (Anderson, 1988; Massad et al., 2005): (1) Enhancements to the logic and specification of current theories and concepts relating to disease transmission; (2) Generation of new testable hypotheses through computer program-based (so-called *in silico*) experiments or simulation processes; (3) Prediction of the future course of an epidemic and/or the impact of preventive measures; and (4) Identification of types of epidemiological data needed to refine understanding of disease epidemiology and/or make better predictions.

On the basis of the particular aims of the exercise, models are sometimes applied retrospectively to interpret historical epidemiological data and are sometimes used prospectively to generate predictions. In practice, retrospective analysis often provides the basis for predictive modeling (see Box 2-4). Examples of retrospective or historical modeling of emerging zoonoses include analysis of both the recent past (e.g., modeling analysis of recent Ebola outbreaks [Chowell et al., 2004; Ferrari et al., 2005; Legrand et al., 2007]) and the more distant past (e.g., modeling of 1918 influenza pandemic [Mills et al., 2004; Sertsou et al., 2006; Vynnycky et al., 2007]).

Predictive modeling is used to evaluate future scenarios as more or less likely, and to explore the possible benefits and/or risks of alternative realities. These alternatives could include alternative disease surveillance efforts (e.g., increased testing of live cattle for *M. bovis* or increased efforts to detect bovine tuberculosis in slaughtered cattle); various possible culling policies designed to reduce future disease incidence; and alternative policies aimed at controlling or eradicating disease. Predictive modeling for disease emergence is a difficult and complex challenge. Although some data on which to model do exist, the biological and ecological characteristics needed for an outbreak to occur are unknown. Therefore, to improve the science behind any effort in modeling emergence, particularly of pathogens, it seems axiomatic that hypotheses need to be generated and data gathered to either strengthen and support or refute and abandon the premise being studied. The prospects of successfully predicting emergence events would be greatly enhanced by systematic data collection on the patterns of presence and prevalence of infectious agents in animal populations, which means developing and implementing a systematic, ongoing integrated disease surveillance program that is global in scope. A longitudinal study of the underlying factors driving disease emergence, including those associated with animal production systems and climate change, could provide valuable information to such a program. To inform such a study, the pairing of complex mathematical models with remote sensing data could be useful to correlate environment with disease outbreaks and more accurately predict future disease events (Ford et al., 2009).

Mathematical models have also been developed and deployed during ongoing epidemics to help advise control policies. Such "real-time" modeling presents a number of challenges, including rapid collection and communication of input data, validating the process of model development, and generating formal estimates of model parameters from initially sparse data, noting that rigorous methods to fit such models to data are more complex and computationally burdensome than those required for traditional statistical models. Even so, real-time modeling can inform the management of an epidemic: Examples include the 2001 epidemic of foot-and-mouth disease in the UK, the 2003 global SARS epidemic, and the 2009 influenza A(H1N1) pandemic.

Projecting into the Future

Projections are defined as "the numerical consequences of the assumptions chosen. The numbers are conditional on the assumptions being fulfilled" (Keyfitz, 1972, p. 347). In the context of infectious disease, these assumptions could take the form of "if" circumstances: a closed population of a particular size, number of people encountering (and potentially infecting) each other randomly at a particular rate, and the introduction of a person infectious with a disease of a certain transmissibility into a fully susceptible population. Possible epidemic scenarios could be described, although the resulting incidence of disease on any subsequent day could not be derived with certainty due to random chance. Projections provide useful

BOX 2-4
Simulation of Human Influenza Transmission in Thailand

Using detailed demographic data (including population distribution and household size) and newly derived parameter estimates from reanalysis of historic data (including U.S. and UK 1918 pandemic mortality data), Ferguson and colleagues (2005) simulated human influenza transmission in Thailand to evaluate the potential effectiveness of targeted mass prophylactic use of antiviral drugs and social distancing to contain influenza. Figure **a** shows the time sequence (in days) of an epidemic, with spreading in a single simulation of an epidemic with R_0 = 1.5. Red indicates presence of infected individuals, and green indicates the density of people who recovered from infection or died. Figure **b** shows the daily incidence of infection over time for R_0 = 1.5 in the absence of control measures. Thick blue lines show the average for realizations resulting in a large epidemic; grey shading represents 95 percent confidence limits of the incidence time-series. Multicolored thin lines show a sample of realizations, illustrating a large degree of stochastic variability.

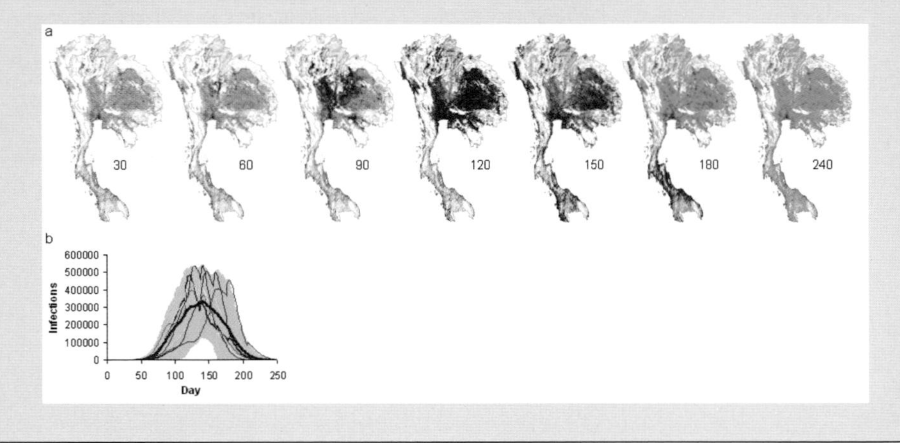

information to predict (or forecast) the future, insofar as the assumptions (from model structure to parameter values) are realistic.

Traditional statistical methods (most often time-series and regression) are sometimes used to provide short-term predictions of infectious disease incidence, quantifying past trends, and projecting them forward (see Box 2-5). Temporal, seasonal, and spatial trends were quantified along with temporal correlation to predict the incidence of meningococcal disease in France, and the model was based solely on trends observed in the detailed incidence data available (Knorr-Held and Richardson, 2003). An alternative approach is to predict incidence based on risk factors previously observed to be associated with incidence rates. For example, having previously

A similar study was published simultaneously in *Science* (Longini et al., 2005). The World Health Organization issued the following statement:

> The models provide additional information which will help WHO and public health officials in our Member States to improve pandemic influenza preparedness planning. . . . Several countries have already purchased stockpiles of antiviral drugs and WHO has taken steps to establish an international stockpile. . . . If we have a chance to reduce the scale of a pandemic with antivirals and other public health measures, the success of these interventions will depend on effective disease surveillance and early reporting in risk-prone countries. Before any stockpile can be used effectively, both must be strengthened. (WHO, 2005a)

These influenza studies offered the authors' most plausible set of transmission scenarios in order to inform policymakers, along with other available evidence. The next decisions are how much effort and what type to invest in planning for a serious future human and animal health crisis.

Surveillance data are critical to underpin estimation of key epidemiological parameters, which in turn determine which transmission scenarios are most plausible.

SOURCE: Ferguson et al. (2005); WHO (2005a).

shown an association between weather conditions and the presence of St. Louis encephalitis hemagglutination inhibition antibodies in wild birds, a hydrology model and a logistic regression model were combined to predict the incidence of human cases of St. Louis encephalitis, and these predictions were found to perform well looking 2 to 4 months ahead (Shaman et al., 2003, 2006).

The predictions from transmission models under different scenarios can be compared to inform debate about the potential consequences (both risks and benefits) of alternative courses of action. In this context, mathematical models have the advantage of transparency, since the basis for making predictions (for example, about the impact of control measures) is

BOX 2-5
Predicting an Outbreak

Anyamba and colleagues (2006, 2007) observed that sea surface temperatures in the equatorial east Pacific ocean increased anomalously during July to October 2006, indicating El Niño conditions. Such conditions previously had been associated with excess rainfall in East Africa. Such rainfall was predicted to give rise to the normalized difference vegetation index (NDVI), and a Rift Valley fever (RFV) model, based on the NDVI data, indicated that in October to December 2006 there would be an elevated risk of RVF in northern Kenya, central Somalia, and subsequently Tanzania.

Based on these results, early warning advisories were issued by the Food and Agriculture Organization of the United Nations and the World Health Organization to alert countries' authorities in early November 2006 of the elevated risk of RVF outbreaks (WHO, 2007c; Anyamba et al., 2009). On this basis "the [U.S.] Department of Defense–Global Emerging Infections Surveillance and Response System and the Department of Entomology and Vector-borne Disease, United States Army Medical Research Unit–Kenya initiated entomological surveillance in Garissa, Kenya, in late November 2006, weeks before subsequent reports of unexplained hemorrhagic fever in humans in this area" (Anyamba et al., 2009, p. 957).

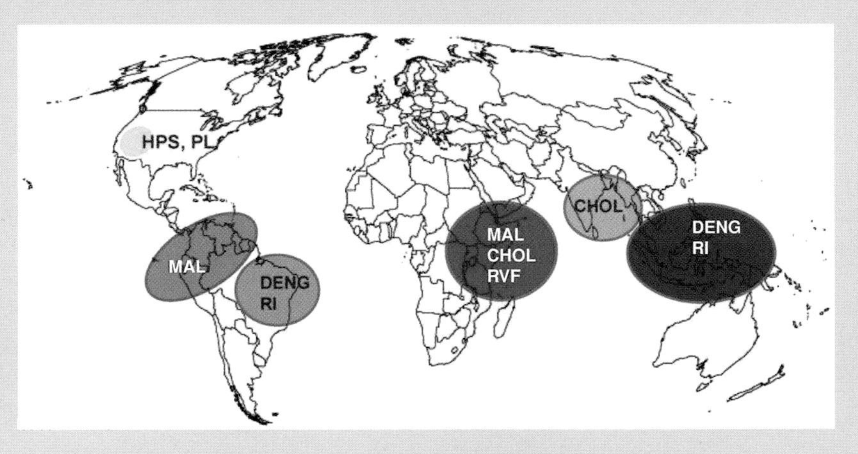

Hotspots of potential elevated risk for disease outbreaks under El Niño conditions, 2006–2007.

available for inspection, criticism, and change (Woolhouse, 2008b). Often, models will be the best evidence we have to inform decisionmaking. Models can also be used to gain insight into situations where an intervention was implemented and an unexpected result was obtained. As with any modeling exercise (other factors being equal), a model that has been shown to

SOURCE: Anyamba et al. (2006).

The index case was a patient in Kenya who experienced the onset of symptoms on November 30, 2006 (CDC, 2007; WHO, 2007c), and the Kenyan cases peaked in late December. From November 30, 2006, to March 12, 2007, 684 cases were reported in Kenya, including 155 deaths; 114 cases were reported in Somalia, including 51 deaths; and 290 cases were reported in Tanzania, including 117 deaths (WHO, 2007c).

The model's successful prediction of the epidemic enabled the affected countries to be forewarned of the increased risk (Kaplan, 2007). "The early warning enabled the government of Kenya, in collaboration with the World Health Organization, the United States Centers for Disease Control and Prevention, and the Food and Agriculture Organization of the United Nations to mobilize resources to implement disease mitigation and control activities in the affected areas, and prevent its spread to unaffected areas" (Anyamba et al., 2009, p. 957).

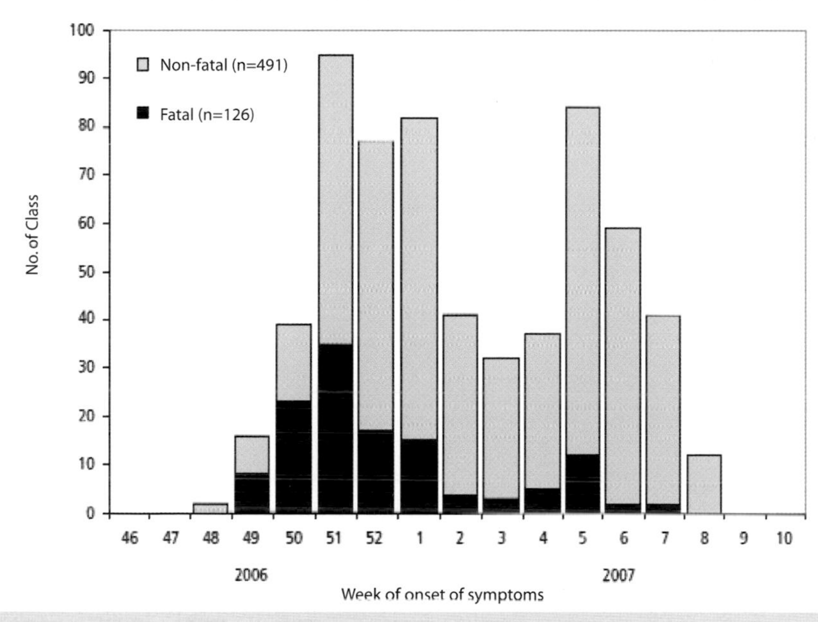

Cases of Rift Valley fever meeting inclusion criteria by date of onset of symptoms, Kenya, December 2006–February 2007 (n = 617).

produce accurate predictions has increased credibility compared with one that has only been shown to fit data well retrospectively.

Dynamical mathematical models of disease transmission, in contrast to statistical models of trend or association, are better suited to longer term predictions and predictions of new and emerging threats. They also have the potential to explore usefully "what if" scenarios, such as sudden changes in control policies or human behavior (e.g., due to travel restrictions). There are considerable challenges posed by such studies. Infectious disease incidence depends directly on various factors of the particular disease under study (e.g., population size, weather, or risk behaviors). Thus, making accurate predictions requires both accurately incorporating the roles of important drivers into the transmission model and making accurate predictions of how these drivers will behave in the future. Mathematical models are valuable tools for policymakers, but are best used as one component of the decisionmaking process, which should draw on all kinds of evidence available.

INTERNATIONAL AND NATIONAL SUPPORT IS CRITICAL

Zoonotic diseases can transcend boundaries and affect multiple countries, thus the support of both the national and international community is critical for effectively responding to them. The control of HPAI H5N1 at the international and national levels has provided insight into how different actors cooperate and collaborate on zoonotic disease concerns. The experiences reported here are based mainly on independent evaluation reports from FAO, United Nations System Influenza Coordinator (UNSIC), and the World Bank.

International Level

At the international level, WHO, FAO, and OIE are the main players in the international HPAI H5N1 arena.[6] According to their respective mandates, WHO focuses on the human health aspects and FAO on the implementation of the standards and strategies that OIE sets for animal health. The scope and mode of operation of these three agencies is quite different. WHO has a significant country presence, which enables it to more directly affect national decisionmaking. FAO has a much-limited presence at field level, normally without any animal health expertise in its country offices. Finally, OIE has a 40-person staff, a limited number of regional representatives, and no specific country representation. These organizations (without the United Nations Children's Fund, or UNICEF) cooperated well in the *Codex Alimentarius* Committee, which sets food safety standards. This committee was established by WHO and FAO, and now also has close relations with OIE. It was described by the recent Independent External

[6]To support communication about HPAI H5N1 and its control, UNICEF was added as an additional technical agency, although its role and mandate were never clearly articulated.

Evaluation panel of FAO as an example of an effective partnership among international organizations (FAO, 2007).

The start-up of the collaboration among the international agencies in addressing the HPAI H5N1 threat was difficult and slow, however. The first outbreaks of the current H5N1 strain of HPAI occurred in December 2003, with major outbreaks in 10 East Asian countries in January 2004. The first WHO strategy (2005b), without any discernable FAO or OIE input, was prepared in early September 2005. A joint FAO/OIE strategy prepared in collaboration with WHO was prepared by November 2005, or nearly 2 years after the outbreak (FAO et al., 2005). The reasons for the delays were caused by a lack of understanding of the mission of the involved agencies, lack of understanding on the epidemiology of the disease, difference of opinions among the agencies on how to respond, and the slow pace of resource mobilization.

This delay led to a rather fragmented approach that was arguably one of the main factors in the slow donor response in providing financial support, which caused donors to get involved in a bilateral fashion, based on the advice of their own technicians. The overwhelming number of missions of the technical agencies with large numbers of expatriate specialists and the complexity of procedures were also frequently mentioned at the country level as important issues (FAO, 2007).

Starting in mid-2005, and in particular leading up to and following the Beijing Conference, the International Finance Institutions (IFIs) and the World Bank also became directly involved in the HPAI H5N1 campaign, although the World Bank had supported Vietnam with an earlier emergency loan. This opened a new set of constraints, which affected the implementation of the campaign, especially administrative and procedural aspects. These constraints became especially apparent in the cooperation between WHO, FAO, and the IFIs, where the respective roles of these United Nation's agencies as cooperators for technical expertise and as contractors for services led to conflicts with the procurement rules of the IFIs. These administrative differences were exacerbated by differences in fiduciary requirements between the technical agencies and the IFIs (Willitts-King et al., 2008).

The cooperation among WHO, FAO, OIE, and to some extent UNICEF significantly improved over time because of the major increase in funding, the strong pressure from donors, and the excellent coordination role of UNSIC. There are now weekly conference calls, and there is a stronger cooperation in the preparation of the strategy updates. The institutions work together in the preparation of Integrated National Action Plans. A mutual trust between the main day-to-day decisionmakers in these organizations has emerged. However, even now, the cooperation is mainly concerned with strategy development and planning, yet there are few joint activities on implementing disease surveillance and control. The relationship between UNSIC and the technical agencies—especially WHO, which sees itself as the lead technical agency in human health and pandemic preparedness—is still a challenge (Willitts-King et al., 2008).

At the *individual level*, the three agencies provided a rapid reaction. For example, FAO, with input from OIE, organized an international workshop in East Asia on HPAI H5N1 only 3 weeks after the first outbreak. FAO became involved quite early with its Special Fund for Emergency and Rehabilitation Activities in the implementation of control measures. This flexible tool, with much lighter administrative requirements than normally demanded in FAO, provided FAO with the flexibility to respond early to the disease outbreaks. The lack of funds, however, caused the initial support that FAO provided in the affected and at-risk countries to be limited and restricted to strengthening disease surveillance systems, providing protective gear, and supporting epidemiological studies. Funding included almost no support in containing the disease, such as support of public administrations to be able to enforce movement control, compensation for culling, and vaccination. Similarly, WHO focused on the stocking of antivirals, although it could have used its much greater country presence to raise greater awareness and train local staff in the epidemiology and control of HPAI H5N1.

National Level

At the national level, cooperation among the respective ministries of health, agriculture, and the environment in many countries is cumbersome at best. They often have separate human and animal disease reporting procedures and communication channels during a disease outbreak. Environmental agencies are the weakest in the public sector, and efforts to bring them together are often confronted with major transaction costs, bureaucratic delays, and competency issues. The main lessons learned from the HPAI H5N1 campaign point to the importance of political support for disease control and the existence of an institutional framework.

Political support is crucial for disease control. The picture, which emerges from the reviews, shows ownership and political will at the highest levels to effectively plan and implement HPAI H5N1 campaigns. In several countries, this lack of ownership has led to inadequate interministerial collaboration; grossly insufficient national funding for human, veterinary, and wildlife services; and reluctance to share animal disease incidence information. These trends will severely affect the sustainability of future HPAI H5N1 activities.

The institutional framework is another critical element. Key observations that emerge from the reviews concern these factors: (1) the hierarchical place of HPAI H5N1 campaigns in government, and experience in the current campaigns seems to indicate that placement at a higher level (deputy prime minister, ministry of finance) than the line ministries of health or agriculture gives better results[7]; (2) decentralization, which, with some exceptions,[8] severely obstructs lines of command[9]; (3) the limited simulation testing and the general neglect in the preparation of most national preparedness and Integrated National Action Plans; and (4) the limited involvement of the private sector and, in particular, the nearly complete lack of use of private service providers (private veterinarians and paraveterinarians) under a sanitary mandate.

Lessons Learned

In an early phase of an emerging outbreak, countries need to define a mutually agreed-upon strategy with the international organizations concerned and with other relevant institutions. As was the case with the HPAI H5NI control campaign, it is important to collaborate early on with institutions specialized in environmental health and wildlife. This could be the function of the current UNSIC, whose current mandate expires in December 2010 and would have to be extended. Many developing countries lacked funding for investment in the surveillance of and response to HPAI H5N1. To avoid lack of funds to control an emerging disease at an early stage, sustainable funding is needed for highly infectious zoonotic diseases. To foster cooperation at the national level, governments need to establish special permanent, functional cross-sector coordination mechanisms, either through the exchange of memorandums of agreement between the different ministries and agencies involved, or a coordinating authority (e.g., special task force) above the sectoral human health, veterinary, and environmental agencies (e.g., the prime minister or deputy prime minister). In the case of an emerging disease outbreak, such institutions would define the control strategy, prepare contingency plans, and oversee their implementation; an option would be to let such a task force evolve into an independent agency. Finally, they need to cultivate a new style of leadership that promotes cooperation, teambuilding, and mentoring. This would need to be achieved through education and underpinned by incentive systems, which recognizes achievements in these areas rather than the current performance systems that often promote single department goals and individual achievements.

---

[7]Other disease control campaigns (HIV/AIDS) find that strengthening line ministries might be more efficient.

[8]For example, in India, where the identification of HPAI H5N1 was a national priority, with upfront government financial support and technical assistance from the central level, the full cooperation of the states was secured.

[9]At the local level, early communication between the human and animal health authorities may reduce the likelihood of the spread of disease from animals to humans.

CONCLUSION

Recent human outbreaks of zoonotic diseases have unavoidably resulted in increased attention to their impacts on national economies, international trade, household livelihoods, and human morbidity and mortality. Recent socioeconomic changes and the increase in international trade have also been critical drivers of zoonotic disease emergence and spread.

Disease surveillance is critical for detecting the emergence of zoonotic pathogens in human populations, preventing their spread between animal populations, and preventing transmission to human populations. The earlier an emerging pathogen can be detected and eliminated or controlled, the smaller the emergency response and cost will be. In addition, models of disease transmission have been successful in predicting future zoonotic disease outbreaks and trends. They have been used to make informed decisions on the relative risks and benefits of preventive measures aimed at managing the risk at low levels prior to infection. Data from surveillance systems are necessary for more accurately predicting future disease outbreaks. Accurately predicting or anticipating a disease outbreak enables local human and animal health authorities to implement prevention and control efforts, averting the need for costly emergency responses. Accurate prediction is important for preventing an outbreak altogether, decreasing an outbreak's duration, and lessening its impact on national and household economies and on human health.

The case for systematic and sustainable zoonotic disease surveillance, as presented in this chapter, is based on the committee's conclusion that conditions promoting the driving forces for zoonotic disease emergence are intensifying (further discussed in Chapter 3), that technologies and approaches that could be employed to develop a global system are available, and that the socioeconomic and health consequences for humans and animals are too enormous for inaction.

REFERENCES

Anderson, R. M. 1988. Epidemiological models and predictions. *Trop Geogr Med* 40(3): S30–S39.

Andrews, N. J. 2009. *Incidence of variant Creutzfeldt-Jakob disease diagnoses and deaths in the UK January 1994–December 2008.* Statistics Unit, Centre for Infections, U.K. Health Protection Agency (HPA). London, UK: HPA. http://www.cjd.ed.ac.uk/cjdq60.pdf (accessed March 25, 2009).

Anyamba, A., J. P. Chretien, J. Small, C. J. Tucker, and K. J. Linthicum. 2006. Developing global climate anomalies suggest potential disease risks for 2006–2007. *Int J Health Geogr* 5(60):60.

Anyamba, A., J. P. Chretien, J. Small, C. J. Tucker, P. B. Formenty, J. H. Richardson, S. C. Britch, and K. J. Linthicum. 2007. *Forecasting the temporal and spatial distribution of a Rift Valley fever outbreak in East Africa: 2006–2007.* Paper presented at the American Society of Tropical Medicine and Hygiene 2007 Conference, Philadelphia, PA, November 4–8.

Anyamba, A., J. P. Chretien, J. Small, C. J. Tucker, P. B. Formenty, J. H. Richardson, S. C. Britch, D. C. Schnabel, R. L. Erickson, and K. J. Linthicum. 2009. Prediction of a Rift Valley fever outbreak. *Proc Natl Acad Sci U S A* 106(3):955–959.

AusAID (Australian Agency for International Development). 2009. *AusAID assistance to combat avian influenza and other emerging and resurging zoonotic diseases—Total commitments since 2003.* http://www.ausaid.gov.au/keyaid/avian/assistance.pdf (accessed April 11, 2009).

Barry, J. M. 2005. *The great influenza: The epic story of the deadliest plague in history.* London, UK: Penguin Books.

Bernoulli, D. 1766. Essai d'une nouvelle analyse de la mortalité causée par la petite vérole et des avantages de l'inoculation pour la prévenir. In *Histoire de l'académie royale des sciences, avec mémoires de mathématique et de physique.* Paris, France. Pp. 1–40.

Brahmbhatt, M. 2006. *Economic impacts of avian influenza propagation.* Presented during the First International Conference on Avian Influenza in Humans, Institut Pasteur, Paris, France, June 29.

Brahmbhatt, M., and A. Dutta. 2008. *On SARS type economic effects during infectious disease control.* Policy research working paper 4466. Washington, DC: The World Bank.

Burkom, H. S. 2003. Biosurveillance applying scan statistics with multiple, disparate data sources. *J Urban Health* 80(2 Suppl 1):i57–i65.

Burns, A., D. van der Mensbrugghe, and H. Timmer. 2008. *Evaluating the economic consequences of avian influenza.* Washington, DC: The World Bank.

Cash, R. A., and V. Narasimhan. 2000. Impediments to global surveillance of infectious diseases: Consequences of open reporting in a global economy. *Bull World Health Organ* 78(11):1358–1367.

Caspari, C., M. Christodoulou, and E. Monti. 2007. *Prevention and control of animal diseases worldwide: Economic analysis—Prevention versus outbreak costs.* Final report (Part I). Berlin, Germany: Agra CEAS Consulting.

CDC (U.S. Centers for Disease Control and Prevention). 1994. International notes update: Human plague—India, 1994. *MMWR* 43(41): 761–762.

CDC. 2007. Rift Valley fever outbreak—Kenya, November 2006–January 2007. *MMWR* 56(4):73–76.

CDC. 2008. West Nile virus activity in the United States from 1999–2008. Statistics, Surveillance, and Control. Atlanta, Georgia: CDC. http://www.cdc.gov/ncidod/dvbid/westnile/surv&control.htm (accessed January 26, 2009).

CDC. 2009. *Fact Sheet: Variant Creutzfeldt-Jakob disease.* http://www.cdc.gov/ncidod/dvrd/vcjd/factsheet_nvcjd.htm#surveillance (accessed February 13, 2009).

Chomel, B. B., A. Belotto, and F. X. Meslin. 2007. Wildlife, exotic pets, and emerging zoonoses. *Emerg Infect Dis* 13(1):6–11.

Chowell, G., N. W. Hengartner, C. Castillo-Chavez, P. W. Fenimore, and J. M. Hyman. 2004. The basic reproductive number of Ebola and the effects of public health measures: The cases of Congo and Uganda. *J Theor Biol* 229(1):119–126.

Collins, K. 2007. *Statement of Keith Collins, chief economist, U.S. Department of Agriculture, before the U.S. House of Representatives Committee on Agriculture.* October 18. http://www.usda.gov/oce/newsroom/archives/testimony/2007a/Housetstoutlook10_17_07r2.doc (accessed April 6, 2009).

CRS (U.S. Congressional Research Service). 2008a. *International illegal trade in wildlife: Threats and U.S. policy*, edited by L. S. Wyler and P. A. Sheikh. Washington, DC: Library of Congress.

CRS. 2008b. *Global health: Appropriations to USAID programs from FY2001 through FY2008*, edited by T. Salaam-Blyther. Washington, DC: Library of Congress.

Darby, P. M. 2003. *The economic impact of SARS.* Special Briefing, Canadian Tourism Research Institute. Ottawa, Ontario: The Conference Board of Canada. http://sso.conferenceboard.ca/documents.aspx?did=539 (accessed January 26, 2009).

de Haan, C., T. J. van Veen, B. Brandenburg, J. Gauthier, F. Le Gall, R. Mearns, and M. Siméon. 2001. *Livestock development: Implications for rural poverty, the environment, and global food security.* Washington, DC: The World Bank.

Defra (UK Department for Environment, Food and Rural Affairs). 2009a. *Summary of passive surveillance reports in Great Britain.* http://www.defra.gov.uk/vla/science/docs/sci_tse_stats_gboverview.pdf (accessed April 11, 2009).

Defra. 2009b. *BSE statistics—Schemes.* http://www.defra.gov.uk/animalh/bse/statistics/schemes.html (accessed April 11, 2009).

Dobson, A., and J. Foufopoulos. 2001. Emerging infectious pathogens of wildlife. *Philos Trans R Soc Lond B Biol Sci* 356(1411):1001–1012.

DTZ Pieda Consulting. 1998. *Economic impact of BSE on the UK economy.* Report to United Kingdom's agricultural departments and Her Majesty's Treasury. Edinburgh, UK: DTZ Pieda Consulting.

Easterbrook, J. D., J. B. Kaplan, N. B. Vanasco, W. K. Reeves, R. H. Purcell, M. Y. Kosoy, G. E. Glass, J. Watson, and S. L. Klein. 2007. A survey of zoonotic pathogens carried by Norway rats in Baltimore, Maryland. *Epidemiol Infect* 135(7):1192–1199.

Ebrahim, M., and J. Solomon. 2006. Exotic pets in the U.S. may pose health risk. The Associated Press, November 27.

Einsweiler, S. 2008. *Monitoring the wildlife trade.* Presentation, Fifth Committee Meeting on Achieving Sustainable Global Capacity for Surveillance and Response to Emerging Diseases of Zoonotic Origin, Washington, DC, December 1–2.

Elci, C. 2006. The impact of HPAI of the H5N1 strain on economies of affected countries. In *Proceedings of the Conference on Human and Economic Resources.* First International Conference on Human and Economic Resources, Izmir, Turkey, May 24–25. Izmir, Turkey: Izmir University of Economics. Pp. 104–117.

Engler, M., and R. Parry-Jones. 2007. *Opportunity of threat—The role of the European Union in global wildlife trade.* Brussels, Belgium: Traffic Europe.

FAO (Food and Agriculture Organization of the United Nations). 2002. *Japanese encephalitis/Nipah outbreak in Malaysia.* FAO/WHO Global Forum on Food Safety Regulators, Marrakech, Morocco, January 28–30. Rome, Italy: FAO. http://www.fao.org/DOCREP/MEETING/004/AB455E.HTM, (accessed January 26 2009).

FAO. 2006. *Impacts of animal disease outbreaks on livestock markets.* Committee on Commodity Problems. 21st Session of the Intergovernmental Group on Meat and Dairy Products, November 14.

FAO. 2007. *The challenge of renewal.* Report of the Independent External Evaluation of the Food and Agriculture Organization of the United Nations (FAO) submitted to the Council Committee for the Independent External Evaluation of FAO (CC-IEE), September. Rome, Italy: FAO. ftp://ftp.fao.org/docrep/fao/meeting/012/k0827e02.pdf (accessed July 19, 2009).

FAO. 2009. FAOSTAT, FAO Statistical Division. Rome, Italy: FAO.

FAO and APHCA (Food and Agriculture Organization of the United Nations Regional Office for Asia and Animal Production and Health Commission for Asia and the Pacific). 2002. *Manual on the diagnosis of Nipah virus infection in animals.* Bangkok, Thailand: FAO.

FAO, OIE, and WHO (Food and Agriculture Organization of the United Nations, World Organization for Animal Health, and World Health Organization). 2005. *A global strategy for the progressive control of highly pathogenic avian influenza (HPAI)*. http://www.fao.org/avianflu/documents/HPAIGlobalStrategy31Oct05.pdf (accessed July 19, 2009).

FAO, OIE, WHO, UNSIC, UNICEF (Food and Agriculture Organization of the United Nations, World Organization for Animal Health, World Health Organization, United Nations System Influenza Coordinator, United Nations Children's Fund), and World Bank. 2008. *Contributing to one world, one health: A strategic framework for reducing risks of infectious diseases at the animal–human–ecosystems interface*. Consultation document prepared for the Inter-ministerial Meeting on Avian and Pandemic Influenza, Sharm-el-Sheikh, Egypt, October 14.

Ferguson, N. M., D. A. Cummings, S. Cauchemez, C. Fraser, S. Riley, A. Meeyai, S. Iamsirithaworn, and D. S. Burke. 2005. Strategies for containing an emerging influenza pandemic in Southeast Asia. *Nature* 437(7056):209–214.

Ferrari, M. J., O. N. Bjornstad, and A. P. Dobson. 2005. Estimation and inference of R_0 of an infectious pathogen by a removal method. *Math Biosci* 198(1):14–26.

Fèvre, E. M., R. W. Kaboyo, V. Persson, M. Edelsten, P. G. Coleman, and S. Cleaveland. 2005. The epidemiology of animal bite injuries in Uganda and projections of the burden of rabies. *Trop Med Int Health* 10(8):790–798.

Ford, T. E., R. R. Colwell, J. B. Rose, S. S. Morse, D. J. Rogers, and T. L. Yates. 2009. Using satellite images of environmental changes to predict infectious disease outbreaks. *Emerg Infect Dis* 15(9). http://www.cdc.gov/EID/content/15/9/1341.htm (accessed August 28, 2009).

Gerson, H., B. Cudmore, N. E. Mandrak, L. D. Coote, K. Farr, G. Baillargeon. 2008. Monitoring international wildlife trade with coded species data. *Conserv Biol* 22(1):4–7.

Grants.gov. 2009a. *RFI Avian and Pandemic Influenza and Zoonotic Disease Program PREDICT.* file:///N:/Zoonotic%20Diseases/Public%20Access%20File/USAID%20Summary.RFI.Predict.4.16.09.htm (accessed September 24, 2009).

Grants.gov. 2009b. *RFI Avian and Pandemic Influenza and Zoonotic Disease Program RESPOND.* file:///N:/Zoonotic%20Diseases/Public%20Access%20File/USAID%20Summary.RFI.Predict.4.16.09.htm (accessed September 24, 2009).

Gubler, D. J. 2001. Silent threat: Infectious diseases and U.S. biosecurity. *Georgetown J of International Affairs* II(2):15–23.

Hanna, D., and Y. Huang. 2004. The impact of SARS on Asian economies. *Asian Econ Pap* 3(1):102–112.

Holland, P. W. 1986. Statistics and causal inference. *J Am Stat Assoc* 81(396):945–960.

IOM and NRC (Institute of Medicine and National Research Council). 2008. *Achieving sustainable global capacity for surveillance and response to emerging disease of zoonotic origin: Workshop report.* Washington, DC: The National Academies Press.

Jambiya, G., S. H. A. Milledge, and N. Mtango. 2007. *"Night time spinach": Conservation and livelihood implications of wild meat use in refugee situations in Northwestern Tanzania.* Dar es Salaam, Tanzania: TRAFFIC East/Southern Africa.

Jenkins, P. T., K. Genovese, and H. Ruffler. 2007. *Broken screens: The regulation of live animal importation in the United States.* Washington DC: Defenders of Wildlife. http://www.defenders.org/resources/publications/programs_and_policy/international_conservation/broken_screens/broken_screens_report.pdf (accessed April 11, 2009).

Jewell, N. P. 2003. *Statistics for epidemiology.* Boca Raton, Florida: Chapman & Hall/CRC Press.

Jones, K. E., N. G. Patel, M. A. Levy, A. Storeygard, D. Balk, J. L. Gittleman, and P. Daszak. 2008. Global trends in emerging infectious diseases. *Nature* 451(7181):990–993.

Kaplan, K. 2007. *Model successfully predicts Rift Valley fever outbreak*. Agricultural Research Service, U.S. Department of Agriculture. http://www.ars.usda.gov/is/pr/2007/070216.htm (accessed January 29, 2009).

Keyfitz, N. 1972. On future population. *J Am Stat Assoc* 67 (338):347–363.

Kimball, A. M., and R. Davis. 2006. Costs of epidemics in APEC economies. In *Plagues, power, and politics: Infectious disease and international policy*, edited by A. Price-Smith. Toronto, Canada: Palgrave Publishers.

Kimball, A. M., and K. Taneda. 2004. Emerging infections and global trade: A new method of gauging impact. *Rev Sci Tech* 23(3):753–760.

King, L. J. 2004. Introduction—Emerging zoonoses and pathogens of public health concern. *Rev Sci Tech* 23(2):429–430.

Knobel, D. L., S. Cleaveland, P. G. Coleman, E. M. Fevre, M. I. Meltzer, M. E. Miranda, A. Shaw, J. Zinsstag, and F. X. Meslin. 2005. Re-evaluating the burden of rabies in Africa and Asia. *Bull World Health Organ* 83(5):360–368.

Knorr-Held, L., and S. Richardson. 2003. A hierarchical model for space-time surveillance data on meningococcal disease incidence. *Appl Stat* 52(2):169–183.

Koonse, B. 2008. *Regulators and aquaculture certification: Can we use it?* Presentation, FAO Aquaculture Certification Workshop, Silver Spring, Maryland, May 29–30. http://library.enaca.org/certification/washington08/presentation-koonse.pdf (accessed January 22, 2009).

Legrand, J., R. F. Grais, P. Y. Boelle, A. J. Valleron, and A. Flahault. 2007. Understanding the dynamics of Ebola epidemics. *Epidemiol Infect* 135(4):610–621.

Lembo, T., K. Hampson, D. T. Haydon, M. Craft, A. Dobson, J. Dushoff, E. Ernest, R. Hoare, M. Kaare, T. Mlengeya, C. Mentzel, and S. Cleaveland. 2008. Exploring reservoir dynamics: A case study of rabies in the Serengeti ecosystem. *J Appl Ecol* 45(4):1246–1257.

Longini, I. M., Jr., A. Nizam, S. Xu, K. Ungchusak, W. Hanshaoworakul, D. A. Cummings, and M. E. Halloran. 2005. Containing pandemic influenza at the source. *Science* 309(5737):1083–1087.

Marano, N. 2008. *CDC's role in preventing introduction of zoonotic diseases via animal importation*. Presentation, Fifth Committee Meeting on Achieving Sustainable Global Capacity for Surveillance and Response to Emerging Diseases of Zoonotic Origin, Washington, DC, December 1–2.

Marano, N., P. M. Arguin, and M. Pappaioanou. 2007. Impact of globalization and animal trade on infectious disease ecology. *Emerg Infect Dis* 13(12):1807–1809.

Massad, E., M. N. Burattini, L. F. Lopez, and F. A. Coutinho. 2005. Forecasting versus projection models in epidemiology: The case of the SARS epidemics. *Med Hypotheses* 65(1):17–22.

McKibbin, W., and A. Sidorenko. 2006. *Global macroeconomic consequences of pandemic influenza*. Sydney, Australia: Lowy Institute.

McMichael, A. J. 2004. Environmental and social influences on emerging infectious diseases: Past, present and future. *Philos Trans R Soc Lond B Biol Sci* 359(1447):1049–1058.

Mills, C. E., J. M. Robins, and M. Lipsitch. 2004. Transmissibility of 1918 pandemic influenza. *Nature* 432(7019):904–906.

Morens, D. M., G. K. Folkers, and A. S. Fauci. 2004. The challenge of emerging and re-emerging infectious diseases. *Nature* 430(6996):242–249.

Murphy, F. A. 2008. Emerging zoonoses: The challenge for public health and biodefense. *Prev Vet Med* 86(3–4):216–223.

OAU-IBAR (Organization of African Unity-Interafrican Bureau for Animal Resources). 2009. *Pan-African programme for the control of epizootics*. http://www.au-ibar.org/ach_animhealth/pace.html (accessed April 11, 2009).

Odiit, M., P. G. Coleman, W. C. Liu, J. J. McDermott, E. M. Fèvre, S. C. Welburn, and M. E. Woolhouse. 2005. Quantifying the level of under-detection of *trypanosoma brucei rhodesiense* sleeping sickness cases. *Trop Med Int Health* 10(9):840–849.

OECD and WHO (Organisation for Economic Co-operation and Development and World Health Organization). 2003. *Food borne disease in the OECD countries: Present state and economic costs.* Paris, France: OECD.

Osterholm, M. T. 2005. Preparing for the next pandemic. *Foreign Af* 84(4):24–37.

Otte, J., D. Roland-Holst, and D. Pfeiffer. 2006. *HPAI control measures and household incomes in Vietnam.* Rome, Italy: FAO Pro Poor Livestock Policy Initiative.

Plowright, R. K., S. H. Sokolow, M. E. Gorman, P. Daszak, and J. E. Foley. 2008. Causal inference in disease ecology: Investigating ecological drivers of disease emergence. *Front Ecol Environ* 6(8):420–429.

Pomareda, C. 2001. *Propuesta de programa hemisférico de sanidad agropecuaria e inocuidad de alimentos.* Presentada por el IICA a consideración de Organismos Internacionales de Financiamiento de Desarrollo, Agencias de Cooperación Bilateral. San José, Costa Rica.

Presanis, A. M., D. De Angelis, D. J. Spiegelhalter, S. Seaman, A. Goubar, and A. E. Ades. 2008. Conflicting evidence in a Bayesian synthesis of surveillance data to estimate human immunodeficiency virus prevalence. *J R Stat Soc Ser A* 171(4):915–937.

Price-Smith, A. T. 1998. Contagion and chaos: Infectious disease and its effects on global security and development. *University of Toronto Centre for International Studies Working Paper 1998–001.* Toronto, Ontario: Centre for International Studies.

Sertsou, G., N. Wilson, M. Baker, P. Nelson, and M. G. Roberts. 2006. Key transmission parameters of an institutional outbreak during the 1918 influenza pandemic estimated by mathematical modelling. *Theor Biol Med Model* 3(38):38.

Shaman, J., J. F. Day, and M. Stieglitz. 2003. St. Louis encephalitis virus in wild birds during the 1990 south Florida epidemic: The importance of drought, wetting conditions, and the emergence of *Culex nigripalpus* (diptera: Culicidae) to arboviral amplification and transmission. *J Med Entomol* 40(4):547–554.

Shaman, J., J. Day, M. Stieglitz, S. Zebiak, and M. Cane. 2006. An ensemble seasonal forecast of human cases of St. Louis encephalitis in Florida based on seasonal hydrologic forecasts. *Clim Change* 75(4):495–511.

Singer, R. S., L. A. Cox, Jr., J. S. Dickson, H. S. Hurd, I. Phillips, and G. Y. Miller. 2007. Modeling the relationship between food animal health and human foodborne illness. *Prev Vet Med* 79(2–4):186–203.

Singh, S. P., D. C. Reddy, M. Rai, and S. Sundar. 2006. Serious underreporting of visceral leishmaniasis through passive case reporting in Bihar, India. *Trop Med Int Health* 11(6):899–905.

Stephenson, J. 2003. Monkeypox outbreak a reminder of emerging infections vulnerabilities. *JAMA* 290(1):23–24.

Taylor, L. H., S. M. Latham, and M. E. Woolhouse. 2001. Risk factors for human disease emergence. *Philos Trans R Soc Lond B Biol Sci* 356(1411):983–989.

The BSE Inquiry. 2000. Economic impact and international trade. In *The BSE Inquiry: The inquiry into BSE and variant CJD in the United Kingdom.* Vol. 10. London, UK: Stationery Office. http://www.bseinquiry.gov.uk/report/volume10/chapted3.htm#273979 (accessed January 14, 2009).

The end of cheap food. 2007. *The Economist* 385(8558):11–12.

Theile, S., A. Steiner, and K. Kecse-Nagy. 2004. *Expanding borders: New challenges for wildlife trade controls in the European Union.* Brussels, Belgium: TRAFFIC Europe.

Turner, R. M., D. J. Spiegelhalter, G. C. Smith, and S. Thompson. 2009. Bias modelling in evidence synthesis. *J R Stat Soc Ser A* 172(1):21–47.

UNWTO (United Nations World Tourism Organization). 2009. Quick overview of key trends. *UNWTO World Tourism Barometer* 7(1):1–9. http://unwto.org/facts/eng/pdf/barometer/UNWTO_Barom09_1_en_excerpt.pdf (accessed March 26, 2009).

U.S. House of Representatives, Committee on Natural Resources. 2008. *Poaching American security: Impacts of illegal wildlife trade.* 110th Cong., March 5.

USITC (U.S. International Trade Commission). 2008. *Global beef trade: Effects of animal health, sanitary, food safety, and other measures on U.S. beef exports.* Investigation no. 332-488. Washington, DC: USITC. http://hotdocs.usitc.gov/docs/pubs/332/pub4033.pdf (accessed February 18, 2009).

van Zwanenberg, P., and E. Millstone. 2002. Mad cow disease 1980s–2000: How reassurances undermined precaution. In *Late lessons from early warnings: The precautionary principle 1896–2000.* Environmental issue report No 22. Copenhagen, Denmark: European Environment Agency.

Vorou, R. M., V. G. Papavassiliou, and S. Tsiodras. 2007. Emerging zoonoses and vector-borne infections affecting humans in Europe. *Epidemiol Infect* 135(8):1231–1247.

Vynnycky, E., A. Trindall, and P. Mangtani. 2007. Estimates of the reproduction numbers of Spanish influenza using morbidity data. *Int J Epidemiol* 36(4):881–889.

Wells, G. A., A. C. Scott, C. T. Johnson, R. F. Gunning, R. D. Hancock, M. Jeffrey, M. Dawson, and R. Bradley. 1987. A novel progressive spongiform encephalopathy in cattle. *Vet Rec* 121(18):419–420.

WHO (World Health Organization). 2003. *Consensus document on the epidemiology of severe acute respiratory syndrome (SARS).* WHO/CDS/CSR/GAR/2003.11. Geneva, Switzerland: WHO.

WHO. 2004. *Summary of probable SARS cases with onset of illness from 1 November 2002 to 31 July 2003.* Geneva, Switzerland: WHO.

WHO. 2005a. *WHO welcomes pandemic influenza response modelling papers, August 3.* http://www.who.int/mediacentre/news/statements/2005/s08/en/index.html (accessed March 26, 2009).

WHO. 2005b. *Responding to the avian influenza pandemic threat: Recommended strategic actions.* Epidemic and Pandemic Alert and Response. WHO/CDS/CSR/GIP/2005.8 Geneva, Switzerland: WHO.

WHO. 2007a. *World health report 2007.* Geneva, Switzerland: WHO.

WHO. 2007b. *Rift Valley fever in Kenya, Somalia and the United Republic of Tanzania.* Disease Outbreak News Alerts, May 9. Geneva, Switzerland: WHO. http://www.who.int/csr/don/2007_05_09/en/index.html (accessed February 10, 2009).

WHO. 2007c. Outbreaks of Rift Valley fever in Kenya, Somalia and United Republic of Tanzania, December 2006–April 2007. *Wkly Epidemiol Rec* 82(20):169–178. http://www.who.int/wer/2007/wer8220.pdf (accessed March 26, 2009).

WHO. 2009. *Cumulative number of confirmed human cases of avian influenza A/(H5N1) reported to WHO, 23 April 2009 update.* Geneva, Switzerland: WHO. http://www.pandemicflunews.us/cumulative-number-of-confirmed-human-cases-of-avian-influenza-ah5n1-reported-to-who-5/ (accessed April 23, 2009).

Willitts-King, B., A. Smith, and L. Sims. 2008. *Evaluation of United Nations System Influenza Coordination (UNSIC)—Final report.* http://www.undg.org/docs/9411/UNSIC-evaluation-final-report-submitted-July-22-2008.pdf (accessed March 25, 2009).

Wolfe, N. D., C. P. Dunavan, and J. Diamond. 2007. Origins of major human infectious diseases. *Nature* 447(7142):279–283.

Woolhouse, M. E. 2008a. Epidemiology: Emerging diseases go global. *Nature* 451(7181):898–899.

Woolhouse, M. 2008b. Exemplary epidemiology. *Nature* 453(7191):34.

Woolhouse, M., and R. Antia. 2008. Emergence of new infectious diseases. In *Evolution in Health and Disease*, 2nd ed., edited by S. C. Stearns and J. K. Koella. Oxford, UK: Oxford University Press. Pp. 215–228.

Woolhouse, M. E., and E. Gaunt. 2007. Ecological origins of novel human pathogens. *Crit Rev Microbiol* 33(4):231–242.

Woolhouse, M. E., and S. Gowtage-Sequeria. 2005. Host range and emerging and reemerging pathogens. *Emerg Infect Dis* 11(12):1842–1847.

Woolhouse, M. E., L. H. Taylor, and D. T. Haydon. 2001. Population biology of multihost pathogens. *Science* 292(5519):1109–1112.

Woolhouse, M. E. J., R. Howey, E. Gaunt, L. Reilly, M. Chase-Topping, and N. Savill. 2008. Temporal trends in the discovery of human viruses. *Proc R Soc B* 275(1647):2111–2115.

World Bank. 2005. *East Asia update: Countering global shocks*. November. Washington, DC: The World Bank. http://web.worldbank.org/WBSITE/EXTERNAL/COUNTRIES/EASTASIAPACIFICEXT/EXTEAPHALFYEARLYUPDATE/0,,contentMDK:20708543~menuPK:550232~pagePK:64168445~piPK:64168309~theSitePK:550226,00.html (accessed October 17, 2008).

World Bank. 2008a. *World development report 2008: Agriculture for development*. Washington, DC: The World Bank.

World Bank. 2008b. *What's driving the wildlife trade? A review of expert opinion on economic and social drivers of the wildlife trade and trade control efforts in Cambodia, Indonesia, Lao PDR, and Vietnam*. Washington, DC: The World Bank.

World Resources Institute, United Nations Environment Programme, United Nations Development Programme, and the World Bank. 1996. Box 2.3 The Black Death revisited: India's 1994 plague epidemic. In *World Resources 1996–97: The urban environment*. New York: Oxford University Press.

WTO (World Trade Organization). 2008. *International trade statistics*. Geneva, Switzerland: WTO.

WTO. 2009. International Trade Statistics retrieved May 27, 2009, from WTO Statistics Database. Geneva, Switzerland: WTO.

Zohrabian, A., M. I. Meltzer, R. Ratard, K. Billah, N. A. Molinari, K. Roy, R. D. Scott II, and L. R. Petersen. 2004. West Nile virus economic impact, Louisiana, 2002. *Emerg Infect Dis* 10(10):1736–1744.

3

Drivers of Zoonotic Diseases

> *"A transcendent moment nears upon the world for a microbial perfect storm. Unlike the meteorological perfect storm—happening just once in a century—the microbial perfect storm will be a recurrent event. The two events share a common feature; a combination of factors is the driving force behind each."*
>
> *—Microbial Threats to Health: Emergence, Detection, and Response (Institute of Medicine, 2003)*

Zoonotic disease emergence is a complex process. A series of external factors, or *drivers*, provide conditions that allow for a select pathogen to expand and adapt to a new niche. The drivers for the most part are ecological, political, economic, and social forces operating at local, national, regional, and global levels. Regions where these factors are most densely aggregated, most highly prevalent, and where risk of a disease event are most intense can be considered zoonotic disease "hotspots." In this chapter, the committee reviews many of the drivers underlying this process of disease emergence and reemergence. Though not an exhaustive review, it reveals the multiplicity and the complexity of their inter-relationships.

OVERVIEW OF ZOONOTIC DISEASE EMERGENCE AND REEMERGENCE

Zoonotic disease emergence often occurs in stages, with an initial series of spillover events, followed by repeated small outbreaks in people, and then pathogen adaptation for human-to-human transmission. Each stage might have a different driver, and therefore a different control measure. As mentioned in Chapter 2, human immunodeficiency virus-1 (HIV-1) emerged from chimpanzees in Africa, spilling over to humans repeatedly before its global spread (Hahn et al., 2000). This initial phase of emergence was driven by bushmeat hunting and was the primary driver of its emergence. A second phase of emergence was driven by increased urbanization and road expansion in Central Africa beginning in the 1950s, and dispersal of index cases harboring prototype HIV-1 infections that were transmissible from person to person. The virus then entered the rapidly expanding global air travel network and became pandemic, with its emergence in North America, Europe, and Asia, accelerated by changes in sexual behavior, drug use, trade in blood derivatives, and population mobility.

Nipah virus is another example of a recently discovered paramyxovirus with fruit bat reservoir hosts. It caused a large-scale outbreak in Malaysian pig farmers in 1998. It is a growing threat due to its broad host range, wide geographical distribution, high case fatality, reports of human-to-human transmission, and the lack of vaccines or effective therapies (CDC, 1999; Eaton et al., 2006; Gurley et al., 2007). A recent analysis of food-animal production data from the index site—a commercial pig farm in Malaysia—before and during the outbreak shows that the emergence was likely caused by repeated introduction of Nipah virus from the wildlife reservoir into an intensively managed, commercial pig population site planted with mango trees (Daszak et al., 2006). This repeated introduction led to changes in infection dynamics in the pigs and a long-term, within-farm persistence of virus that would otherwise have died out. This causative mechanism has been previously proposed as a driver of highly pathogenic avian influenza (HPAI) H5N1 dynamics in poultry and the emergence of other pathogens (Pulliam et al., 2007).

An overview of how certain factors lead to disease emergence and reemergence is outlined in Figure 3-1. There is currently a great deal of interest in studying the underlying drivers of emerging diseases, from the proximal to the primary, to better target control programs.

THE HUMAN–ANIMAL–ENVIRONMENT INTERFACE

Historical Perspective on the Human–Animal Interface

The hunter-gatherer lifestyle supported early human societies for millennia, and this lifestyle could support an estimated 4 million people worldwide. About 10,000 years ago, hunter-gatherers began to settle, planting crops and husbanding wild animals to the point of domestication. This pattern continued more or less uninterrupted until the end of the 17th century when Thomas Malthus wrote in *An Essay on the Principle of Population* that human growth would soon outstrip the ability of the world to feed it. Fortunately, Malthusian predictions proved untrue, largely because of the change in agricultural systems from extensive to intensive. This change was accelerated by the growth of large urban centers and the invention of the railway, allowing food to move more freely from the farm to the table.

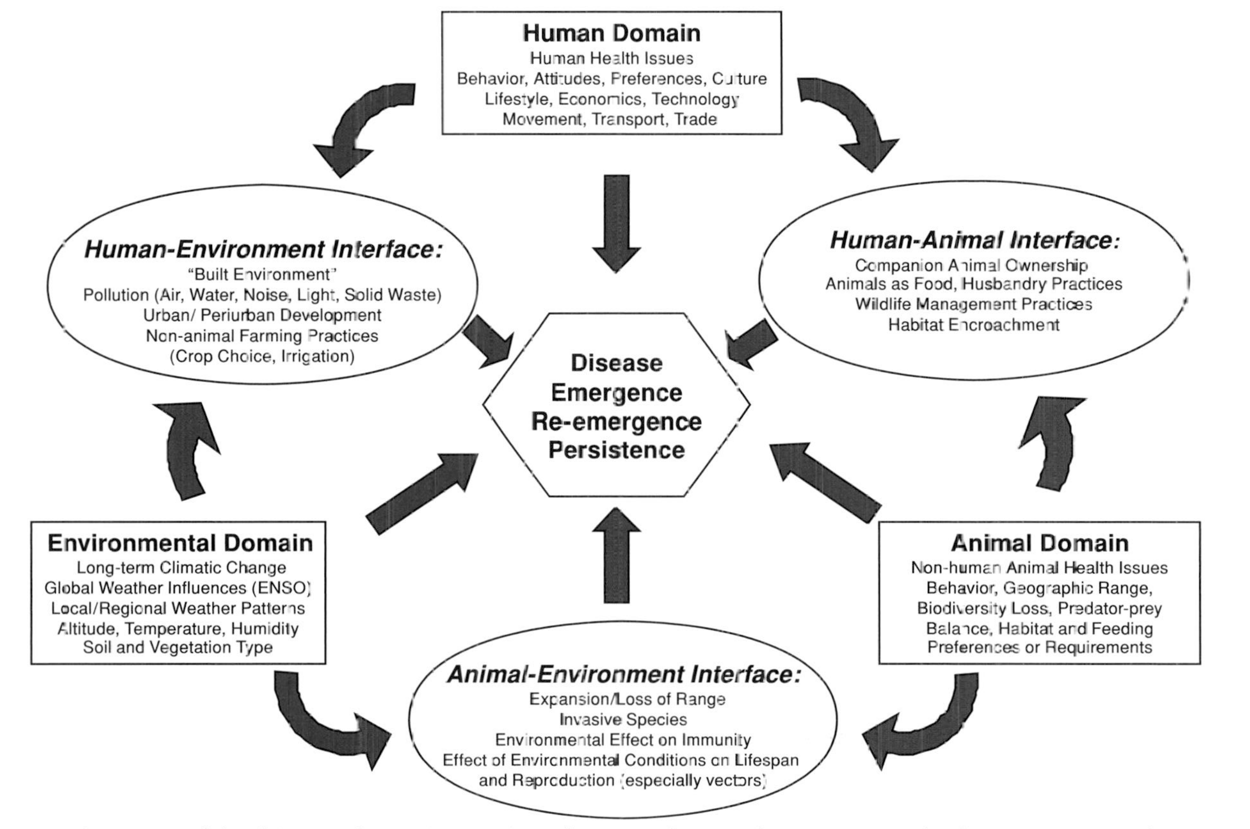

FIGURE 3-1 Overview of the driver-pathogen interactions that contribute to the emergence of infectious zoonotic diseases.
SOURCE: Treadwell (2008).

The "Green Revolution"[1] further increased crop yields and the separation of humans from the source of their food.

Since the 1960s, the production of food animals has grown phenomenally. Global milk production has doubled, meat production has tripled, and egg production has increased four-fold. Part of this is due to greater numbers of animals. However, genetic enhancement has also played a role, leading to higher overall production per animal.

Current Trends in Animal Protein Production

World demand for animal protein is increasing, and projections for consumption are staggering. Between 2000 and 2030, global meat production is expected to increase by approximately 2 percent per annum until 2015 and then slightly more than 1 percent per annum until 2030 (Steinfeld, 2004). Most of this demand is expected to come from the developing world, where rapid population expansion and higher per-capita incomes will drive people to change from a diet of rice, beans, and corn to one that incorporates more animal protein, a phenomenon known as the "nutrition transition" (Delgado, 2003). How will this demand be met? Most recent growth in intensive agriculture and projected growth for the next 30 years is mostly in the developing world, where intensive food-animal production facilities are being set up. These facilities are almost entirely based on feed grain, and in Asia, feed grain is imported from other parts of the world (see discussion later in this chapter on Global Food Systems and Food Safety). These collective changes in agricultural production and distribution, referred to as the "Livestock Revolution," are driven by globalization and the developing world's emerging middle class. The Livestock Revolution is characterized by vertical integration, the introduction of large supermarkets in developing countries, regional concentrations of animals, and a move to locate production facilities geographically at the farthest reaches permitted by regulations (Steinfeld, 2004).

Fueled by a growing population, rising incomes, and related urbanization, the consumption of meat and milk in the developing world grew slightly more than 3 and 2 percent per year, respectively, from 1992 to

[1]The term "Green Revolution" was coined by the director of the U.S. Agency for International Development in 1968 to describe the phenomenal growth in production of rice and wheat. The Rockefeller and Ford foundations made research investments to improve breeding varieties combined with expanded use of fertilizers, other chemical inputs, and irrigation. This led to dramatic yields of these grains, particularly in Asia and Latin America, in the late 1960s. Although heralded as a major achievement in establishing levels of national food security for developing countries, it is also criticized for causing environmental damage, including polluting waterways with chemicals, affecting the health of farm workers, and killing beneficial insects and wildlife (IFPRI, 2002).

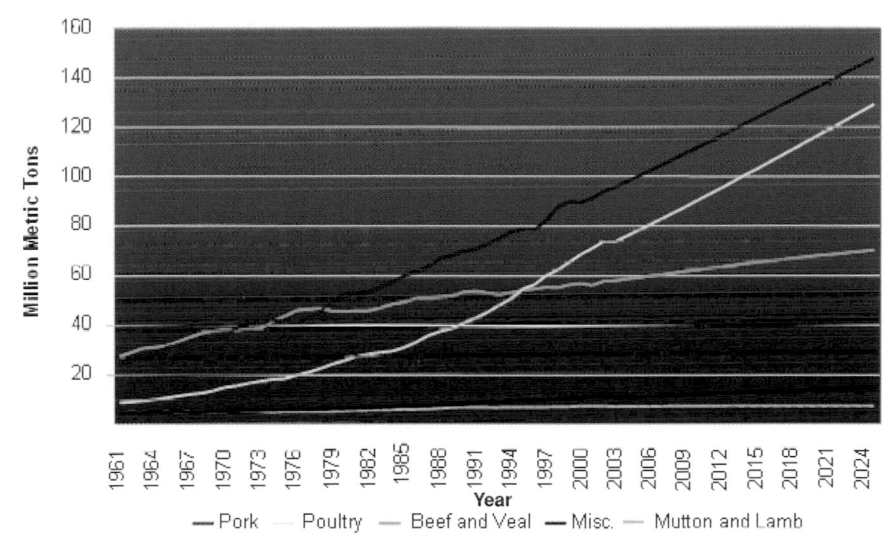

FIGURE 3-2 Projected production of animal meat by species, 1961–2025.
SOURCE: Newcomb (2004). Reprinted with permission from Bio Economic Research Associates, LLC (bio-era™). All rights reserved.

2002.[2] Growth was particularly strong in China, where over that same period meat and milk consumption grew by nearly 6 and 8 percent per year, respectively. Most of the growth occurred in poultry and swine; beef consumption grew at a much lower rate (see Figure 3-2). In contrast, per-capita total meat consumption in the developed world remained practically static in the same period, although there has been a slight shift from beef to chicken.

This strong expansion and resulting concentration of meat and milk production in the developing world has consequences for global human and animal health, which is explored in more detail later in this chapter. The shift of production to the developing world transfers the industry to a region with generally weak public services and regulatory oversight

[2]The underlying quantitative parameters driving this growth over the period 1992–2002 are (1) population increases of 1.7 percent per year in the developing world versus 0.4 in the developed world; (2) per-capita gross domestic product increase of 3.9 percent in the developing world versus 0.4 percent in the developed world; and (3) expenditure elasticity (percentage increase in expenditure on an item with a 1 percent increase in total expenditure) for meat in low-income countries of 0.78 percent, in middle-income countries of 0.64 percent, and in high-income countries of 0.36 percent (Searle et al., 2003).

mechanisms, which were unprepared for fast growth and major structural changes.

Intensified food-animal production has epidemiological consequences (see Box 3-1). Natural herds often have a low rate of reproduction and production. Humans have domesticated animals to ensure a more regular, safer, and convenient food supply. The objective of husbandry is to reach a natural balance between the host and its parasites while promoting efficient and economical production. Any increase in production must be matched with a refinement of management and disease control strategies. Although the factors listed under a "man-made ecosystem" (Box 3-1) are caused by influences of human intervention, their adjustments or maintenance are not necessarily under human control, and could lead to higher levels of disease risk. But at the same time, the level of risk could be reduced through more intensively managed and maintained factors with respect to animal health and well-being.

BOX 3-1
Epidemiological Factors Comparing Natural and Man-made Ecosystems

Natural Ecosystem	Man-made Ecosystem
• Wandering herds grazing extensive areas	• Herds are permanently housed (zero grazing)
• Intermingled species so that mixed grazing occurs	• Mixed herds have become single species
• Different species unaffected by the parasites of others	• Excreted pathogens are available to others of the same species
• In the open air, expiratory droplet infections are of little importance	• Animals are crowded on limited land
• Natural avoidance distances minimize direct contact	• Crowding allows closer contact
• Predators remove diseased animals early in the course of the disease	• Predators are eliminated; sick are helped to survive while excreting pathogens
• Hosts and parasites reach a balance so that both live with little harm	• Balance is upset as new niches are created
• Epidemics occur only when populations increase past a certain point	• Increased risk of disease

DRIVERS INFLUENCING EMERGING AND REEMERGING ZOONOSES

Human Population Growth and Distribution

Global Population Growth

The second half of the 20th century was a time of unprecedented population growth. According to United Nations (UN) estimates and forecasts, the world population more than doubled from an estimated 2.5 billion in 1950 to more than 6.5 billion in 2005 (see Figure 3-3), an annual average growth rate of 1.72 percent (United Nations, 2007). Although growth rates peaked in the late 1960s at slightly more than 2 percent and had declined to slightly more than 1 percent in the first 5 years of the 21st century, annual population increments continued to increase in the late 1980s and were projected to peak at about 8 billion by 2050. The UN's medium variant forecast, based on the assumption of continued fertility declines in low-income countries, shows the world population continuing to increase to slightly more than 9 billion by 2050.

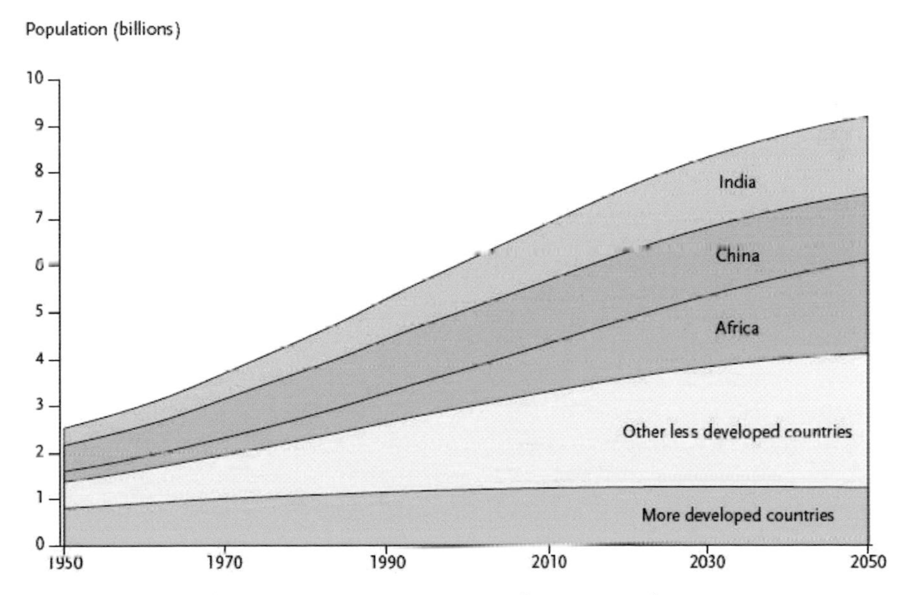

FIGURE 3-3 World population projections, median variant forecasts.
SOURCE: United Nations (2007). Reprinted with permission from the Population Reference Bureau.

Population growth has been unevenly distributed around the globe and is expected to become even more so in the next few decades. The developed countries—essentially Europe, North America, Australia, Japan, and New Zealand—represented nearly a third of the total growth in 1950, a proportion that had declined to less than 19 percent by 2005 (see Figure 3-3). Sub-Saharan Africa has shown the highest growth rates, averaging nearly 3 percent per annum in the late 1980s. The bulk of the absolute population increments have occurred in Asia, with annual increases reaching 57 million around 1985, declining only to slightly less than 50 million by 2005. More than half of these annual increases are now accounted for by South Central Asia, predominantly India, Pakistan, and Bangladesh. The bulk of future population growth is expected to occur in developing countries. The share of world population of the developed countries is forecast to decline to less than 14 percent by 2050, while sub-Saharan Africa is forecast to increase to nearly 20 percent. By 2050, of the global annual increment of 37 million, 22 million will occur in sub-Saharan Africa, whose population will still be increasing by more than 1 percent per annum, and 12 million will occur in South Central Asia (United Nations, 2007).

Population Mobility

Once a zoonotic disease has emerged, its spread in the human population is likely to be facilitated by population movements. Migration, also called long-term population resettlement, is likely to spread diseases that have a long period of latency or duration of infectiousness, whereas short-term mobility for periods of days or weeks, typical of "travel" patterns, may rapidly spread diseases with short resolution periods. The latter is illustrated by the spread of severe acute respiratory syndrome (SARS) from Hong Kong to Toronto within weeks in spring 2003, and the spread of the influenza A(H1N1) virus from Mexico to New York in April 2009.

Measurement of both intra- and international migration is poor, with most estimates coming from census data on birthplace. The global count of foreign-born persons now living in a different country has increased moderately, from about 75 million in 1965 to about 175 million in 2000 (United Nations, 2002). This growth is somewhat misleading, however, because a portion of the increase resulted from the break-up of the Soviet Union. About half of the world's international migrants have moved between developing countries. As of 1990, the United Nations (2002) estimated that about 13 percent of international migrants were living in Africa, 36 percent in Asia, 21 percent in Europe, 20 percent in North America, 6 percent in Latin America, and 4 percent in Oceania.

Population displacements as a result of conflict or natural disaster are likely to create conditions of crowding and poor sanitation that are highly conducive to the spread of infectious diseases. As of 2007, the Office of the United Nations High Commissioner for Refugees reported a total of 16 million refugees, under its or the United Nations Relief and Works Agency mandates, 26 million persons reported internally displaced as a result of conflict, and 25 million reported internally displaced as a result of natural disasters (UNHCR, 2009). Given the caveat that definitions and data collection procedures have varied over time, the numbers of refugees and internally displaced persons have not changed dramatically over several decades.

Human travel associated with tourism, business, and other moves not associated with changing residence have increased rapidly over the past 50 years and are projected to continue to increase. As shown in Figure 3-4, the revenue passenger kilometers represent the total number of passengers traveling globally multiplied by the number of kilometers they commercially fly, illustrating the increasing number of people and goods that are traveling farther and faster around the globe.

Human movement has significant implications for human and animal health. Not only are travelers (tourists, businesspeople, and other workers) at risk of contracting communicable diseases when visiting tropical countries, but they also can act as vectors for delivering infectious diseases to a different region or potentially around the world, as in the case of SARS. Refugees have become impoverished and more exposed to a wide range of health risks because of their status (Toole and Waldman, 1997), and

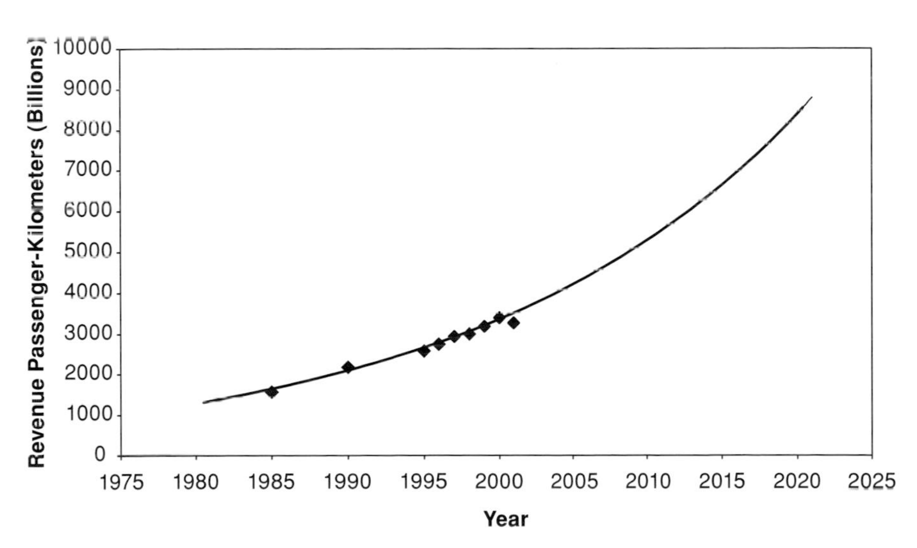

FIGURE 3-4 Volume of global air traffic, 1985–2001, and projection of future trends, 2001–2021.
SOURCE: Adapted from Daszak and Cunningham (2003).

their populations have been reported to harbor hepatitis B, tuberculosis, and various parasitic diseases (Loutan et al., 1997). Immigrants may come from nations where infectious diseases such as tuberculosis and malaria are endemic, and refugees may come from situations where crowding and malnutrition create ideal conditions for the spread of diseases such as cholera, shigellosis, malaria, and measles (CDC, 1998).

Urbanization

Populations in urban areas are typically less exposed to animal contact than rural populations, depending on the market structures and production systems of live food animals, but urbanites may also live in more crowded conditions conducive to disease transmission. The increase of global population over the past 50 years has been roughly paralleled by an increase in the level of urbanization. In 2005, the world's population was nearly 50 percent urbanized, a figure forecast to rise to nearly 70 percent by 2050 (United Nations, 2008). Developing countries as a whole, and South Central Asia and sub-Saharan Africa in particular, are somewhat less urbanized than the global average, though the differences have narrowed over time. By contrast, in all regions except sub-Saharan Africa, the rural population is forecast to be declining by 2050, and has probably been declining since the early 1990s in Latin America. Of course, cities grow in part by encroaching on surrounding farmland. The combination of reduced population increments and declining rural populations is likely to increase pressures on land resources in the future.

Human Behavior and Cultural Factors

Researchers have identified several social and cultural factors as drivers of emerging zoonotic diseases (Mayer, 2000; Patz et al., 2000; Daszak et al., 2001; Macpherson, 2005). Changing demographics and unprecedented population movement, as well as increased global flow of people, goods, food-animals, food products, and domestic and wild animals, all affect "microbial traffic" and emerging viral, bacterial, and parasitic zoonoses (Morse, 1993; Mayer, 2000). Social changes resulting in altered land and water-use patterns, intensified agricultural practices, deforestation and reforestation, and human and domestic animal encroachment on wildlife habitats also affect the movement of pathogens. These factors contribute to cross-species pathogen transmission and the emergence of new epidemic diseases that affect humans and animals, including the transmission of zoonotic diseases to humans and the anthropogenic movement of pathogens into new geographic spaces affecting the health of wildlife (Daszak et al., 2001).

Natural and Built Environments

The built environment—environments made, modified, and used by humans—is characterized by a sense of cultural aesthetics that influences how buildings, gardens, ponds, and parks are constructed. Environments are modified not only for aesthetic reasons, but also for utilitarian needs to provide a larger, general population with access to a public good or utility, such as dams for hydroelectric power or canal-building for transportation. Built environments have provided breeding sites for disease vectors such as *Aedes aegypti*, the mosquito which transmits dengue fever.

Culture, society, and religion influence the kinds of foods people eat, how foods are prepared, and the demand for foods at particular times (Shanklin, 1985). For example, each year 2–4 million Muslims from more than 140 countries make the pilgrimage to Mecca in Saudi Arabia for the Hajj or for Umrah (year-long lesser religious rites). During the religious festivals of Eid al-Adha,[3] up to 10–15 million small ruminants or 64 percent of the global trade of live sheep (Shimshony and Economides, 2006) are ritually slaughtered in various countries, including Saudi Arabia where Mecca is located, but even outside urban areas such as Washington, DC, to feed an estimated 12–15 million people. Most of these animals are shipped alive to the Arabian peninsula from countries across the Red Sea in East Africa and the Horn of Africa, where diseases that affect both humans and animals, such as the mosquito-borne disease Rift Valley fever (RVF), are endemic (Ahmed et al., 2006; Davies, 2006). Because animals are dispatched rapidly to preserve their value and the incubation period of diseases such as RVF is days longer than the transport time, conditions are ripe for disease spread. In 2000–2001, RVF was reported in Saudi Arabia (CDC, 2000) and has the potential to become an epidemic if not carefully monitored. Challenges to disease surveillance include not only heavy human and animal traffic and crowded conditions in ports and pilgrimage sites, but also political instability in the region and lack of cooperation among countries, which undermines the reporting of sick animals.

Food Preferences

Taste is a cultural phenomenon that influences food preparation and is also a driver of zoonotic disease transmission and infection. Globalization has also fostered the taste for foods from other cultures that contain raw meat or fish (e.g., sushi), and this can facilitate a number of parasitic zoonoses (Macpherson, 2005). In both Indonesia and China, a preference for

[3]Eid al-Adha (Arabic for "Festival of the Sacrifice") is a major Islamic festival that takes place at the end of the Hajj observed by Muslims throughout the world to commemorate the faith of Ibrahim.

the consumption of freshly slaughtered local chicken draws people to "wet markets" that vend live poultry (as well as other animals) for slaughter either onsite or at the buyer's home (Liu, 2008; Padmawati and Nichter, 2008). Local chickens in Indonesia are considered better tasting, resistant to disease, and strength-enhancing when consumed. Local chickens also fetch a higher price in the market and are trucked to major cities from the countryside to meet demand (Diwyanto and Iskandar, 1999). This practice puts consumers in contact with live fowl and freshly killed wild animals (primates, reptiles, bats, etc.) as well as domesticated animals (e.g., dogs, civets, pigs) and their feces, which may be infected with pathogens and contribute to the transmission of zoonotic diseases such as SARS and HPAI H5N1. Consumer preference for fresh products of wet markets is a complicating factor for health authorities that are trying to reduce health risks.

Bushmeat consumption, especially of primates, has been tied to zoonotic diseases such as HIV and Ebola (Peeters et al., 2002; Chapman et al., 2005; Daszak, 2006). Bushmeat may either be consumed as an inexpensive source of protein or as a sought-after delicacy, according to cultural value related to taste, wealth, and cultural significance. Bushmeat has cultural significance in not only religious rites, which increase demand for meat (Adeola, 1992), but also ethnic identity, nostalgia, and social memory (Holtzman, 2006). The demand for bushmeat is driven by cultural factors as well as wild game availability, poverty, food insecurity,[4] and an increased demand for protein. Increases in household wealth, however, appear to shift preference from bushmeat to the meat of domesticated animals (Schmink and Wood, 1992; Stearman and Redford, 1995) or narrow the range of bushmeat species consumed (Hames, 1991; Layton et al., 1991).

Most bushmeat is not taken in a simple subsistence manner, that is, directly from the forest to the table. An estimated 90 percent of all bushmeat consumed moves through a distinct and well-organized market chain, with numerous nodes along the supply chain where the meat changes hands multiple times between the animal's death and its presence on the dinner table (de Merode and Colishaw, 2006). The exchangers in this process include, among others, hunters, porters, bicycle traders, wholesalers, market-stall owners, and food preparers. Each person handling the meat or carcasses is

[4]In 2007, more than 900 million people suffered from malnutrition due to chronic food insecurity, an increase of 75 million in 1 year (FAO, 2009a). Recent events such as increased farming for use in biofuels, high world oil prices, and escalating consumer demand in emerging economies such as India and China have caused major fluctuations in food security, particularly for the urban poor, raising the number of people who are at least periodically food insecure to 2 billion (FAO, 2009a). Globally, bushmeat forms an important part of the diet for many poor households (de Merode et al., 2004). As prices of imports increase or strife breaks down international market chains, the consumption of bushmeat increases (Karesh et al., 2005).

exposed to the normal flora as well as any pathogens present. Additional sources of infection include the remnants and wastes from the carcasses, which could be scavenged and taken to even more new hosts.

Repeated transmission of viruses to humans, most of which do not result in human-to-human transmission, is termed "viral chatter" (Wolfe et al., 2005). For example, simian foamy viruses are known to infect bushmeat hunters regularly, but to date there has been no evidence of human-to-human transmission (Wolfe et al., 2004). More bushmeat means more viral chatter, which will increase the incidence of human infections, increase the number of pathogens that may infect humans, and increase the probability of eventual human-to-human transmission of one of these agents. As food insecurity increases, the bushmeat market becomes more essential and more lucrative, creating more opportunities for transmission of pathogens to humans.

The consumption of wild-animal products is also driven by cultural dietetic practices related to health promotion and disease treatment, known as zootherapeutics. Animal products are deemed to have medicinal value, and when consumed, play an important role in ethnomedical systems to increase strength as well as enhance virility (Afolayan and Yakubu, 2009) or to treat illness in humans and domestic animals (Martin et al., 2001; Mathias and McCorkle, 2004; Kakati et al., 2006; Mahawar and Jaroli, 2008; Soewu, 2008).

Companion Animals

The popularity of companion animals is a cultural phenomenon subject to social and economic contingencies. These include animals kept for display as well as animals for which humans develop a special relationship that extends beyond the animals' value for work, substance, or sale. For example, despite the risk of HPAI H5N1, backyard chickens are allowed in the kitchen and treated as companion animals by some Indonesians the same way an American might care for a dog or cat. Fighting cocks are groomed and handled daily by their owners who express considerable affection for them. Primates are kept as pets in parts of the Cameroon where high rates of simian immunodeficiency virus have been recorded (Peeters et al., 2002). Pastoralists in Africa and Hindus in India have special relationships with cattle that extend beyond their monetary or exchange value. Dogs and cats are the most popular companion animals (found in 63 percent of American homes) and are at once associated with positive health benefits ranging from physical health (e.g., lower blood pressure and cholesterol, increased exercise) to mental health (e.g., improved psychological coping with stress, decreased psychotropic medication use among the elderly). At the same time pet ownership increases the chances of zoonotic infection from several different types of diseases (e.g., salmonellosis and *Giardia*, *Cryptosporidium* and toxoplasmosis, rabies). The transnational trade in exotic animals from birds to nontraditional companion animals (e.g., prairie dogs that carry monkeypox in the United States) is growing and creating new challenges for both human and animal health professionals and demands their closer collaboration (Pickering et al., 2008).

Global Food Systems and Food Safety

The livestock production system,[5] farm and market structure, and farm geography are major variables that define the emergence and consecutive spread of a zoonotic disease.

Production Systems

Seré and Steinfeld (1996), who prepared the standard work on livestock production systems, distinguished two groups of farming systems. The first are the pure animal production systems, in which less than 10 percent of the total value of outputs comes from non-livestock farming activities, can be further differentiated into pure grassland-based systems and landless (or industrial) systems, which buy at least 90 percent of their feed from other enterprises. The second are the mixed farming systems, where livestock farming is associated with cropping. Globally, the mixed farming system is the most important producer of beef and milk. The production of pork is about equally distributed over mixed and industrial systems, whereas the industrial system is the dominant origin of poultry meat. The future will probably see a stagnation of the grazing system, a slight decrease in the mixed farming system, and a continuation of the strong increase in industrial swine and poultry production units (Steinfeld et al., 2006).

Farm and Marketing Structure

Projections suggest that farm size will increase in about one-half of the world, and shrink in the other. Economies of scale in production, and in particular in meeting stricter food safety and environmental standards and the low, marginal returns to labor in the food-animal production sector,

[5] A production system clusters production units (herds, farms, ranches), which, because of the similar environment in which they operate, can be expected to produce according to similar production functions. This similar environment can be characterized by the physical (climate, soils, and infrastructure) and biological environments (plant biomass production, food-animal species composition) and economic and social conditions (prices, population pressure and markets, human skills, and access to technology and other services) and policies (land tenure, trade, and subsidy policies) (Seré and Steinfeld, 1996).

will drive the process of the increase in size and scale in the industrialized world. For example, in the United States, the share of the value of pork production from farms with sales of $500,000 increased from 14 percent in 1989 to 64 percent in 2002, and for poultry meat, from 40 to 68 percent over the same period (McDonald et al., 2006). On the other hand, in most developing countries, population pressure has led to an increase in the number of farm holdings and a subsequent decrease in farm size. For example, in India, the number of farm holdings increased from 70 million in 1970–1971 to nearly 98 million in 1985–1986. Farm holdings further increased to approximately 105 million in 1990–1991, with a major shift to landless and marginal farm holdings (AERC, 2005).

Balance of Food Production and Its Ecological Impacts

Livestock production is strongly linked to land. Livestock production uses nearly 4 billion, generally intensively managed hectares (ha) of land, of which 0.5 billion are for feed crops such as corn and soya (33 percent of the total cropland); slightly more than 1 billion are for pasture with relatively high productivity, and the remaining 2 billion ha are extensive pastures with relatively low productivity (Steinfeld et al., 2006). Expansion of demand for food-animal products can be met by intensifying land use, increasing the yield per unit area, or expanding the area under feed crops or grassland. Until the 1960s, increasing the livestock population and expanding the area under feed and fodder crops have been the main trends. As a result, the conversion of natural habitats to pastures and crop land has been rapidly growing. More land has been converted for the growing of crops between 1950 and 1980 than in the preceding 150 years (MEA, 2005). There are major regional differences, however, with continuing strong crop-land area expansion in Asia and Latin America, but a reduction of agricultural land-use in North America and Europe. These trends are expected to continue, with a stronger accelerating conversion of natural habitat into crop land in sub-Saharan Africa (Steinfeld et al., 2006).

Recent trends show a tendency toward intensification, with higher yields per area of feed crops and per animal, and lower feed inputs per unit of production. For example, global corn yields increased from 31,542 hectograms (hg) per ha in 1980 to 50,102 hg per ha in 2005 (FAO, 2009b), and the amount of feed required to produce 1 kilogram (kg) of poultry meat decreased in the United States from 1.92 kg in 1957 to 1.62 kg in 2001 (Havenstein et al., 2003). This increase in productivity has been achieved through a greater use of capital and technology, mainly through purchased goods (e.g., feed and pharmaceutical inputs) and services (e.g., animal health and expert advice).

Parallel Evolution of Marketing Systems and Production Geography

Production and marketing systems develop to supply demand for animal products most efficiently while reducing production and delivery costs. The marketing system differs depending on the pattern of food-animal production. In relatively simple production systems, distances to markets are short, and most products are marketed on foot or fresh in wet markets. Unsold stock or products, after having been in contact with live or fresh material from other origins, are often taken outside the market, thus increasing the chance of disease spread. As economic development progresses further, and distances between producer and consumer lengthen, supermarket chains with more stringent standards emerge. Their share in total sales is rapidly increasing, in particular in East Asia and Latin America (Reardon et al., 2003).

These trends have major implications for the emergence of zoonotic diseases. In countries where consumption and production grow most, which cover a large part of the developing world, there is still a high density of smallholders, together with an emerging, often poor biosecure industrial sector. This was described as a high-risk situation in the emergence of HPAI H5N1 (Slingenbergh et al., 2004). Moreover, the concentration of the larger industrial operations around the urban centers results in major environmental pressures on soil and water. This presents another set of conditions favorable for the emergence of new zoonotic pathogens, although if they are professionally managed and adopt highly integrated production compartments with strict biosecurity measures, they actually reduce the animal-human interface and can reduce the disease pressure. Finally, these risks are further exacerbated by the open market system.

The Case of Poultry Production in Southeast Asia

Smallholder poultry keeping, also known as "backyard poultry," has been advocated for decades by the Food and Agriculture Organization of the United Nations as a strategy for poverty reduction. The greatest density of poultry is in East and Southeast Asia (see Figure 3-5). Along with wet market supply of fresh poultry, there has been an increasing urbanization of smallholder poultry keeping. As previously mentioned, urbanization of the human population has been rapid, and the migration of people has been accompanied by the migration of their animals. For example, the global distribution of swine appears to be heavily concentrated in East and Southeast Asia, along with poultry (see Figure 3-6). This can present public health concerns and challenges, given that pigs can play a crucial role in influenza ecology and epidemiology because of their susceptibility to both human and avian viruses; scientists consider them a "potential 'mixing

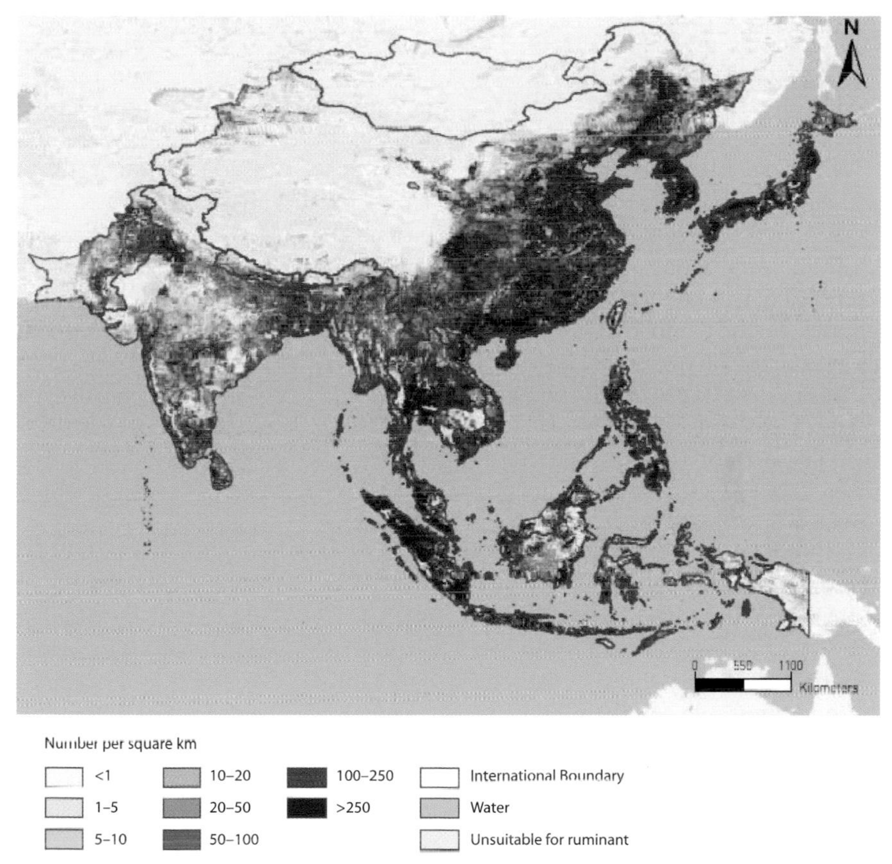

FIGURE 3-5 Distribution of poultry in East and Southeast Asia.
SOURCE: FAO (2007). Reprinted with permission from FAO.

vessel' for influenza viruses, from which reassortants may emerge" (Capua and Alexander, 2008, p. 4).

Two major trends have occurred in poultry agriculture in the region since the 1960s. First, intensive poultry agriculture was introduced into Thailand in the late 1960s through a strategic partnership between the Charoen Pokphand Corporation (known as "CP Corp") and Arbor Farms in the United States. This was a core technology that was adopted to create the first fully vertically integrated approach (seeds for animal feed, and animals purposed for fast food) in Asia. In 1978, CP Corp registered as corporation #001 in the People's Republic of China and introduced the first barns containing more than 5,000 birds into that country. By the 1990s, CP Corp was the largest chicken producer in Asia (Horn, 2004), and by

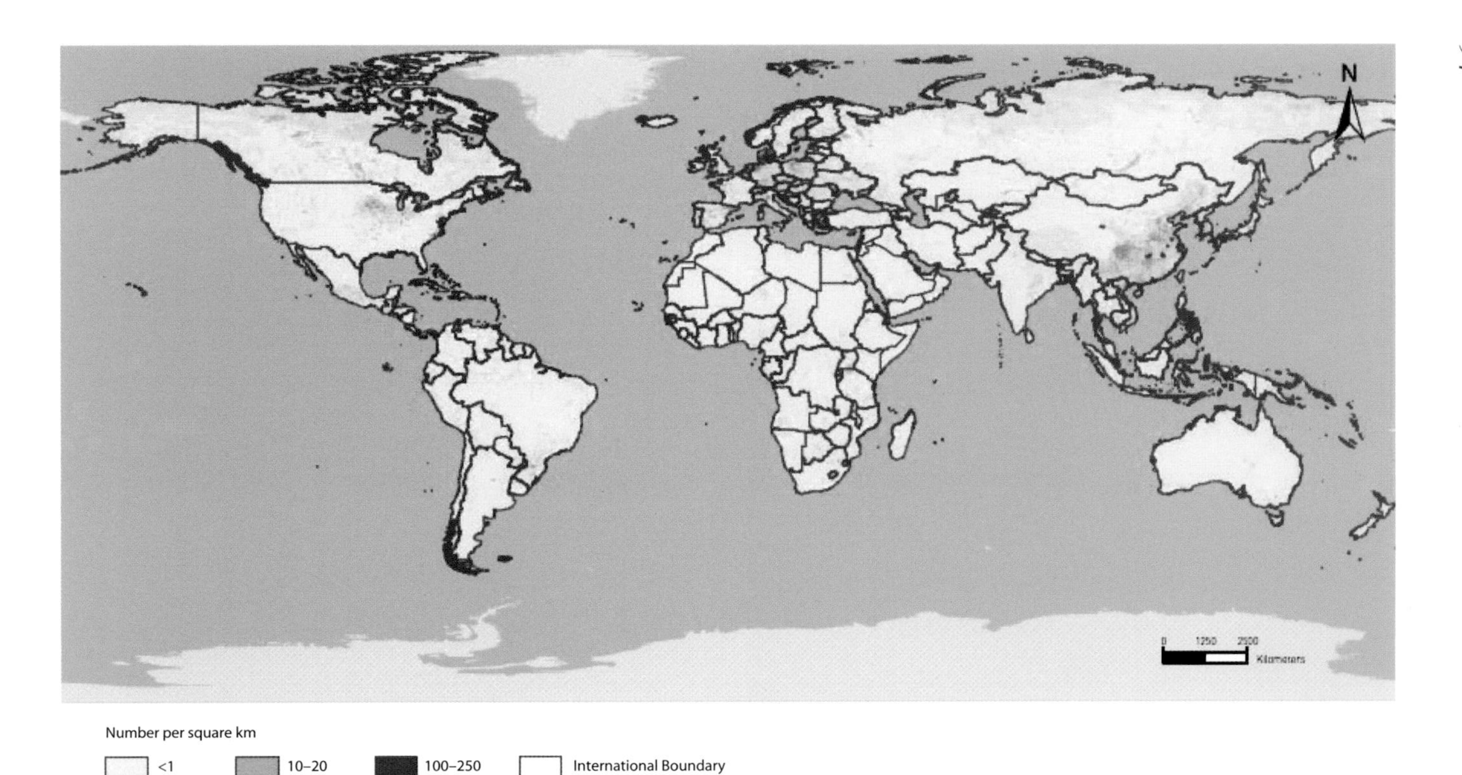

FIGURE 3-6 Global swine distribution.
SOURCE: FAO (2007). Reprinted with permission from FAO.

2003, Thailand was the third largest producer of poultry in the world. It is not known to what extent these coincidental trends may have "set the stage" for the avian influenza outbreaks that have ravaged the region since 2003 (Kimball, 2006). Other issues unique to Southeast Asia confound the control of avian influenza. Waterfowl are asymptomatic reservoirs for HPAI H5N1. Thus, the traditional practice of free-range raising of ducks serves to disseminate infection among vulnerable poultry flocks.

Legal and Illegal Trade

Legal Wildlife Trade

Few reliable estimates can quantify the global illegal trade in wildlife[6] or its value, but some estimates are in the billions of dollars annually. Some analysts identify the United States, the People's Republic of China, and the European Union as the areas with greatest demand, driven by the need for specific animal parts to use in zootherapeutics (e.g., powdered rhino horn), for human consumption (e.g., bushmeat), as symbols of wealth (e.g., hunting trophies), and as exotic pets (e.g., black palm cockatoos). The United States purchases nearly 20 percent of all legal wildlife products on the global market (CRS, 2008). Source countries of both legal and illegal exports tend to include developing countries with rich biological diversity (CRS, 2008).

The World Organization for Animal Health (OIE), acting through the World Trade Organization, deals almost entirely with a series of diseases listed as "notifiable," which are of importance to agriculture and trade. High-impact diseases that are present in introduced wildlife and that do not affect human or food-animal health are rarely the subject of legislation, even though OIE has the authority to list wildlife diseases as notifiable due to their impact on wildlife and the environment. In both developed and developing countries, the legislative authority and responsibility over human and ecosystem health impacts of the wildlife trade are unclear or poorly coordinated.

A recent study by the Consortium for Conservation Medicine showed that more than half a million shipments containing more than 1 billion live animals were imported into the United States between 2000 and 2006

[6]Illegal trade in wildlife is defined as "Illicit procurement, transport, and distribution—internationally and domestically—of animal parts and derivatives thereof, in contravention of laws, foreign, and domestic, and treaties. Illegal wildlife trade ranges in scale from single-item, local bartering to multi-ton, commercial-sized consignments shipped all over the world. Wildlife contraband may include live pets, hunting trophies, fashion accessories, cultural artifacts, ingredients for traditional medicines, wild meat for human consumption, and other products" (CRS, 2008, p. 1).

(Smith et al., 2009). Nearly all of these shipments were designated for commercial purposes (e.g., pet and food trade), and nearly 80 percent contained animals from wild populations (Smith et al., 2009). Annual shipments of live animals traded by the United States increased significantly over the time period of the study, as did the number of individual animals traded.

In the United States, the U.S. Department of Agriculture is responsible for health inspection of wildlife shipments, but only for those animals used in food production. Thus, when wildlife are imported into the United States, they are inspected by the U.S. Fish and Wildlife Service (USFWS)[7] to examine their CITES (The World Conservation Union's Convention on International Trade in Endangered Species of Wild Fauna and Flora) status, but minor clinical signs are unlikely to be reported. Wildlife reservoirs of zoonotic pathogens often show no clinical signs, so they would likely be missed in the USFWS screenings of shipments.

Furthermore, the focus of the agency is conservation, not disease prevention and detection. In 2007, the USFWS processed 188,000 wildlife shipments worth more than $2 billion, conducted 14,000 investigations, and recorded a total of more than 200 million live wildlife legally imported into the United States (Einsweiler, 2008). By the fourth quarter of 2008, USFWS had 114 inspectors stationed at 38 ports of entry/exit and 201 special agents stationed around the country. Even with these resources, the agency physically inspects an average of only 25 percent of all wildlife shipments (Einsweiler, 2008).

The U.S. Centers for Disease Control and Prevention (CDC) also plays a role in regulating and monitoring the U.S. importation of animals used for nonfood production purposes.[8] A recent example is the 2003 U.S. outbreak of human monkeypox, a zoonosis harbored by African rodents imported into the United States for the pet trade. After 215 CDC employees spent 65 person-days investigating 72 human cases and confirming 37 of the cases (Marano, 2008), CDC and the Food and Drug Administration (FDA) jointly used emergency powers to ban importation of this pathogen's species

[7]The U.S. Fish and Wildlife Service (USFWS) conserves, protects, and enhances fish, wildlife, and plants and their habitats and is responsible for ensuring that imports meet international CITES (The World Conservation Union's Convention on International Trade in Endangered Species of Wild Fauna and Flora) requirements. The USFWS collaborates with the U.S. Department of Agriculture, the U.S. Food and Drug Administration, the U.S. Centers for Disease Control and Prevention, and U.S. Customs. Globally, it collaborates with the INTERPOL Wildlife Crime Working Group and the Association of Southeast Asian Nations Wildlife Law Enforcement Network (Einsweiler, 2008).

[8]The U.S. Centers for Disease Control and Prevention's Division of Global Migration and Quarantine enforces Department of Health and Human Services' authority at 20 ports of entry to protect human health and has the authority to restrict importation of animals and products if they pose threats to human health. These may include dogs, cats, turtles, tortoises, terrapins, nonhuman primates, etiologic hosts, vectors, agents, African rodents, persons, carriers, and things (IOM, 2006; Marano, 2008).

reservoirs and restrict interstate movement of African rodents and prairie dogs. In September 2008, FDA lifted its restrictions on the interstate movement of prairie dogs, but the CDC national importation ban remained in place. A CDC official noted, however, that the ban on African rodents also resulted in an increase in the U.S. importation of rodents for the commercial pet trade from other continents, especially Asia (Marano, 2008).

The trade in wildlife has led to the introduction of pathogens that threaten human and animal health, agricultural production, and biodiversity. The human-mediated introduction of infectious disease and vectors, termed "pathogen pollution" (Daszak et al., 2000), is expected to continue to rise via future expansion of global travel and trade (Cunningham et al., 2003; Daszak and Cunningham, 2003). There appears to be a growing awareness of this impact by the wildlife trade, particularly following SARS and human monkeypox. This adds pressure to deal with the welfare and conservation impact of the trade, in particular the repeated introduction of invasive species (Eterovic and Duarte, 2002; Reed, 2005; Fowler et al., 2007).

ENVIRONMENTAL FACTORS

Emerging infectious diseases are by definition in a process of flux, either rising in incidence, expanding in host or geographic range, or changing in pathogenicity, virulence, or some other factor. It is increasingly clear that large-scale, often anthropogenic, environmental changes are among the most important drivers of emerging zoonoses. These drivers include land-use changes (e.g., deforestation, agricultural encroachment, and urban sprawl), climate change, and more subtle products of anthropogenic change such as biodiversity loss (IOM, 1992; Krause, 1992, 1994; Morse, 1993; Daszak et al., 2000, 2001; Anderson et al., 2004). These drivers often act via complex pathways that are poorly understood. For example, fragmentation, which may be due to suburban expansion of housing developments, generally leads to loss of biodiversity; this has been linked to heightened Lyme disease risk in the northeastern United States (Ostfeld and Keesing, 2000; Allan et al., 2003; LoGiudice et al., 2003).

Unraveling this complexity will require long-term field research to account for annual variation in environmental or other factors. For example, it has taken more than a decade to demonstrate the mechanistic interaction of biodiversity changes and Lyme disease risk in the United States, and the link between El Niño-Southern Oscillation (ENSO), rainfall, and hantavirus pulmonary syndrome in the Southwest desert (Mills et al., 1999). However, these studies have key value to human and animal health in that they demonstrate causative links that can be used, for example, to predict climate-linked outbreaks of vector-borne diseases (Linthicum et al., 1987, 1999).

Deforestation

Rates of deforestation have increased exponentially since the beginning of the 20th century. Although reforestation has been conducted in some developed countries (e.g., parts of Europe and the United States), 2–3 percent of global forests continue to be lost each year with the majority of losses in tropical countries. Deforestation and processes that lead to it have a number of ecosystem consequences. Deforestation decreases the overall habitat available for wildlife species. It also modifies the structure of environments, for example, by fragmenting habitats into smaller patches separated by agricultural activities or human populations. Increased "edge effect" (from a patchwork of varied land uses) can further promote interaction among pathogens, vectors, and hosts. This edge effect has been well documented for Lyme disease (Glass et al., 1995). Similarly, increased activity in forest habitats (through human behavior or occupation) appears to be a major risk factor for leishmaniasis (Weigle et al., 1993). Evidence is mounting that deforestation and ecosystem changes have implications for the distribution of many other microorganisms and the health of human, domestic animal, and wildlife populations.

Deforestation, with subsequent changes in land-use and human settlement patterns, has coincided with an upsurge of malaria and its vectors in Africa (Coluzzi et al., 1979; Coluzzi, 1984, 1994), in Asia (Bunnag et al., 1979), and in Latin America (Tadei et al., 1998). When tropical forests are cleared for human activities, they are typically converted into agricultural or grazing lands. This process is usually exacerbated by road construction, which causes erosion and allows previously inaccessible areas to become colonized by people (Kalliola and Flores, 1998). Cleared lands and culverts that collect rainwater are in some areas far more suitable for larvae of malaria-transmitting Anopheline mosquitoes than are intact forests (Tyssul Jones, 1951; Cruz Marques, 1987; Charlwood and Alecrim, 1989). Deforestation and logging often result in exposure of small groups of people and food-animals to new pathogens, particularly where bushmeat hunting occurs (Wolfe et al., 2000). Finally, land-use changes drive some of these pathogen introductions and migrations, and those changes increase the vulnerability of habitats and populations to these introductions. Human migrations also drive land-use changes that, in turn, drive infectious disease emergence.

Habitat Fragmentation

One of the key products of anthropogenic land-use change is the fragmentation of wildlife habitat, which alters the composition of host species in an environment and the fundamental ecology of microorganisms. Top predators and other species at higher trophic levels usually exist at low-population density and are sensitive to changes in food availability. The smaller patches left after fragmentation reduce sufficient prey populations, causing local extinction of predators and a subsequent increase in the density of their prey species. Smaller fragments in North American forests have fewer small mammal predators and higher densities of white-footed mice, a highly competent reservoir of the Lyme disease pathogen *Borrelia burgdorferi* (Ostfeld and Keesing, 2000). In these fragments, the risk of Lyme disease infection in people is higher; in less modified habitats, increasing diversity of alternative and less competent reservoirs dilute this risk (Ostfeld and Keesing, 2000). Therefore increasing diversity provides a "dilution effect"—a buffer against disease risk that is lost when habitat is fragmented (Schmidt and Ostfeld, 2001).

Agriculture

Crop Irrigation and Breeding Sites

Agriculture occupies most of the world's arable land and uses more than two-thirds of the world's fresh water (Horrigan et al., 2002). The subsequent increase in irrigation reduces water availability for other uses and increases breeding sites for disease vectors. Irrigation development in the southern Nile Delta following construction of the Aswan High Dam has caused a rapid rise in mosquito populations and an increase in the *Culex*-borne disease, Bancroftian filariasis (Harb et al., 1993; Thompson et al., 1996). Onchocerciasis and trypanosomiasis are further examples of vector-borne parasitic diseases that may be triggered by changing land-use and water management patterns. In addition, large-scale use of pesticides has had other deleterious health effects on farm workers, including poisoning, hormone disruption, and cancer (Blair et al., 2005; Bretveld et al., 2006; Calvert et al., 2008).

Food-Borne Diseases

The expansion of international food trade has led to a series of disease outbreaks and the emergence of some novel agents. U.S. importation of strawberries from Mexico, raspberries from Guatemala, carrots from Peru, and coconut milk from Thailand have caused recent outbreaks. Some recent outbreaks of food-borne diseases in meat and vegetables can also be attributed to domestically produced food. Food safety is an important factor in human health. Food-borne disease accounts for an estimated 76 million illnesses, 325,000 hospitalizations, and 5,200 deaths in the United States each year (CDC, 2005). Other dangers include antibiotic-resistant organisms, such as Cyclospora, *Escherichia coli* O157:H7, and other pathogenic *E. coli* associated with hemolytic uremic syndrome in children (Dols et al., 2001).

Secondary Effects

There are secondary health effects associated with agricultural production. Examples include the emerging microbial resistance from antibiotics in animal waste that is found in groundwater fed from farm run-off, and the introduction of microdams for irrigation in Ethiopia that resulted in a seven-fold increase in malaria (Ghebreyesus et al., 1999).

Encroachment into Wildlife Habitat

Alterations of ecosystems and natural resources contribute to the emergence and spread of infectious disease agents. Human encroachment on wildlife habitat has broadened the interface between wildlife and humans, resulting in increased opportunities for both the emergence of novel or re-emergence of known infectious diseases in wildlife and their transmission to people. Rabies is an example of a zoonotic disease carried by animals that has become habituated to urban environments. Bats colonize buildings; skunks and raccoons scavenge human refuse; and in many countries, feral dogs in the streets are common and a major source of human infection (Singh et al., 2001).

Infectious diseases can also pass from people to wildlife. Nonhuman primates have acquired measles from ecotourists (Wallis and Lee, 1999). Also, drug resistance in gram-negative enteric bacteria of wild baboons with limited human contact is significantly less common than in baboons near urban or semi-urban human settlements (Rolland et al., 1985).

Climate Change

Climate models for greenhouse warming predict that geographic changes will take place in a number of water-borne (e.g., cholera) and vector-borne (e.g., malaria, yellow fever, dengue, leishmaniasis) diseases. These changes will be driven largely by increases in precipitation leading to favorable habitat availability for vectors, intermediate and reservoir hosts, or warming that leads to expansion of ranges in low latitudes, oceans, or mountain regions. Two phenomena indicate that climate change will likely have a heightened impact on key human diseases. First, a strong link exists between ENSO and outbreaks of RVF, cholera, hantavirus, and a range of emergent diseases (Colwell, 1996; Bouma and Dye, 1997; Linthicum et al., 1999; Anyamba et al., 2009). If ENSO cycles become more intense, as they are predicted to do under climate change scenarios, these events may become more extensive and have greater impact. Secondly, recent expansion of *Culicoides* species, the vector species that spreads the diseases bluetongue and African Horse Sickness, into Northern Europe, has led to outbreaks of bluetongue there as recently as 2006, and has put Europe on alert for the potential introduction of African Horse Sickness. The recent geographic expansion of this vector species has been hypothesized to have a climate-change link, although this remains a controversial point (Purse et al., 2005; Wilson et al., 2008).

TECHNOLOGICAL CHANGES LEADING TO DISEASE EMERGENCE

Disease Diagnosis and Detection

Routine disease diagnosis has a central role in disease surveillance. Although it is not a direct driver of disease emergence, differences in laboratory diagnostic approaches and diagnostic goals between the human and animal health fields, variable levels of communication, and limited comparison of microbial populations in humans and animals can hinder early recognition of an emerging zoonotic disease event. These factors can delay intervention and response with consequent amplification of the impact in both human and animal populations.

The laboratory infrastructure and approach is quite different in resource-constrained countries. Although some point-of-care assays for targeted diseases such as avian influenza are available for animals, few are actually deployed in laboratories at the district or community level. Assays for zoonotic diseases such as brucellosis—which are simple, commonly used in developed countries, and easily deployed—are not uniformly available in developing countries. Routine infectious disease diagnosis in animals is virtually nonexistent in sub-Saharan Africa and in much of the Near and Far East, where expertise that is on par with most state diagnostic laboratories is simply not available. Diagnosis of animal diseases is often established in the field through familiarity of field personnel, such as veterinarians or community animal health paraprofessionals, with clinical presentations for transboundary infectious diseases of importance to the country for trade and disease-free status. Confirmatory diagnosis is made in national laboratories when possible, and OIE reference laboratories when not. Some of these diseases will be zoonotic (e.g., RVF), while many are not. As a result, diagnosis of zoonotic diseases in developing countries is most often first made in humans. However, diagnosis of zoonotic disease agents is also quite limited in resource-constrained countries except at the national level.

Exceptions can be found, most often supported by a combination of national, donor nation, and nongovernmental organization funding. Examples include the CDC International Emerging Infections program at the Kenya Medical Research Institute, the Uganda Virus Research Unit in Entebbe, and the International Center for Diarrheal Diseases Research, Bangladesh. In general, however, the challenges of routine diagnosis of and communication about zoonotic agents found in developed countries are exponentially amplified in the developing world by a nearly universal lack of sustained laboratory infrastructure for disease diagnosis. As a result, the majority of infectious diseases remain undiagnosed in much of the developing world. The threat of pandemic influenza and other emerging diseases has stimulated donor support to develop the ability to diagnose specific agents in humans. Unfortunately, the animal disease diagnostic infrastructure has not been included in this enhanced donor support in most resource-constrained countries. Additionally and with few exceptions, communication between the human and animal health sectors remains limited.

Early recognition and intervention in an emerging infectious zoonotic disease event is essential to limit spread, whether it involves a novel agent such as the SARS virus or an adaptation of a routinely recognized pathogen such as influenza virus. Limitations in conventional approaches to diagnosis of infectious diseases in humans and animals, while not directly driving emerging disease events, can contribute to spread within the population. Differential diagnoses for unusual disease events need to be expanded to include the unknown or not-yet-discovered pathogen. Recognition of these limitations will help inform a strategic approach toward effective zoonotic disease surveillance.

Farm Management

As identified earlier, the most remarkable trend in farm management over the past 30 years has been toward intensification, which has its origins in the United States. The ready availability of inexpensive grain and the rapid growth of an efficient transportation system have made it possible to supply large concentrations of animals with sufficient feed. As shown in Box 3-1, large-scale facilities in manmade ecosystems permit the production of more units of consumable nutrients produced per unit of input than other systems. Intensive agriculture has since spread to all parts of the world, and it has both advantages and disadvantages (see Box 3-2).

Disease Management for Food-Producing Animals

As previously mentioned, food-producing animals are economic entities. Disease treatment is not administered to individual animals; instead, the entire herd is monitored. Although it might seem easy to protect the human population from serious zoonotic diseases (e.g., anthrax or *Brucella*)

BOX 3-2
Advantages and Disadvantages of Intensive Agriculture Related to Zoonotic Diseases

Advantages

- **Increased ease of monitoring.** With animals congregated and the focus on profit, avoiding disease is important in minimizing losses. Consequently, there are good incentives for continual health and disease observations, as well as working quickly to stamp out any disease that emerges in order to maximize profits.
- **Improved food security.** Any losses due to disease are decreased in intensive production systems, and thus more animal protein is produced. Intensive systems are more efficient than extensive or household systems, so overall increased animal protein is available.
- **Increased ease of biosecurity.** In the agricultural context, biosecurity means the protection of animals from external diseases and pests. In a large operation that is well maintained, controlling access in and out is easily accomplished. There are incentives to biosecurity as maximal production requires optimal health, and therefore increased biosecurity is more profitable. Biosecurity in extensive or household operations is extremely difficult.

Disadvantages

- **Increased probability of the spread of a novel agent.** The likelihood of a pathogen spreading is greater as a function of having a dense herd or flock. This was seen with Nipah virus spread in Malaysia in 1998–1999 among pigs, farmers, and bats (Daszak et al., 2006).
- **Increased concentration of environmental degradation.** Controlling waste products from an intensive operation is challenging. Although the same waste products are generated from the equivalent number of animals kept under extensive conditions, the waste products with intensive production systems are concentrated. In most developed countries, strict environmental regulations are in place regarding disposal and treatment of waste from these concentrated operations. However, in less regulated environments, the waste can be dispensed inappropriately into areas that might allow for transmission of intestinal pathogens into humans.

through vaccination of all at-risk animals, in practice, food-animals are only vaccinated against diseases as a matter of cost–benefit if there is a concern regarding the health of the herd or a high probability of human health risk.

Although the topic of antibiotic resistance is beyond the charge of the committee and is in itself the topic of other major studies, the committee recognized the importance of the issue to make a few observations. Antibiotics are commonly used in food-animals as a prophylactic measure, as "growth promoters," and as a treatment in a very minor proportion. The use of antibiotics for growth promotion began in the 1940s when the poultry industry discovered that the use of tetracycline-fermentation by-products resulted in improved performance (Stokstad et al., 1949), although the mechanisms for improved performance are not completely understood. Research has suggested that growth promotion works by affecting changes in intestinal tract microorganisms, resulting in better absorption of nutrients and consequently improvements in weight gain (Stock and Mader, 1984; Preston, 1987; Elam and Preston, 2004). Poultry and swine production systems account for most of the use of antibiotics in feed, with 44 and 42 percent of all growth-promotant antibiotics used in these two species, respectively. Beef production is responsible for the remaining 14 percent (Mellon et al., 2001). The discontinued use of fluoroquinolones and macrolides in U.S. broiler production could predispose people to greater health risks as a result of increased illness rates in animals, greater microbial loads in servings from affected animals, and hence increased potential for human illness (Cox and Popken, 2006).

Other investigators have found direct links between the feeding of antibiotics and the presence of resistant bacteria in the vicinity, with potential spread to humans. Tetracycline resistance was found in 77 percent and 68 percent of *E. coli* and *Enterococci* isolated from samples obtained at a swine concentrated animal feeding operation (CAFO) in the United States (Stine et al., 2007). In a Danish study (Smith et al., 2002), the application of pig manure as fertilizer for farmland resulted in the detection of elevated occurrences of tetracycline-resistant bacteria in the soil immediately after pig manure slurry was spread. Gibbs and colleagues (2006) evaluated the air plume downwind from a CAFO and found a greater concentration of antibiotic-resistant bacteria within and downwind of the swine facility than upwind. Some reports have postulated an association between human and animal health, food-animal antibiotic resistance, and antibiotic resistance in clinical isolates (Teuber, 2001; Smith et al., 2002). Clearly there is concern regarding low-level antibiotic use in food-producing animals, and more scientific data are needed to develop meaningful policies and procedures to protect both human and animal health while optimizing food-animal production.

Biotechnology and Lack of Biosecurity

Biotechnology has precipitated disease emergence in three ways: (1) through medical innovations; (2) as a result of laboratory escapes; and (3) through personal contact with laboratory animals or biological agents in a research setting. A further area of concern is bioterrorism and the manipulation of microbiological agents to make them more readily contagious or infectious among humans.

Medical Innovations

In recent years, transplantation has resulted in several cases of zoonotic diseases infecting transplant recipients. Perhaps the most widely cited instance was an organ donor who was infected with rabies. His organs subsequently infected and killed four transplant recipients (Burton et al., 2005). A second instance involved two clusters of unusual disease in transplant recipients, in which lymphocytic choriomeningitis virus was eventually diagnosed, with seven or eight transplant recipients dying. The organ donor kept a pet hamster that had a strain identical to those isolated from some of the transplant recipients (Fischer et al., 2006).

Xenotransplantation is the transplantation of living organs, tissues, or cells from one species to another, and is considered by some as a solution to the shortage of human organs and tissues. In the late 1990s, several companies were working with pigs that were genetically modified to have a human gene to help decrease the organ rejection response. These pigs were bred to fill the supply–demand gap for human organ transplantation. However, the discovery of an endogenous porcine retrovirus slowed the enthusiasm for this developing field because it proved extremely difficult to create a population of pigs without this retrovirus. The retrovirus is present in the genome in multiple copies. Researchers feared the virus could emerge from porcine-origin cells in intimate apposition within the circulation of the recipient human and adapt to create a transmissible epidemic (Boneva et al., 2001).

Laboratory Escapes

The SARS virus was grown and studied in numerous laboratories around the world. Spread outside of the laboratory has occurred on several occasions, including accidents in Taiwan, Singapore, and China. The incident in China was particularly worrisome as it resulted in three cycles of person-to-person transmission (Lim et al., 2006). Perhaps the most notable and devastating example of laboratory escape is the 1979 incident at Sverdlovsk, Russia, where anthrax spores were disseminated within a population due to inadequate biosecurity and failure to change filters in a timely and adequate manner. This escape resulted in nearly 70 human deaths (National Security Archive, 2001).

Laboratory Animals or Biological Agents in Research

As biotechnology grows and studies in animals continue, there is always the possibility of zoonotic disease occurring in the scientific staff who are responsible for the care of the animals, or in laboratory workers engaged in microbiological aspects of the disease. There have been numerous instances of humans becoming infected with a zoonotic agent within a laboratory, either through contact with animals or working with the infectious agent. To date none of these has resulted in subsequent person-to-person spread. Examples include glanders, tularemia, Q fever, Venezuelan equine encephalitis, and herpes B (Hall et al., 1982; CDC, 2000; Rusnak et al., 2004).

Bioterrorism

Though intentional release and use of pathogens to threaten a nation's security is also beyond the scope of this study, it is important to mention that it as a driver for zoonotic diseases. In fact, many of the CDC Category A, B, and C bioterrorism agents—such as anthrax, plague, tularemia, brucellosis, and cryptosporidium—are zoonoses. Much has been written about the potential of biotechnology to create a "superbug," an organism that could pass rapidly through the population, causing massive morbidity and mortality. To date there is little scientific evidence that this is easily achievable, but the threat remains.

INADEQUATE GOVERNANCE

Inadequate governance systems at the local, national, and international levels are another driver. For purposes of this report, "governance" refers to the structures, rules, and processes that societies individually and collectively use to organize themselves to prevent, prepare for, and respond to human and animal health threats. Each driver analyzed in this chapter raises its own set of governance issues within countries and in the relations between nations. The most effective way to prevent zoonotic disease threats is to bring the various drivers of such threats under better control. However, increasing fears of zoonotic disease emergence and spread underscore the lack of confidence in the legal, regulatory, and enforcement mechanisms established by nations to address the political, economic, and cultural trends that exacerbate zoonotic threats.

Poor governance that undermines a country's ability to prevent zoonoses from emerging and to control the harm their spread might cause flows from many factors. These include the absence of needed regulatory authority, antiquated rules, uncoordinated policy and governmental capacities, lack of resources to devote to addressing difficult health, social, and economic problems, and the speed and scale of globalization.

Governance capacities are crucial to fund, organize, and operate the rules, personnel, laboratory capabilities, information networks, and response interventions needed to identify zoonotic threats early and to act swiftly against them. Crafting and sustaining integrated human and animal health governance capacities locally, nationally, and globally proves difficult for many reasons ranging from complacency in developed countries to the debilitating effects of widespread poverty in least-developed nations. Despite these difficulties, these capabilities need support by strong governance strategies and mechanisms because they serve national interests for human and animal population health; thus governmental bodies need to take responsibility for disease prevention, surveillance, and response. Failure to do so not only contributes to the emergence and spread of zoonotic pathogens, but also creates a blind spot in any attempts to establish a global system of disease surveillance, prevention, and control. Chapter 7 provides an in-depth discussion about the governance challenges facing countries and the international community.

CONCLUSION

The drivers of zoonotic disease can be quite complex—individually and collectively. Although some of these drivers may be understood in isolation or in their simpler, temporal interactions with each other (e.g., food insecurity for workers in a logging or mining camp in Africa, leading to increased hunting and consumption of bushmeat), the complex ways in which they change over time (sometimes in lengthy intervals as with HIV) and how they interact are not well understood. Constant with the coexistence of humans on the planet are the challenges that the drivers present for when, how, and where zoonotic diseases will emerge.

The committee concludes that there are few efforts for regular or systematic review of the scientific information about these drivers. Such a review is needed to inform strategic action that can mitigate the consequences of drivers by national and global policymakers or international donors dedicated to global development and poverty reduction. The efforts are also minimal when governments or governance entities negotiate international treaties for activities or interests not specifically geared toward protecting human and animal health, but which may impact them. The committee also concludes that dedicated attention and resources to improve our recognition of and comprehension about these factors is a significantly noticeable gap in global zoonotic disease surveillance, reporting, and response efforts.

REFERENCES

Adeola, M. O. 1992. Importance of wild animals and their parts in the culture, religious festivals, and traditional medicine of Nigeria. *Environ Conser* 19:125–134.

AERC (Agro-Economic Research Centre, Visvabharati). 2005. *Analysis of trends in operational holdings: Consolidated report.* Department of Agriculture and Co-operation, Ministry of Agriculture, Government of India. http://agricoop.nic.in/study7.htm (accessed July 1, 2009).

Afolayan, A. J., and M. T. Yakubu. 2009. Erectile dysfunction management options in Nigeria. *J Sex Med* 6(4):1090–1102.

Ahmed, Q. A., Y. M. Aabi, and Z. A. Memish. 2006. Health risks at the Hajj. *Lancet* 367(9515):1008–1015.

Allan, B. F., F. Keesing, and R. S. Ostfeld. 2003. Effect of forest fragmentation on Lyme disease risk. *Conserv Biol* 17(1):267–272.

Anderson, P. K., A. A. Cunningham, N. G. Patel, F. J. Morales, P. R. Epstein, and P. Daszak. 2004. Emerging infectious diseases of plants: Pathogen pollution, climate change and agrotechnology drivers. *Trends Ecol Evol* 19(10):535–544.

Anyamba, A., J. P. Chretien, J. Small, C. J. Tucker, P. B. Formenty, J. H. Richardson, S. C. Britch, D. C. Schnabel, R. L. Erickson, and K. J. Linthicum. 2009. Prediction of a Rift Valley fever outbreak. *Proc Natl Acad Sci U S A* 106(3):955–959.

Blair, A., D. Sandler, K. Thomas, J. A. Hoppin, F. Kamel, J. Coble, W. J. Lee, J. Rusiecki, C. Knott, M. Dosemeci, C. F. Lynch, J. Lubin, and M. Alavanja. 2005. Disease and injury among participants in the Agricultural Health Study. *J Agric Saf Health* 11(2):141–150.

Boneva, R. S., T. M. Folks, and L. E. Chapman. 2001. Infectious disease issues in xenotransplantation. *Clin Microbiol Rev* 14(1):1–14.

Bouma, M. J., and C. Dye. 1997. Cycles of malaria associated with El Niño in Venezuela. *JAMA* 278(21):1772–1774.

Bretveld, R. W., C. M. G. Thomas, P. T. J. Scheepers, G. A. Zielhuis, and N. Roeleveld. 2006. Pesticide exposure: The hormonal function of the female reproductive system disrupted? *Reprod Biol Endocrinol* 4:30–44.

Bunnag, T., S. Sornmani, S. Pinithpongse, and C. Harinasuta. 1979. Surveillance of waterborne parasitic infections and studies on the impact of ecological changes on vector mosquitoes of malaria after dam construction. *Southeast Asian J Trop Med Public Health* 10(4):656–660.

Burton, E. C., D. K. Burns, M. J. Opatowsky, W. H. El-Feky, B. Fischbach, L. Melton, E. Sánchez, H. Randall, D. L. Watkins, J. Chang, and G. Klintmalm. 2005. Rabies encephalomyelitis: Clinical, neuroradiological, and pathological findings in 4 transplant recipients. *Arch Neurol* 62(6):873–882.

Calvert, G. M., J. Karnik, L. Mehler, J. Beckman, B. Morrissey, J. Sievert, R. Barrett, M. Lackovic, L. Mabee, A. Schwartz, Y. Mitchell, and S. Moraga-McHaley. 2008. Acute pesticide poisoning among agricultural workers in the United States, 1998–2005. *Am J Ind Med* 51(12):883–898.

Capua, I., and D. J. Alexander. 2008. Ecology, epidemiology and human health implications of avian influenza viruses: Why do we need to share genetic data? *Zoonoses Public Health* 55(1):2–15.

CDC (U.S. Centers for Disease Control and Prevention). 1998. Preventing emerging infectious diseases: A strategy for the 21st century. Overview of the updated CDC plan. *MMWR* 47(15):1–14.

CDC. 1999. Update: Outbreak of Nipah virus—Malaysia and Singapore, 1999. *MMWR* 48(16):335–337.

CDC. 2000. Laboratory-acquired human glanders—Maryland, May 2000. *MMWR* 49(24): 532–535.

CDC. 2005. *Foodborne illness.* Division on bacterial diseases. http://www.cdc.gov/ncidod/dbmd/diseaseinfo/foodborneinfections_t.htm (accessed July 1, 2009)

Chapman, C. A., T. R. Gillespie, and T. L. Goldberg. 2005. Primates and the ecology of their infectious diseases: How will anthropogenic change affect host-parasite interactions? *Evol Anthropol* 14(4):134–144.

Charlwood, J. D., and W. A. Alecrim. 1989. Capture-recapture studies with the South American malaria vector Anopheles darlingi, Root. *Ann Trop Med Parasitol* 83(6):569–576.

Coluzzi, M. 1984. Heterogeneities of the malaria vectorial system in tropical Africa and their significance in malaria epidemiology and control. *Bull World Health Organ* 62 (Suppl):107–113.

Coluzzi, M. 1994. Malaria and the Afrotropical ecosystems: Impact of man-made environmental changes. *Parassitologia* 36(1–2):223–227.

Coluzzi, M., A. Sabatini, V. Petrarca, and M. A. Di Deco. 1979. Chromosomal differentiation and adaptation to human environments in the *Anopheles gambiae* complex. *Trans R Soc Trop Med Hyg* 73(5):483–497.

Colwell, R. R. 1996. Global climate and infectious disease: The cholera paradigm. *Science* 274(5295):2025–2031.

Cox, L. A., Jr., and D. A. Popken. 2006. Quantifying potential human health impacts of animal antibiotic use: Enrofloxacin and macrolides in chickens. *Risk Anal* 26(1):135–146.

CRS (U.S. Congressional Research Service). 2008. *International illegal trade in wildlife: Threats and U.S. policy*, edited by L. S. Wyler and P. A. Sheikh. Washington, DC: Library of Congress.

Cruz Marques, A. 1987. Human migration and the spread of malaria in Brazil. *Parasitol Today* 3(6):166–170.

Cunningham, A. A., P. Daszak, and J. P. Rodríguez. 2003. Pathogen pollution: Defining a parasitological threat to biodiversity conservation. *J Parasitol* 89(Suppl):S78–S83.

Daszak, P. 2006. Risky behavior in the Ebola zone. *Anim Conserv* 9(4):366–367.

Daszak, P., and A. A. Cunningham. 2003. Anthropogenic change, biodiversity loss and a new agenda for emerging diseases. *J Parasitol* 89(Suppl):S37–S41.

Daszak, P., A. A. Cunningham, and A. D. Hyatt. 2000. Emerging infectious diseases of wildlife—Threats to biodiversity and human health. *Science* 287(5452):443–449.

Daszak, P., A. A. Cunningham, and A. D. Hyatt. 2001. Anthropogenic environmental change and the emergence of infectious diseases in wildlife. *Acta Trop* 78(2):103–116.

Daszak, P., R. Plowright, J. H. Epstein, J. Pulliam, S. Abdul Rahman, H. E. Field, A. Jamaluddin, S. H. Sharifah, C. S. Smith, K. J. Olival, S. Luby, K. Halpin, A. D. Hyatt, A. A. Cunningham, and the Henipavirus Ecology Research Group (HERG). 2006. The emergence of Nipah and Hendra virus: Pathogen dynamics across a wildlife-livestock-human continuum. In *Disease ecology: Community structure and pathogen dynamics*, edited by S. K. Collinge, and C. Ray. Oxford, UK: Oxford University Press. Pp. 186–201.

Davies, F. G. 2006. Risk of a Rift Valley fever epidemic at the Hajj in Mecca, Saudi Arabia. *Rev Sci Tech* 25(1):137 147.

de Merode, E., and G. Cowlishaw. 2006. Species protection, the changing informal economy, and the politics of access to the bushmeat trade in the Democratic Republic of Congo. *Conserv Biol* 20(4):1262–1271.

de Merode, E., K. Homewood, and G. Cowlishaw. 2004. The value of bushmeat and other wild foods to rural households living in extreme poverty in Democratic Republic of Congo. *Biol Conserv* 118(5):573–581.

Delgado, C. L. 2003. Rising consumption of meat and milk in developing countries has created a new food revolution. *J Nutr* 133(11):3907S–3910S.

Diwyanto, K., and S. Iskandar. 1999. Kampung chickens: A key part of Indonesia's livestock sector. In *Livestock industries of Indonesia prior to the financial crisis*. Food and Agriculture Organization of the United Nations Regional Office for Asia and the Pacific. Bangkok, Thailand: FAO. ftp://ftp.fao.org/docrep/fao/004/ab986e/AB986E00.pdf (accessed July 19, 2009).

Dols, C. L., J. M. Bowers, and A. E. Copfer. 2001. Preventing food- and water-borne illnesses. *Am J Nurs* 101(6):24AA–24HH.

Eaton, B. T., C. C. Broder, D. Middleton, and L. F. Wang. 2006. Hendra and Nipah viruses: Different and dangerous. *Nat Rev Microbiol* 4(1):23–35.

Einsweiler, S. 2008. *Monitoring the wildlife trade.* Presentation, Fifth Committee Meeting on Achieving Sustainable Global Capacity for Surveillance and Response to Emerging Diseases of Zoonotic Origin, Washington, DC, December 1–2.

Elam, T. E., and R. D. Preston. 2004. *Fifty years of pharmaceutical technology and its impact on the beef we provide to consumers.* Johnston, IA: Growth Enhancement Technology Information Team. http://www.beeftechnologies.com/documents/whitePaper-summary.pdf (accessed March 4, 2007).

Eterovic, A., and M. R. Duarte, 2002. Exotic snakes in São Paulo City, southeastern Brazil: Why xenophobia? *Biodivers Conserv* 11(2):327–333.

FAO (Food and Agriculture Organization of the United Nations). 2007. *Gridded livestock of the world 2007*, edited by W. Wint and T. Robinson. Rome, Italy: FAO.

FAO. 2009a. *Briefing paper: Hunger on the rise—Soaring prices add 75 million people to global hunger rolls.* Rome, Italy: FAO. http://www.fao.org/newsroom/common/ecg/1000923/en/hungerfigs.pdf (accessed January 28, 2009).

FAO. 2009b. FAOSTAT. FAO Statistical Division. Rome, Italy: FAO.

Fischer, S. A., M. B. Graham, M. J. Kuehnert, C. N. Kotton, A. Srinivasan, F. M. Marty, J. A. Comer, J. Guarner, C. D. Paddock, D. L. DeMeo, W. J. Shieh, B. R. Erickson, U. Bandy, A. DeMaria, Jr., J. P. Davis, F. L. Delmonico, B. Pavlin, A. Likos, M. J. Vincent, T. K. Sealy, C. S. Goldsmith, D. B. Jernigan, P. E. Rollin, M. M. Packard, M. Patel, C. Rowland, R. F. Helfand, S. T. Nichol, J. A. Fishman, T. Ksiazek, and S. R. Zaki. 2006. Transmission of lymphocytic choriomeningitis virus by organ transplantation. *N Engl J Med* 354(21):2235–2249.

Fowler, A. J., D. M. Lodge, J. F. Hsia. 2007. Failure of the Lacey Act to protect U.S. ecosystems against animal invasions. *Front Ecol Environ* 5(7):353–359.

Ghebreyesus, T. A., M. Haile, K. H. Witten, A. Getachew, A. M. Yohannes, M. Yohannes, H. D. Teklehaimanot, S. W. Lindsay, and P. Byass. 1999. Incidence of malaria among children living near dams in northern Ethiopia: Community based incidence survey. *BMJ* 319(7211):663–666.

Gibbs, S. G., C. F. Green, P. M. Tarwater, L. C. Mota, K. D. Mena, and P. V. Scarpino. 2006. Isolation of antibiotic-resistant bacteria from the air plume downwind of a swine confined or concentrated animal feeding operation. *Environ Health Perspect* 114(7):1032–1037.

Glass, G. E., B. S. Schwartz, J. M. Morgan, III, D. T. Johnson, P. M. Noy, and E. Israel. 1995. Environmental risk factors for Lyme disease identified with geographic information systems. *Am J Public Health* 85(7):944–948.

Gurley, E. S., J. M. Montgomery, M. J. Hossain, M. Bell, A. K. Azad, M. R. Islam, M. A. Molla, D. S. Carroll, T. G. Ksiazek, P. A. Rota, L. Lowe, J. A. Comer, P. Rollin, M. Czub, A. Grolla, H. Feldmann, S. P. Luby, J. L. Woodward, and R. F. Breiman. 2007. Person-to-person transmission of Nipah virus in a Bangladeshi community. *Emerg Infect Dis* 13(7):1031–1037.

Hahn, B. H., G. M. Shaw, K. M. De Cock, and P. M. Sharp. 2000. AIDS as a zoonosis: Scientific and public health implications. *Science* 287(5453):607–614.

Hall, C. J., S. J. Richmond, E. O. Caul, N. H. Pearce, and I. A. Silver. 1982. Laboratory outbreak of Q fever acquired from sheep. *Lancet* 1(8279):1004–1006.

Hames, R. B. 1991. Wildlife conservation in tribal societies. In *Biodiversity, culture, conservation and ecodevelopment*, edited by M. L. Oldfield, and J. B. Alcorn. Boulder, CO: Westview Press. Pp. 172–199.

Harb, M., R. Faris, A. M. Gad, O. N. Hafez, R. Ramzy, and A. A. Buck. 1993. The resurgence of lymphatic filariasis in the Nile delta. *Bull World Health Organ* 71(1):49–54.

Havenstein, G. B., P. R. Ferket, and M. A. Qureshi. 2003. Growth, livability, and feed conversion of 1957 versus 2001 broilers when fed representative 1957 and 2001 broiler diets. *Poult Sci* 82(10):1500–1508.

Holtzman, J. D. 2006. Food and memory. *Annu Rev Anthropol* 35(1):361–378.

Horn, R. 2004. The families that own Asia: Thailand. *TIME Asia Magazine*, February 23.

Horrigan, L., R. S. Lawrence, and P. Walker. 2002. How sustainable agriculture can address the environmental and human health harms of industrial agriculture. *Environ Health Perspect* 110(5):445–456.

IFPRI (International Food Policy Research Institute). 2002. *Green revolution: Curse or blessing?* Washington, DC: IFPRI.

IOM (Institute of Medicine). 1992. *Emerging infections: Microbial threats to health in the United States*, edited by J. Lederberg, R. E. Shope, and S. C. J. Oakes. Washington, DC: National Academy Press.

IOM. 2003. *Microbial threats to health: Emergence, detection, and response*. Washington, DC: The National Academies Press.

IOM. 2006. *Quarantine stations at ports of entry: Protecting the public's health*. Washington, DC: The National Academies Press.

Kakati, L. N., B. Ao, and V. Doulo. 2006. Indigenous knowledge of zootherapeutic use of vertebrate origin by the Ao tribe of Nagaland. *J Hum Ecol* 19(3):163–167.

Kalliola, R., and P. S. Flores. 1998. *Geoecología y desarollo Amazónico: Estudio integrado en la zona de Iquitos, Perú*. Sulkava, Perú: Finnreklama Oy.

Karesh, W. B., R. A. Cook, E. L. Bennett, and J. Newcomb. 2005. Wildlife trade and global disease emergence. *Emerg Infect Dis* 11(7):1000–1002.

Kimball, A. M. 2006. *Risky trade: Infectious disease in the era of global trade*. Aldershot, UK: Ashgate Press.

Krause, R. M. 1992. The origin of plagues: Old and new. *Science* 257(5073):1073–1078.

Krause, R. M. 1994. Dynamics of emergence. *J Infect Dis* 170(2):265–271.

Layton, R., R. Foley, and E. Williams. 1991. The transition between hunting and gathering and the specialized husbandry of resources: A socio-ecological approach. *Curr Anthropol* 32(3):255–274.

Lim, W., K. C. Ng, and D. N. Tsang. 2006. Laboratory containment of SARS virus. *Ann Acad Med Singapore* 35(5):354–360.

Linthicum, K. J., C. L. Bailey, F. G. Davies, and C. J. Tucker. 1987. Detection of Rift Valley fever viral activity in Kenya by satellite remote sensing imagery. *Science* 235(4796): 1656–1659.

Linthicum, K. J., A. Anyamba, C. J. Tucker, P. W. Kelley, M. F. Myers, and C. J. Peters. 1999. Climate and satellite indicators to forecast Rift Valley fever epidemics in Kenya. *Science* 285(5426):397–400.

Liu, T-S. 2008. Custom, taste and science: Raising chickens in the Pearl River delta region, South China. *Anthropology & Medicine* 15(1):7–18.

LoGiudice, K., R. S. Ostfeld, K. A. Schmidt, and F. Keesing. 2003. The ecology of infectious disease: Effects of host diversity and community composition on Lyme disease risk. *Proc Natl Acad Sci U S A* 100(2):567–571.

Loutan, L., D. Bierens de Haan, and L. Subilia. 1997. The health of asylum seekers: From communicable disease screening to post-traumatic disorders [in French]. *Bull Soc Pathol Exot* 90(4):233–237.

Macpherson, C. N. L. 2005. Human behaviour and the epidemiology of parasitic zoonoses. *Int J Parasitol* 35(11–12):1319–1331.

Mahawar, M. M., and D. P Jaroli. 2008. Traditional zootherapeutic studies in India: A review. *J Ethnobiol and Ethnomed* 4(1):17.

Marano, N. 2008. *CDC's role in preventing introduction of zoonotic diseases via animal importation.* Presentation, Fifth Committee Meeting on Achieving Sustainable Global Capacity for Surveillance and Response to Emerging Diseases of Zoonotic Origin, Washington, DC, December 1–2.

Martin, M., E. Mathias, and C. M. McCorkle. 2001. *Ethnoveterinary medicine: An annotated bibliography of community animal healthcare.* London, UK: ITDG Publishing.

Mathias, E., and C. M. McCorkle. 2004. Traditional livestock healers. *Rev Sci Tech* 23(1): 277–284.

Mayer, J. D. 2000. Geography, ecology and emerging infectious diseases. *Soc Sci Med* 50(7–8): 937–952.

McDonald, J., R. Hoppe, and D. Banker. 2006. *Growing farm size and the distribution of farm payments.* U.S. Department of Agriculture, Economic Research Service (ERS). Economic Brief no. 6. Washington, DC: ERS.

MEA (Millennium Ecosystem Assessment). 2005. *Ecosystems and human well-being: Biodiversity synthesis.* Washington, DC: World Resources Institute.

Mellon, M., C. Benbrook, and K. L. Benbrook. 2001. *Hogging it: Estimates of antimicrobial abuse in livestock.* Cambridge, MA: Union of Concerned Scientists. http://www.ucsusa.org/index.html (accessed June 16, 2009).

Mills, J. N., T. G. Ksiazek, C. J. Peters, and J. E. Childs. 1999. Long-term studies of Hantavirus reservoir populations in the southwestern United States: A synthesis. *Emerg Infect Dis* 5(1):135–142.

Morse, S. S. 1993. Examining the origins of emerging viruses. In *Emerging viruses*, edited by S. S. Morse. New York: Oxford University Press. Pp. 10–28.

National Security Archive. 2001. Volume V: Anthrax at Sverdlovsk, 1979—U.S. intelligence on the deadliest modern outbreak. In *The September 11th Sourcebooks*, edited by R. A. Wampler and T. S. Blanton. National Security Archive Electronic Briefing Book N. 61. http://www.gwu.edu/~nsarchiv/NSAEBB/NSAEBB61/ (accessed July 19, 2009).

Newcomb, J. 2004. *Economic analysis of impacts of Avian Influenza, SARS, FMD and other infectious disease outbreaks.* Presentation, One World—One Health Workshop, Beyond zoonoses: The threat of emerging diseases to human security and conservation and the implications for public policy, Bangkok, Thailand, November 15.

Ostfeld, R. S., and F. Keesing. 2000. Biodiversity and disease risk: The case of Lyme disease. *Conserv Biol* 14(3):722–728.

Padmawati, S., and M. Nichter. 2008. Community response to avian flu in central Java, Indonesia. *Anthropology & Medicine* 15(1):31–51.

Patz, J. A., T. K. Graczyk, N. Geller, and A. Y. Vittor. 2000. Effects of environmental change on emerging parasitic diseases. *Int J Parasitol* 30(12–13):1395–1405.

Peeters, M., V. Courgnaud, B. Abela, P. Auzel, X. Pourrut, F. Bibollet-Ruche, S. Loul, F. Liegeois, C. Butel, D. Koulagna, E. Mpoudi-Ngole, G. M. Shaw, B. H. Hahn, and E. Delaporte. 2002. Risk to human health from a plethora of simian immunodeficiency viruses in primate bushmeat. *Emerg Infect Dis* 8(5):451–457.

Pickering, L. K., N. Marano, J. A. Bocchini, F. J. Angulo, and the Committee on Infectious Diseases. 2008. Exposure to nontraditional pets at home and to animals in public settings: Risks to children. *Pediatrics* 122(4):876–886.

Preston, R. L. 1987. The role of animal drugs in food animal production. In *Proceedings of the symposium on animal drug use—Dollars and sense.* Rockville, MD: Center for Veterinary Medicine. Pp. 127–134.

Pulliam, J. R., J. G. Dushoff, S. A. Levin, and A. P. Dobson. 2007. Epidemic enhancement in partially immune populations. *PLoS ONE* 2(1):e165.

Purse, B. V., P. S. Mellor, D. J. Rogers, A. R. Samuel, P. P. Mertens, and M. Baylis. 2005. Climate change and the recent emergence of bluetongue in Europe. *Nat Rev Microbiol* 3(2):171–181.

Reardon, T., C. P. Timmer, C. B. Barrett, and J. Berdegu. 2003. The rise of supermarkets in Africa, Asia, and Latin America. *Am J Agric Econ* 85(5):1140–1146.

Reed, R. N. 2005. An ecological risk assessment of nonnative boas and pythons as potentially invasive species in the United States. *Risk Anal* 25(3):753–766.

Rolland, R. M., G. Hausfater, B. Marshall, and S. B. Levy. 1985. Antibiotic-resistant bacteria in wild primates: Increased prevalence in baboons feeding on human refuse. *Appl Environ Microbiol* 49(4):791–794.

Rusnak, J. M., M. G. Kortepeter, R. J. Hawley, A. O. Anderson, E. Boudreau, and E. Eitzen. 2004. Risk of occupationally acquired illnesses from biological threat agents in unvaccinated laboratory workers. *Biosecur Bioterror* 2(4):281–293.

Schmidt, K. A., and R. S. Ostfeld. 2001. Biodiversity and the dilution effect in disease ecology. *Ecology* 82(3):609–619.

Schmink, M., and C. H. Wood. 1992. *Contested frontiers in Amazonia*. New York: Columbia University Press.

Searle, J., A. Regmi, and J. Bernstein. 2003. *International evidence on food consumption patterns*. U.S. Department of Agriculture, Economic Research Service (ERS), Technical Bulletin no. 1904. Washington, DC: ERS.

Seré, C., and H. Steinfeld. 1996. *World livestock production systems: Current status, issues and trends*. FAO Animal Production and Health Paper 127. Rome, Italy: FAO.

Shanklin, E. 1985. Sustenance and symbol: Anthropological studies of domesticated animals. *Annu Rev Anthropol* 14:375–403.

Shimshony, A., and P. Economides. 2006. Disease prevention and preparedness for animal health emergencies in the Middle East. *Rev Sci Tech* 25(1):253–269.

Singh, J., D. C. Jain, R. Bhatia, R. L. Ichhpujani, A. K. Harit, R. C. Panda, K. N. Tewari, and J. Sokhey. 2001. Epidemiological characteristics of rabies in Delhi and surrounding areas, 1998. *Indian Pediatr* 38(12):1354–1360.

Slingenbergh, J., M. Gilbert, K. de Balogh, and W. Wint. 2004. Ecological sources of zoonotic diseases. *Rev Sci Tech* 23(2):467–484.

Smith, D. L., A. D. Harris, J. A. Johnson, E. K. Silbergeld, and J. G. Morris, Jr. 2002. Animal antibiotic use has an early but important impact on the emergence of antibiotic resistance in human commensal bacteria. *Proc Natl Acad Sci U S A* 99(9):6434–6439.

Smith, K. F., M. Behrens, L. M. Schloegel, N. Marano, S. Burgiel, and P. Daszak. 2009. Ecology. Reducing the risks of the wildlife trade. *Science* 324(5927):594–595.

Soewu, D. A. 2008. Wild animals in ethnozoological practices among the Yorubas of Southwestern Nigeria and the implications for biodiversity conservation. *Afr J Agric Res* 3(6):421–427.

Stearman, A. M., and K. H. Redford. 1995. Game management and cultural survival: The Yuqui ethnodevelopment project in lowland Bolivia. *Oryx* 29(1):29–34.

Steinfeld, H. 2004. The livestock revolution: A global veterinary mission. *Vet Parasitol* 125(1–2):19–40.

Steinfeld, H., P. Gerber, T. Wassenaar, V. Castel, M. Rosales, and C. de Haan. 2006. *Livestock's long shadow, issues and options*. Rome, Italy: FAO.

Stine, O. C., J. A. Johnson, A. Keefer-Norris, K. L. Perry, J. Tigno, S. Qaiyumi, M. S. Stine, and J. G. Morris, Jr. 2007. Widespread distribution of tetracycline resistance genes in a confined animal feeding facility. *Int J Antimicrob Agents* 29(3):348–352.

Stock, R., and T. Mader. 1984. *Feed additives for beef cattle*. Publication G85-761-A. Lincoln, NE: University of Nebraska Extension.

Stokstad, E. L., T. H. Jukes, J. Pierce, A. C. Page, Jr., and A. L. Franklin. 1949. The multiple nature of the animal protein factor. *J Biol Chem* 180(2):647–654.

Tadei, W. P., B. D. Thatcher, J. M. Santos, V. M. Scarpassa, I. B. Rodrigues, and M. S. Rafael. 1998. Ecologic observations on anopheline vectors of malaria in the Brazilian Amazon. *Am J Trop Med Hyg* 59(2):325–335.

Teuber, M. 2001. Veterinary use and antibiotic resistance. *Curr Opin Microbiol* 4(5): 493–499.

Thompson, D. F., J. B. Malone, M. Harb, R. Faris, O. K. Huh, A. A. Buck, and B. L. Cline. 1996. Bancroftian filariasis distribution and diurnal temperature differences in the southern Nile delta. *Emerg Infect Dis* 2(3):234–235.

Toole, M. J., and R. J. Waldman. 1997. The public health aspects of complex emergencies and refugee situations. *Annu Rev Public Health* 18:283–312.

Treadwell, T. 2008. *Convergence of forces behind emerging and reemerging zoonoses, and future trends in zoonoses.* Presentation, Institute of Medicine/National Research Council Workshop on Sustainable Global Capacity for Surveillance and Response to Emerging Zoonoses, Washington, DC, June 25–26.

Tyssul Jones, T. W. 1951. Deforestation and epidemic malaria in the wet and intermediate zones of Ceylon. *Indian J Malarial* 5:135–161.

UNHCR (United Nations High Commission for Refugees). 2009. *2008 global trends: Refugees, asylum-seekers, returnees, internally displaced and stateless persons.* Geneva, Switzerland: UNHCR.

United Nations. 2002. *International migration report 2002.* Department of Economic and Social Affairs, Population Division. New York: United Nations.

United Nations. 2007. *World population prospects: The 2006 revision, medium variant.* Department of Economics and Social Affairs, Population Division. New York: United Nations.

United Nations. 2008. *World urbanization prospects: The 2007 revision.* Department of Economics and Social Affairs, Population Division. New York: United Nations.

Wallis, J., and D. R. Lee. 1999. Primate conservation: The prevention of disease transmission. *Int J Primatol* 20(6):803–826.

Weigle, K. A., C. Santrich, F. Martinez, L. Valderrama, and N. G. Saravia. 1993. Epidemiology of cutaneous leishmaniasis in Colombia: Environmental and behavioral risk factors for infection, clinical manifestations, and pathogenicity. *J Infect Dis* 168(3):709–714.

Wilson, A., K. Darpel, and P. S. Mellor. 2008 Where does bluetongue virus sleep in the winter? *PLoS Biol* 6(8):e210.

Wolfe, N. D., M. N. Eitel, J. Gockowski, P. K. Muchaal, C. Nolte, A. T. Prosser, J. N. Torimiro, S. F. Weise, and D. S. Burke. 2000. Deforestation, hunting and the ecology of microbial emergence. *Global Change and Human Health* 1(1):10–25.

Wolfe, N. D., W. M. Switzer, J. K. Carr, V. B. Bhullar, V. Shanmugam, U. Tamoufe, A. T. Prosser, J. N. Torimiro, A. Wright, E. Mpoudi-Ngole, F. E. McCutchan, D. L. Birx, T. M. Folks, D. S. Burke, and W. Heneine. 2004. Naturally acquired simian retrovirus infections in central African hunters. *Lancet* 363(9413):932–937.

Wolfe, N. D., P. Daszak, A. M. Kilpatrick, and D. S. Burke. 2005. Bushmeat hunting, deforestation, and prediction of zoonoses emergence. *Emerg Infect Dis* 11(12):1822–1827.

4

Achieving an Effective Zoonotic Disease Surveillance System

"Surveillance for emerging diseases contributes to global security. If basic surveillance and laboratory capacities are compromised, will health authorities catch the next SARS [severe acute respiratory syndrome], or spot the emergence of a pandemic virus in time to warn the world and mitigate the damage?"

—Dr. Margaret Chan
Director-General of the World Health Organization
Address at the 23rd Forum on Global Issues
(March 18, 2009)

Given recent experiences of rapidly spreading global outbreaks across borders and continents, an effective emerging zoonotic disease surveillance system will need to be global in scope and effort. A global, integrated zoonotic disease surveillance system needs to detect disease emergence in human or animal populations anywhere in the world at the earliest time possible. Early detection is essential to trigger a timely disease outbreak investigation. Multidisciplinary teams of professionals that have relevant expertise and field experience would identify populations at risk and causes and risk factors for infection, and then rapidly and widely disseminate this information so that immediate and longer-term disease prevention and control interventions can be implemented. The goal of these interventions would be to control the size and geographic scope of the outbreak and to minimize morbidity, mortality, and economic losses in both human and animal populations.

No matter how effective a surveillance and response system is, the increasing prevalence of drivers creates a situation where zoonotic disease pathogens will continue to emerge in human and animal populations, and thus it will be impossible to prevent all disease outbreaks and zoonotic diseases from occurring. However, a global zoonotic disease surveillance system provides great benefits by conveying critical data to inform evidence-based responses, therefore minimizing the opportunity for zoonotic disease emergence, transmission, and spread in both human and animal populations.

This chapter first defines disease surveillance, discusses elements of an effective zoonotic disease surveillance system, and describes how such a system would need to be executed. It then presents an overview of existing emerging zoonotic disease surveillance systems and capacity-building programs for creating the needed workforce. From this overview, the chapter identifies important existing gaps and challenges in the current state of global surveillance.

DEFINING DISEASE SURVEILLANCE

The principal purpose of disease surveillance is to "keep one's finger on the pulse of disease in a population." Successful disease surveillance detects increases in disease occurrence over expected levels early so that effective and timely disease control interventions can be introduced and appropriately targeted to reduce morbidity, mortality, and economic loss. Though several definitions of disease surveillance have been used by human and animal health agencies and experts (Thacker and Berkelman, 1988; Teutsch and Churchill, 2000; IOM, 2007; WHO, 2007a; OIE, 2008a), the committee chose to adopt more appropriately integrated definitions for this report (see Box 4-1).

Disease surveillance strategies were developed to address different sur-

BOX 4-1
Definitions of Surveillance

Zoonotic disease surveillance: The ongoing systematic and timely collection, analysis, interpretation, and dissemination of information about the occurrence, distribution, and determinants of diseases transmitted between humans and animals. Zoonotic disease surveillance reaches its full potential when it is used to plan, implement, and evaluate responses to reduce infectious disease morbidity and mortality in human and animal populations through a functionally integrated human and animal health system.

Surveillance system: The total system of surveillance comprising the components of collection and reporting of disease outcome data from populations at risk, confirmation of the etiological agent by laboratory scientists, and mechanisms and pathways of data analysis, interpretation, reporting feedback, and communication of information to those who will use the data at local, provincial, national, regional, or international levels for response.

Integrated emerging zoonotic disease surveillance system: A system that brings together and links data collection, collation, analysis, presentation/reporting, and dissemination components to provide linked human and animal clinical, epidemiological, laboratory, and risk behavior information on unusual occurrences of emerging zoonotic diseases in both human and animal populations. The information brought to both human and animal health officials by human and animal health authorities would be used for early detection and timely response at local, provincial, national, regional, and international levels.

veillance goals and objectives, diverse ways information would be used, and varying human and financial resources available to support and operate the system. These disease surveillance strategies and systems employ different methods to collect information, and they include active, passive, sentinel, syndromic, risk-based, informal, and rumor-based disease surveillance (Teutsch and Churchill, 2000).

ELEMENTS OF AN EFFECTIVE ZOONOTIC DISEASE SURVEILLANCE SYSTEM

An effective global, integrated zoonotic disease surveillance system requires effective surveillance at national, regional, and international levels, because information from outbreak investigations is used by human and animal health officials at all levels to implement response measures and to evaluate the effectiveness of those responses. A surveillance system is comprised of cyclical elements that provide critical pieces of information, as seen in Figure 4-1. For disease surveillance to be comprehensive, surveillance will need to be planned and conducted across human and animal populations

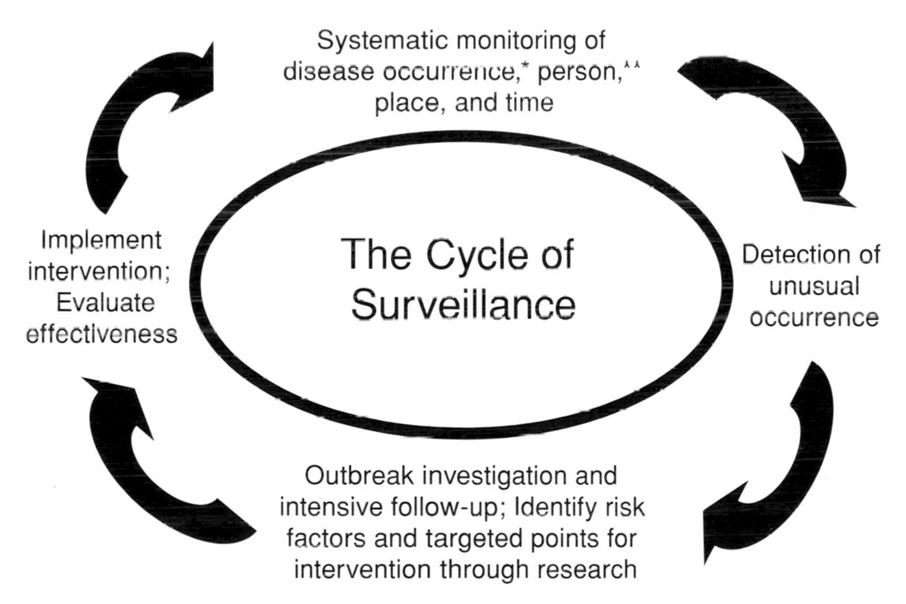

FIGURE 4-1 The cycle of elements comprising an effective infectious disease surveillance system.
NOTES: *disease by clinical signs or detection and confirmation of pathogen or antibody by laboratory diagnoses; **attributes of person would include demographic variables and risky behaviors.

(i.e., domesticated livestock, poultry, and companion animals, and aquatic and terrestrial wildlife), and information transfer will need to be facilitated between the human, animal, and environmental health sectors.

Disease Surveillance System Framework and Components

Designing a disease surveillance system requires decisions on various elements. These include (1) identifying clear objectives; (2) agreeing on a well-defined disease surveillance case definition based on the person (or in the case of animal populations, based on the animal, herd, or flock), place, and time, that can include suspect or probable cases based on clinical and/or epidemiological data, as well as laboratory-confirmed cases; (3) clarifying what information is needed to achieve the objective, and the frequency with which the information is needed; (4) determining the type of disease surveillance system (i.e., active, passive, sentinel, syndromic, etc.); (5) identifying the sources of data and information (clinical, epidemiological, laboratory, and social and anthropological data); (6) determining methods and channels of information dissemination and alerting; and (7) designating clear roles and responsibilities of those who use the information for action (Teutsch and Churchill, 2000). Figure 4-2 shows the further steps

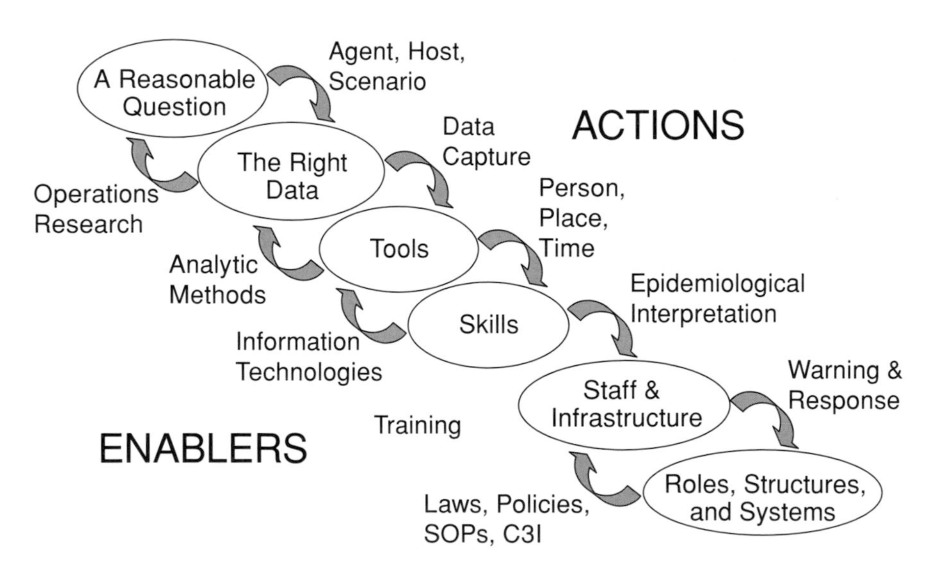

FIGURE 4-2 System requirements for comprehensive human and animal health surveillance.
NOTES: SOPs = standard operating procedures, C3I = communications, command, control, and intelligence.
SOURCE: IOM (2007).

required to design a comprehensive disease surveillance system for human and animal health.

A goal of disease surveillance is to be useful at all levels of the human and animal health systems. In order for the disease surveillance system to be useful, information needs to flow back and forth easily among international, national, and local levels; be timely in detection and laboratory confirmation; include risk factors as a component; and be specific and reliably detect and report disease. Furthermore, the surveillance system will need to be robust under adverse conditions; ensure that information on individual patients and food-animal production owners or industries is secure and remains confidential; be flexible to use innovative information technology for data collection, collation, analysis, presentation, and dissemination; and be compatible for data to be electronically collected and stored across systems.

EXECUTING AN EFFECTIVE ZOONOTIC DISEASE SURVEILLANCE SYSTEM

Identifying, Gathering, Analyzing, and Disseminating Information

The earlier an emerging zoonotic disease can be detected, the timelier the response can be, thereby minimizing transmission and spread and ultimately reducing morbidity and mortality. Data sources need to correctly distinguish an abnormal disease pattern from a typical or expected one. As data are collected, they need to be transmitted for analysis, and such analyses need to be presented in user-friendly, easy-to-understand formats so that decisionmakers can properly interpret and use the information (Mandl et al., 2004a,b). Given the technology available today, these elements are certainly possible to achieve, yet the current system falls short from the target.

Sources of Data

Multiple sources of data from traditional and nontraditional sources have potential use in an integrated disease surveillance system (see Box 4-2). Data can be collected in several ways: by interviewing patients, animal owners, community members, or healthcare providers; administering a questionnaire by mail or phone; searching electronic disease records of established surveillance systems; or searching records from human and animal diagnostic laboratories. Biological samples are collected on site, then safely transported to a laboratory performing requisite tests for laboratory confirmation. Some national monitoring and disease surveillance programs use mail and interview questionnaires as well as a collection of biological

BOX 4-2
Summary of Data Types and Sources for Human and Animal Health Disease Events

Human Health: Traditional Sources

Emergency department chief complaints
Hospital/clinic medical records
Text-based notes
Diagnostic laboratory data
Radiological reports
Physician reports
Emergency Medical Services activity
WHO reports

Human Health: Nontraditional Sources

Digital detection systems
Short Message Service technology
Syndromic surveillance data
Records on pharmaceutical purchases
Patient self-reports
Absenteeism data
Telephone survey results

Animal Health

Diagnostic laboratory data
Farm worker observations
Hospital/clinic medical records
Reportable diseases
Abattoir monitoring programs
Active surveillance programs
Companion animal owner reports
Electronic record systems
Syndromic surveillance

samples for laboratory testing (Traub-Dargatz et al., 2000; USDA, 2000; Wagner et al., 2001).

Screening medical and laboratory records (paper files or electronic databases) for specific entries, or biological sample banks for specific pathogens or lesions, could be part of the active data collection and monitoring system for a disease surveillance system. Pathogen phenotypes and genotypes are routinely submitted to global databanks where they are readily accessible for comparison among laboratories examining outbreak samples. The use of such reference databanks facilitates the rapid identification of unsuspected linked outbreaks even if widely spread by global trade. These types of data retrieval methods are routinely performed in many developed countries, such as for testing suspect cases for rabies, bovine spongiform encephalopathy (BSE) screening of fallen livestock and emergency-slaughtered cattle in Europe (Doherr et al., 1999; Doherr and Audigé, 2001) and of "downer cows" in the United States (USDA-APHIS, 2009a), screening of humans and wild birds for ongoing global influenza viruses funded by the U.S. Agency for International Development (USAID), and the genotypic comparison of food and waterborne bacterial isolates by the U.S. Centers for Disease Control and Prevention's (CDC's) PulseNet (Swaminathan et al., 2001).

Role of Information Technology (IT) in Data Collection and Analysis

Evolving IT has led to a number of breakthroughs and new ways to collect and transmit epidemiological, clinical, demographical, and other information in the field. Examples of new technologies include the use of handheld computers, cell phones, remote sensing, and Internet searches (Beck et al., 2000; Lobitz et al., 2000; Google.org, 2008). These technologies are being used to collect and transmit information from even the most remote and resource-challenged countries. Other breakthroughs in IT include data management and decision software and systems, which facilitate the timely analysis, presentation, interpretation, and use of information by decisionmakers.

The increasingly electronic information stream in human healthcare has permitted the emergence of semi- and fully-automated surveillance systems for symptoms and for other indicators (such as healthcare or drug utilization), which are commonly lumped under "syndromic surveillance" (International Society for Disease Surveillance, 2009). With comparable political will and investments, electronic systems in animal production and conservation could be developed for several purposes including early detection of wildlife die-offs; unexpected culling of livestock or poultry; aberrations in veterinary drug purchases; electronic tracking of bar codes along trade pathways; and electronic trace-back and trace-forwarding of animal products.

Informal Data Sources and Use of Rumor-Based Disease Reporting

With greater Internet access and use and 24/7 informal reporting networks, information on disease outbreak occurrences is increasingly being shared at the first indication of an event through unofficial channels. Real-time information about infectious disease outbreaks is increasingly found in web-based data streams, ranging from official human and animal health reporting to informal news coverage to individual accounts on chat rooms and blogs (Brownstein et al., 2008). Systems that use unstructured informal electronic information have been credited with reducing time to outbreak recognition, preventing governments from suppressing outbreak information, and facilitating the ability of the World Health Organization (WHO) and others to respond to outbreaks and emerging diseases (Madoff and Woodall, 2005). In fact, WHO's Global Outbreak Alert and Response

Network (GOARN) relies on web-based data for daily disease surveillance activities (Grein et al., 2000; Heymann and Rodier, 2001). Of major significance, the revised International Health Regulations 2005 (IHR 2005) authorize WHO to act on informal information to issue recommendations to prevent the spread of diseases (see Chapter 7 on Governance) (Wilson et al., 2008). The Program for Monitoring Emerging Diseases (ProMED-mail) and the Global Public Health Intelligence Network (GPHIN) are two early prototypes of such systems (see Box 4-3).

BOX 4-3
Prototypes of Web-Based Data Sources for Surveillance: ProMED-mail and GPHIN

Founded in 1994 by the Infectious Disease Society of America, the Program for Monitoring Emerging Diseases (ProMED-mail) pioneered the use of the Internet for the detection of outbreaks by e-mailing and posting reports, including many gleaned from its readers, with commentary from a staff of expert moderators. ProMED-mail is now one of the largest publicly available emerging disease and outbreak reporting systems, with more than 45,000 subscribers in over 165 countries. An evaluation of the extent to which ProMED-mail reports lead to timely confirmation and human and animal disease prevention and control efforts, nationally or internationally, is currently underway in collaboration with the HealthMap system.

In collaboration with the World Health Organization (WHO), the Public Health Agency of Canada created the Global Public Health Intelligence Network (GPHIN) in 1997. GPHIN's software application retrieves articles from news feed aggregators based on established search queries in 15-minute intervals on a 24/7 basis to provide an early warning of the possibility of a public health emergency. Although automation is a key component, GPHIN also employs trained analysts who provide essential linguistic, interpretive, and analytical expertise. The data are disseminated to various public health agencies, including WHO, that can perform the necessary public health vetting of the informal report. An early achievement of its potential came in December 1998 when GPHIN was the first to provide preliminary information to the public health community about a new strain of influenza in northern China. During the 2003 severe acute respiratory syndrome outbreak, the GPHIN prototype served as an early-warning system by detecting and informing the appropriate authorities (e.g., WHO and the Public Health Agency of Canada) of an unusual respiratory illness outbreak occurring in Guangdong Province, China as early as November 2002. Comprehensive global access to GPHIN is not available because there is a fee required to join GPHIN. This precludes many resource-challenged countries from participating, including many in areas at higher risk of an emerging zoonotic disease occurrence.

SOURCES: Madoff (2004); PHAC (2004); Madoff and Woodall (2005); Cowen et al. (2006); Zeldenrust et al. (2008).

A number of online disease surveillance systems are now delivering real-time intelligence on emerging infectious diseases to diverse audiences on user-friendly, open-access websites, similar to ProMED-mail and GPHIN. One of these is HealthMap, a freely accessible, automated real time system that monitors, organizes, integrates, filters, visualizes, and disseminates online information about emerging diseases (Freifeld et al., 2008). The site pulls data from more than 20,000 sources every hour, many of which come from news aggregators. Similarly, recent efforts using data from Google (Ginsberg et al., 2009) and Yahoo (Polgreen et al., 2008) have shown that search query data can be harnessed as a form of crowd-sourcing where patterns of specific searches mimic and may even predict disease outbreaks.

Statistical Analysis and Disease Modeling

An infectious disease surveillance system needs to have the capacity to detect disease trends and predict outbreaks, allowing human and animal health authorities to respond in a timely and appropriate manner (USAID, 1998). As mentioned in Chapter 2, surveillance data are crucial for modeling infectious diseases to better understand the dynamics of an epidemic, including transmission patterns, to be able to interpret and critically evaluate epidemiological data, and to design treatment and control strategies.

Laboratory Capability, Capacity, and Networks

Specimen collection, analysis, and laboratory confirmation of the etiological cause of emerging zoonotic disease outbreaks are a vital part of any infectious disease surveillance system. Although rapid field tests are available for a select group of infectious agents (such as influenza A), laboratory confirmation is typically required for pathogen characterization, confirmation of infection, and further preventive actions. Given the multiple problems caused by false-positive reports, laboratory-confirmed cases increasingly provide the bulk of actionable alerts (Rodier et al., 2007). When rapid assays are not available, conventional methods of confirmation may result in significant time delays. Because emerging agents are at times previously unknown organisms, the laboratory system needs to have the capacity to know when something is new and different, have logistics in place to move the samples to laboratories with the necessary advanced discovery capacities, and have protocols flexible enough so that laboratory personnel can cooperate and collaborate to quickly identify the agent causing the outbreak. Disease surveillance systems will therefore need to incorporate both sentinel and reference technical capacity organized into networks at national, regional, or global levels.

Standard for Laboratory Practices and Network Operations

Standards for good laboratory practices overlap with standards for good laboratory network operations. Good laboratory practice principles are simply applied to laboratory facilities that meet proper standards for testing, safety, and security; employ a trained and proficiency-tested staff; have standardized operating procedures, validated test protocols, and properly functioning equipment; and use a communication system that relies on common platforms and accurately and reliably reports test results in a timely manner. Communication lines and logistics need to be established before an event occurs. If each disease emergence represents a new problem to solve, the delay will be both unavoidable and unacceptable. Key points for several of these principles are expanded in Box 4-4.

Human Capacity Requirements from Multiple Disciplines

Executing, managing, and evaluating an effective global, integrated emerging zoonotic disease surveillance system will require human and

BOX 4-4
Principles of Good Laboratory Practice and Network Operation

Laboratory accreditation: For network laboratories, a quality assurance system will guide the application of good laboratory practice standards. Laboratory accreditation continues to be the "gold standard" by which laboratories and their quality assurance system are assessed. The quality assurance and laboratory assessment processes ensure continuous quality improvement.

Validated and standardized assays: Just as a case definition is essential in comparing data on disease incidence and prevalence, validated and standardized assays ideally are used in laboratories throughout a network. Validation refers to examination of a laboratory assay to establish whether it is fit for its purpose, and to establish performance characteristics in the laboratory and in populations of naturally infected individuals, whether humans or animals. Standardization (or harmonization) refers to the use of a common procedure for performing an assay in every network laboratory.

Reference standards: Reference standards for assessing ongoing assay performance, laboratory performance, and network function are necessary to validate assays and continuously assess laboratories. Identifying, characterizing, and providing reference standards is labor intensive and expensive and will not be available for emerging agents in a time-sensitive fashion. Nevertheless reference standards are a critical component to a surveillance system. Reference standards are also referred to as reference materials.

Human resources—training and proficiency testing: Trained technical staff are essential to proper performance of a procedure, no matter how much detail is provided animal health personnel from multiple disciplines. This will require professionals who are trained in basic clinical diagnosis of emerging zoonotic diseases, field epidemiology, laboratory sciences, social sciences, information technology, and communications at national, regional, and global levels. In addition, personnel are needed who have leadership and management skills; who have a vision and understand the need for a national and global integrated system; who have the interpersonal skills to work with experts from different disciplines; and who understand public–private partnerships (Pappaioanou et al., 2003; Perry et al., 2007).

Clinical, Field, and Laboratory Competencies

Clinical diagnostic expertise is essential for making a timely "field" diagnosis of an unexpected, emerging zoonotic disease that occurs in human and/or animal populations, whether it is in primary healthcare clinics or on farms. When a diagnosis is not considered and subsequently missed, serious delays can occur in implementing appropriate, necessary, and immediate

in the protocol. Ideally, training programs in a surveillance laboratory network are standardized, and include a "train the trainer" component that facilitates ongoing training of new personnel within individual laboratories by qualified and certified trainers. Once trained, laboratory staff will need to be proficiency tested to ascertain competence to perform an assay.

Laboratory facilities: Zoonotic diseases by their very nature are considered transmissible to humans. The facility will need to provide an environment in which to safely and securely conduct laboratory operations. Levels of biocontainment are commonly referred to as biosafety level (BSL) and are graded from levels 1–4, with the higher number corresponding to the higher degree of containment required to safely work with the agent. Security will also need to be considered in operating a modern laboratory facility. Specific laboratory techniques essential for agent discovery and characterization (e.g., in vitro or in vivo culture and genetic and molecular analysis) require strict environmental control and specimen flow. Only a limited number of BSL-4 laboratories around the world are designed to work with the most dangerous organisms. Ongoing, expensive operational and technical support is critical to ensure the proper function of BSL-4 facilities.

Implementation of new technology: Technology advances often require costly new equipment, maintenance and reagents, and technical capacity. When they provide a significant advance in capability, these technological advances could be employed in reference laboratories and ultimately reengineered for simplicity and reduced cost to disseminate the technology throughout the laboratory surveillance network. Provisions for funding of instrumentation, maintenance and reagent costs, and training and retaining personnel are all requirements for quality and sustainability.

disease control interventions, and in reporting the event to formal health authorities who can offer added resources and assistance. West Nile virus (WNV), severe acute respiratory syndrome (SARS), highly pathogenic avian influenza (HPAI) H5N1 in humans and poultry, and human monkeypox in the United States are all examples of events where diagnoses were missed early in the outbreaks.

Over the years, astute physicians, nurses, veterinarians, animal technicians, and laboratory scientists have been instrumental in the early detection of emerging zoonotic diseases. Examples include the early detection of anthrax by a keen physician during the U.S. anthrax bioterrorism attack (CDC, 2001) and the suspicion of WNV by a perceptive veterinary pathologist at the New York City Bronx zoo and its link between birds and humans (see Appendix B). Clinical training is experiential and best learned by seeing infected individuals or animals with guidance from an experienced clinician. Providing adequate in-person, hands-on training for zoonotic infections in either human or animal populations is essential but difficult for rare and sporadic or new diseases. New educational information technology and curricula may be of help in the future, but further development and evaluation is needed.

Given the importance of outbreak investigations and other aspects of disease surveillance, competencies in epidemiology, statistics, data collection, analysis, dissemination, communication, disease prevention and control, and program management are critical. Knowledgeable and skilled individuals are needed to investigate disease outbreaks in the field, and to identify their causes, sources of infection, and actionable risk factors that place humans and animals at risk of exposure. A skilled and competent workforce is then needed to communicate information learned, and to implement timely and appropriate responses in both the human and animal health sectors, from local to global levels. It is essential to have a trained workforce that has expertise and experience in the areas of infectious diseases in humans and different animal species, agriculture and animal husbandry practices, natural resources (e.g., wildlife, environment, forestry), and education, both within and across both public and private sectors. Equally important, experts need to be knowledgeable about the valuable kinds of contributions from colleagues in different disciplines and sectors and be able to collaborate and work as a multidisciplinary team.

To identify and confirm an infectious disease outbreak's causative agent in real-time, clinical and epidemiological competencies need to be complemented with expertise and experience in laboratory and pathology diagnostics. Laboratory expertise is needed for correctly obtaining samples and data from the field, analyzing those samples in the laboratory, and interpreting and communicating laboratory-based information to others who have responsibility for determining the best response to the threat. Infectious disease experts within or otherwise connected to the laboratory provide expertise that is often not available elsewhere in a country. Laboratory professionals, therefore, are important members of the team charged with planning, conducting, and monitoring infectious disease surveillance programs and responding to disease outbreaks. Given the importance of linking epidemiological data to laboratory specimens for proper interpretation of results, it is vital that epidemiologists and laboratory scientists work closely together during disease outbreak investigations, and laboratory confirmation, interpretation, and reporting. The consequences of inaccurately interpreting the significance of an event or laboratory result can be disastrous: An incorrect laboratory result or interpretation can cause delays in implementing effective disease control interventions, and result in rapid trade losses, social isolation, international stigmatization, huge costs, and waste of scarce human and medical resources.

Social Science Input

Many human behaviors increase the risk of emergence, exposure, transmission, and spread of emerging infectious diseases. Although risk factor disease surveillance has traditionally been used in human chronic disease and injury efforts (CDC, 2009a), it has increasingly been acknowledged as an important component of infectious disease surveillance (WHO and UNAIDS, 2000). For example, failure to wear personal protective equipment has been implicated in SARS virus transmission from infected patients to uninfected hospital staff caring for SARS patients (Seto et al., 2003; Moore et al., 2005). An understanding of the socioeconomic factors that lead to risky behaviors is helpful for developing strategies that can modify or prevent those behaviors from occurring.

The public's perceived risk of disease can change depending on the disease occurrence, the nature of the disease, and the frequency and accuracy of reports in the media. Expertise in the social sciences is therefore needed to monitor and address risk perceptions for accuracy and to carefully read shifts in psychological, cultural, and political responses to disease whenever emerging diseases are reported (Lau et al., 2003; Menon, 2008). Shifts in risk perception are carefully monitored to detect and minimize the dissemination of misinformation, and to promote an accurate understanding of the sources and causes of disease emergence and routes of transmission. This information would then be used for consistent, evidence based messaging and communications on steps that individuals or whole communities can take to minimize exposure and contribute to prevention and control efforts (Nichter, 2008).

The challenge of risk communication is first and foremost ensuring that the messengers are trusted. The next challenge is finding the right balance of evidence-based, credible messages to promote sustained behavior change without needlessly provoking fear or irrational responses, or causing desensitization because warnings are seen as exaggerated or politically motivated. Studies of public response to epidemics of emerging diseases—such as SARS, HPAI H5N1, BSE, and more recently to pandemic H1N1 2009—point to the importance of honesty, transparency, trust in sources of information, credible communicators, and the speed and appropriateness of government response (Kaufman, 2008; Padmawati and Nichter, 2008; Scoones and Forster, 2008; Briggs and Nichter, 2009).

Social scientists therefore are also needed to explore and implement ways to engender trust in disease surveillance systems conducted under varying conditions and in different cultures (Gilson, 2003, 2006), and to train health personnel to deal more effectively with political pressures that can negatively impact disease reporting when trade and/or tourism might be threatened (Palmer et al., 2009). When there is a lack of trust along the continuum—from the individual to the local, national, regional, international, and global communities, and between the public and private sectors—an integrated zoonotic disease surveillance and response system cannot function at an optimal level.

Finally, social science expertise and capacity is needed to further study the responses of health systems to outbreaks of zoonotic disease at all levels, inclusive of factors that positively and negatively affect and influence vertical (within sector) and horizontal (intersectoral) communication and cooperation.

Necessity of Collaborations from Multiple Sectors

Close partnership between public and private sectors and across human health, agriculture, and natural resources are required for effectively planning and executing a comprehensive, integrated zoonotic disease surveillance system. The human health sector includes private and public physicians, public health professionals, village and community health workers, laboratories, hospitals, and nongovernmental organizations (NGOs) focused on health education, communication, and training. The agricultural, livestock, and poultry sector includes private- and public-sector veterinarians, village, and community animal health workers and technicians, animal producers, food systems, animal hospitals, and NGOs that provide development and capacity-building programs for small livestock and poultry holders, and programs for wildlife conservation, management, and disease surveillance (Kimball et al., 2008; Palmer et al., 2009).

Evaluating the Effectiveness of an Emerging Zoonotic Disease Surveillance System

Several attributes are assessed to evaluate the effectiveness of disease surveillance systems. These attributes include timeliness (detection, confirmation, dissemination), simplicity, flexibility, quality and reliability of data, acceptability, sensitivity (i.e., probability of identifying all cases of disease or outbreaks), high-positive predictive value (i.e., given a report of disease or an outbreak, there is high probability that the outbreak is real), representativeness of the population at risk, stability (i.e., robustness of the system under adverse conditions), and usefulness (WHO, 1977; CDC, 2001; Pappaioanou et al., 2003; Salman et al., 2003; Buehler et al., 2004; Salman, 2008; Wilkins et al., 2008). Important criteria associated with the effectiveness of animal disease monitoring and disease surveillance systems include aims, sampling, coordination and awareness, environmental factors, screening and diagnosis, data collection and transfer, data processing and analysis, and information dissemination (Salman et al., 2003; Salman, 2008).

CDC (2001), WHO (1977), the World Organization for Animal Health (OIE), and others have developed guidelines to assess the effectiveness of a disease surveillance system based on those components and attributes deemed essential. Table 4-1 provides a comparison of core capacity requirements for disease surveillance mentioned in IHR 2005 versus the evaluation of quality mentioned in OIE's Performance of Veterinary Services tool.

REVIEW OF EXISTING DISEASE SURVEILLANCE SYSTEMS FOR ZOONOTIC DISEASES

The committee examined several infectious disease surveillance systems already in operation to identify some effective systems, uncover gaps in efforts, and examine important ways that existing systems could improve to achieve the desired global disease surveillance system. In place of an exhaustive review of all disease surveillance systems that have been developed for human infectious diseases by ministries of health, or those developed for animal diseases of importance by ministries of agriculture, a broad spectrum and mix of existing human and animal infectious disease surveillance systems were presented and discussed at a 2-day workshop, *Achieving Sustainable Global Capacity for Surveillance and Response to Emerging Diseases of Zoonotic Origin: Workshop Summary* (IOM and NRC, 2008). It is important to learn from these and other disease surveillance and response programs as efforts are dedicated to building an effective, global emerging zoonotic disease system.

TABLE 4-1 Comparison of Disease Detection and Response Evaluation Standards for Human and Animal Health: International Health Regulations Versus Performance of Veterinary Services Tool

Specification of Capacities Related to Disease Detection and Response	WHO IHR, 2005	OIE PVS Tool, 2008
Requirement for assessments of infrastructure, and support for early disease detection, disease surveillance, and response	Yes, mandatory within 2 years of the date of entry into force (June 2007)	No, assessments are voluntary, no time requirement
Support by state party/country for assessments, planning, and implementation processes made clear and provided	Yes	No, but best practice described in qualitative ratings
Definitions of disease representing an urgent threat provided	Yes	No, but reference to OIE Terrestrial Code provided for details
Explicit criteria and qualitative levels of advancement described for assessing existing structures and resources to meet core capacity requirements	No	Yes
Minimum requirements for disease detection and reporting articulated by level of the health system	Yes, as core capacities in Annex 1	No, but best practice described in qualitative rating; reference to OIE Terrestrial Code provided for details
Definition/specific listing of essential information to be provided in reporting is provided	Yes, but in general categories only	No, reference to Terrestrial Code provided for details
Minimum requirements for response to emergent event, by level of the health system provided	Yes	No, but general best practice described in qualitative rating; reference to OIE Terrestrial Code provided for details
Development of implementation plans based on assessment	Yes, and required	Yes, but not required, best practice described in qualitative rating
Provides awareness and improves understanding of all sectors regarding fundamental components and critical competencies required to function efficiently	Yes	Yes
Support by international agency for the conduct of assessments and development of implementation plans	Yes	Yes

NOTES: IHR = International Health Regulations; OIE = World Organization for Animal Health; PVS = Performance of Veterinary Services; WHO = World Health Organization.
SOURCES: WHO (2007a); OIE (2008b).

Infectious Disease Surveillance Systems in Human Populations and Species-Specific Animal Populations

At the National Level

At the national level, surveillance systems for human and animal infectious diseases are under the auspices of different departments or ministries (hereinafter collectively known as departments). Infectious disease surveillance of humans falls under the departments of health or public health, while surveillance for infectious diseases in food-animals are typically under the auspices of departments of agriculture or livestock. More recently, surveillance of specific infectious livestock diseases has been conducted by private food production companies raising large populations of food-animals for international trade and human consumption. Surveillance for diseases in wildlife is most often under the purview of departments of natural resources, wildlife, or fish and game. Based on the spread of HPAI H5N1 in birds, funding has been given to wildlife conservation NGOs and universities to support their efforts in complementing efforts by departments of natural resources to conduct surveillance in wildlife. In a few instances, disease surveillance of zoonotic diseases in companion animals has been conducted as special studies (Glickman et al., 2006). However, despite the close human-animal contact between humans and companion animals, responsibility for zoonotic disease surveillance and reporting in these animals—with exceptions of rabies and psittacosis surveillance in dogs and pet birds, respectively—has not been placed under the purview of any department in any country.

At the International Level

Infectious disease surveillance efforts in human populations have focused primarily on diseases such as HIV/AIDS, tuberculosis, malaria, cholera, vaccine preventable diseases, and those causing high morbidity and mortality.

Several global disease surveillance systems and their networks have been instituted primarily for detecting either human outbreaks of emerging zoonotic diseases or animal outbreaks of animal diseases. Examples of human disease surveillance systems reviewed in the workshop report include the WHO Global Outbreak and Response Network, and the U.S. Department of Defense's (DoD's) disease surveillance efforts with the Global Emerging Infections Surveillance and Response System (GEIS) and Early Warning Outbreak Recognition System in Asia, Africa, and other high-risk areas for emerging zoonotic infectious diseases (IOM and NRC, 2008). With regard to surveillance in animal populations, the OIE World Animal

Health Information System, and Food and Agriculture Organization of the United Nations (FAO) Emergency Prevention System for Transboundary Animal and Plant Pests and Diseases were also presented and discussed in the workshop summary (IOM and NRC, 2008).

Although human surveillance systems have identified a number of zoonotic disease outbreaks in humans, these global systems have yet to be adequate in detecting infections in animal populations early enough to prevent transmission from animal to human populations. Unfortunately, because disease surveillance efforts in livestock, poultry, and wildlife typically have been under-resourced even more than disease surveillance in human populations, it is frequently the detection of disease outbreaks in humans that have led to the detection of disease outbreaks in animal populations rather than the reverse.

Integrating Disease Surveillance Efforts Across Human, Animal, and Environmental Health Sectors

Surveillance Efforts in the United States

CDC has made efforts to integrate its zoonotic disease surveillance efforts across human and animal health sectors in the United States. This was most evidently demonstrated in 2007, when CDC established the National Center for Zoonoses, Vector-borne, and Enteric Diseases, which brought expertise in human and animal health together in the same administrative unit. In addition, several animal diagnostic laboratories have joined the Laboratory Response Network for Bioterrorism system of laboratories. The national surveillance system for arboviral diseases, ArboNET, is one example of a national surveillance system that has integrated disease surveillance efforts (see Box 4-5). Other efforts include FoodNet and PulseNet for food-borne diseases (CDC, 2008a) and the National Antimicrobial Resistance Monitoring System for detecting changes in antibiotic resistance in food-borne pathogens.

Zoonotic agents comprise more than 80 percent of the CDC-listed biothreat agents of concern (CDC, 2003a; IOM and NRC, 2008). An optimally integrated surveillance system could integrate existing biosurveillance efforts with ongoing surveillance efforts for zoonotic diseases. Biosurveillance efforts include the National Biosurveillance Integration Center established by the Department of Homeland Security (DHS, 2009a); the Integrated Consortium of Laboratory Networks (ICLN, 2009) established by the Department of Homeland Security; the Biosurveillance Indications and Warning Analytic Community established by CDC (CDC, 2008b); and the National Biosurveillance Advisory Subcommittee of the CDC and the National Biosurveillance Strategy for Human Health in response to Home-

BOX 4-5
ArboNET: Example of an Integrated Zoonotic Disease Surveillance System

An example of a currently functioning zoonotic disease surveillance system that is approaching a state of integration across human and animal populations is the national surveillance system for arboviral diseases, or ArboNET, for West Nile virus fever surveillance in the United States. In this system, the results of surveillance in mosquito and bird populations are disseminated to human and animal health authorities at state and national levels to provide data to trigger mosquito control and increased public health alerts when the risk of infection increases, cautioning people to use insect repellent, wear long-sleeved clothing, and take other actions to decrease exposures to the virus through the bites of infected mosquitoes. Although ArboNET is an example of positive progress made in efforts to integrate data from several sources, work remains to be done in including additional veterinary diagnostic laboratories into the system and improving timeliness of information and communication across sectors.

SOURCE: IOM and NRC (2008).

land Security Presidential Directive 21 (DHS, 2009b). It will be important to learn about the challenges, successes, and failures for coordinating these biosurveillance activities, as similar issues may emerge when integrating multiple human and animal epidemiological and laboratory surveillance systems.

Global Surveillance Efforts

WHO Global Salm-Surv The WHO Global Salm-Surv is a growing international surveillance system for *Salmonella* and other major food-borne pathogens. This relatively new initiative is built on a foundation of a global network of institutions and individuals (WHO, 2009a). This program is relevant to the current challenge because diarrhea-causing *Salmonella* serotypes are zoonotic in origin, and because the mission of this initiative is to "promote integrated, laboratory-based surveillance and foster intersectoral collaboration among human health, veterinary, and food-related disciplines through training courses and activities around the world."

GLEWS WHO, FAO, and OIE have recently joined forces to integrate alert mechanisms for emerging zoonotic diseases in the Global Early Warning System (GLEWS) for major animal diseases. GLEWS builds on the added value of combining and coordinating the alert mechanisms of WHO, FAO, and OIE to communicate with the international community and stakeholders on the occurrence of emerging zoonotic diseases; to aid in prediction, prevention, and control efforts; and to deploy joint field missions to assess and control disease outbreaks. Although this system is promising and offers a platform bridging across the human and animal health sectors at the global level for disease surveillance, response, and interdisciplinary training, it is relatively new and unproven. Much remains to be done to achieve a global, integrated zoonotic disease surveillance system with respect to clinical, epidemiological, laboratory, and risk behavior components.

National and Global Surveillance Efforts of HPAI H5N1

One of the best examples of strategic surveillance for influenza viruses in reservoir and other sentinel species is the U.S. Interagency Strategic Plan for the Early Detection of HPAI H5N1. This was implemented in 2006 across national and state agencies of agriculture, natural resources, and human health, and included other institutions essential for conducting HPAI H5N1 surveillance in wild birds (USGS, 2006). This program is funded by the U.S. Department of Agriculture (USDA) and has tested samples from more than 200,000 wild birds and 100,000 environmental samples in multiple flyways in the United States[1] (WDIN, 2009). The purpose of this program is to detect types of influenza virus that are of potential high virulence in poultry, with particular emphasis on HPAI H5N1. The data gathered since its launch has provided a much better assessment of the prevalence of influenza A in wild birds in the United States.

Fueled by the concern for and specter of the "next 1918 influenza pandemic," there have been additional global efforts aimed at integrated disease surveillance of HPAI H5N1 in wild birds, poultry, and human populations. The U.S. Department of Health and Human Services (HHS), USAID, DoD, and private-sector partners have funded disease surveillance for HPAI H5N1 viruses in wild birds in Southeast Asia and East Africa (GAINS, 2009). The U.S. National Institutes of Health has funded animal disease surveillance for avian influenza viruses in Southeast Asia and Africa (NIAID, 2008). Targeted disease surveillance for avian influenza in wild birds has resulted from human spillover (e.g., Indonesia), and impressive international laboratory coordination has occurred in regions of the world believed to be at higher risk of avian and human infections, because of the concern that a pandemic influenza virus would emerge from wild birds.

Not unlike what has happened with surveillance for HPAI H5N1, infectious disease surveillance for human populations in developing countries has typically been funded through vertical programs to monitor specific diseases, such as HIV/AIDS, tuberculosis, malaria, cholera, and vaccine pre-

[1]Seth Swafford, USDA/APHIS/Wildlife Services, personal communication, July 1, 2009.

ventable diseases. The simultaneous existence of so many independent, vertically operated initiatives duplicates costs, slows the reporting of emerging disease events and data analysis and interpretation, and adversely impacts the use of disease surveillance data to guide and evaluate disease prevention and control efforts. The WHO Regional Office for Africa (AFRO) implemented the Integrated Disease Surveillance and Response (IDSR) program in the 1990s to strengthen disease surveillance by using simplified tools for data collection and analysis, providing channels for reporting and feedback, and strengthening district-level capacity to generate and transform disease surveillance data to inform human and animal health action (Nsubuga et al., 2002; CDC, 2009b). Although AFRO was successful in integrating disease surveillance efforts within the human health community, the IDSR is not connected or linked to surveillance for zoonotic diseases in animal populations. The task of integrating disease surveillance for multiple infectious diseases within ministries of health in developing countries has been difficult, but the barriers to successful linkage and integration of human and animal health surveillance systems have proven to be far greater.

CAPACITY-BUILDING PROGRAMS TO CREATE A MULTIDISCIPLINARY, INTEGRATED WORKFORCE

United States

In the United States, more than 3,000 physicians, PhDs, veterinarians, nurses, dentists, and other professions have graduated from the CDC Epidemic Intelligence Service program since its inception in 1951. The 250+ veterinarians who are graduates of this program have played critical roles in serving public health by working in positions across all CDC centers and at state health departments. In addition to fulfilling their public health responsibilities in infectious diseases, environmental and global health, they have played an important role in bridging human and agricultural health concerns. A growing number of U.S. colleges of veterinary medicine and schools of public health offer joint doctoral degrees in veterinary medicine and master's degrees in public health, an important step in achieving integration across human and animal health sectors (Hueston, 2008). For instance, the Center for Food Security and Public Health at Iowa State University posts a variety of web-based educational materials on zoonotic diseases (The Center for Food Security and Public Health, 2009a).

European Union

In the European Union (EU), the European Programme for Intervention Epidemiology Training provides experiential field training in intervention epidemiology at the national centers for surveillance and control of communicable diseases (EPIET, 2009). The program is aimed at EU physicians, public health nurses, microbiologists, veterinarians, and other health professionals with experience in public health that have an interest in population-based disease prevention and control. The program is hosted at the European Centre for Disease Prevention and Control in Stockholm, Sweden.

Global Programs

Since the 1970s, a number of international programs have been launched to strengthen field epidemiological, laboratory, and program management capacity (Pappaioanou et al., 2003). WHO has sponsored large global human capacity-building efforts to implement immunization programs (WHO, 1980), diarrheal disease control (CDC, 1983), integrated management of the sick child (Victora et al., 2006), and tuberculosis control (WHO, 2009b), among many other programs. CDC, WHO, Rotary International, and others have collaborated to support major capacity-building efforts for polio elimination (CDC, 2009c). USAID has funded significant training programs for the IDSR hosted at AFRO (USAID, 2009).

In the animal health sector, FAO, OIE, and USDA—with funding provided by the World Bank, USAID, and other donor agencies—conduct some capacity-building efforts for disease surveillance and response in animal populations. These initiatives have trained hundreds of animal health experts around the world in diagnostic methods and disease prevention and control for diseases of importance, including zoonoses associated with animal trade (Kerwick et al., 2008). The USDA Animal and Plant Health Inspection Service is oriented to U.S. domestic concerns, but offers clinical and laboratory diagnostic training in foreign animal diseases that can be imported into the United States (USDA-APHIS, 2009b). To address the current risk of HPAI H5N1, the USDA Agricultural Research Service and USDA Foreign Agricultural Service, with funding from the Institute for International Cooperation in Animal Biologics, has provided training on avian influenza diagnostics and disease control to poultry experts around the world (The Center for Food Security and Public Health, 2009b). The USDA and Global Livestock Collaborative Research and Support Program, supported by USAID, has funded the University of California–Davis Avian Flu School Program's 8 training modules and 15 training sessions by Colorado State University on all aspects of avian influenza diagnosis, transmission, surveillance, and disease prevention and control of relevance to poultry, wildlife, and human populations for international use (Salman, 2008; UC Davis School of Veterinary Medicine, 2009).

In general, these independent initiatives are funded by public and private sectors, and they address goals and objectives aimed at confronting human and animal health challenges from single diseases with little to no overlap across human and animal populations. They represent a patchwork of systems that lack the necessary integration to create a global system robust enough to cover emerging zoonotic diseases for both humans and animals.

INCLEN

In 1980, The Rockefeller Foundation created and funded the International Clinical Epidemiology Network (INCLEN), a unique global network of clinical epidemiologists, biostatisticians, health social scientists, health economists, and other health professionals. INCLEN is affiliated with 82 key academic healthcare institutions. Through collaborative, interdisciplinary, evidence-based research, they study high-priority health problems and promote equitable healthcare and efficient use of resources. Their program has addressed both communicable and noncommunicable diseases, as well as health system initiatives. To date, they are not yet oriented to integrated human and animal disease surveillance for zoonotic diseases.

FETPs and FELTPs

With funding from HHS, WHO, USAID, and other partners, CDC has collaborated with more than 40 countries to establish Field Epidemiology Training Programs (FETPs) (Thacker et al., 2001). These programs are modeled on CDC's Epidemic Intelligence Service program but are tailored to the needs of host countries. To date, more than 1,000 trainees have graduated from these 40 programs, and more than 500 trainers have been trained.[2]

Laboratory training components have more recently been added to the FETP model, resulting in the Field Epidemiology and Laboratory Training Programs (FELTPs) (CDC, 2009d). These programs have recently begun to accept veterinarians and laboratory scientists along with physicians in their classes. In 1994, the FELTP in Kenya was established and supported by the Ellison Medical Foundation through the CDC Foundation. It uses both human and animal health expertise to conduct surveillance for emerging zoonotic infectious diseases in special at-risk populations in Nairobi. Its graduates are required to be placed in national or local positions focused on disease surveillance, prevention, and control. A major constraint of FELTPs is that in each country program, a maximum of 15 people are accepted per country program each year. Although there have been other attempts to

[2]Dionisio Jose Herrera Guibert, the Training Programs in Epidemiology and Public Health Interventions NETwork, personal communication, May 26, 2009.

train many more health professionals annually through shorter programs, the FELTP model has proven most successful in establishing and building sustained national core capacity for field epidemiology and laboratory experience with regard to infectious and other disease concerns in both developed and developing countries. The 40 FETPs and FELTPs function independently under national control (CDC, 2009d), but have formed an organization—the Training Programs in Epidemiology and Public Health Interventions NETwork—to enhance communication, coordination, collaboration, and networking.

Integrated multidisciplinary disease surveillance and training efforts are beginning to see success in Africa. The African Field Epidemiology Network is supported by USAID and CDC, and it includes FETP and FELTP programs in Kenya, Ghana, Uganda, and South Africa, and associate programs in Nigeria, south Sudan, Tanzania, Burkina Faso, Mali, Niger, and Togo. Several of these programs are taking steps to further integrate human and animal health surveillance systems in these key African countries (AFENET, 2009).

GAPS AND CHALLENGES

Existing surveillance and response programs were compared to components and attributes previously described in this chapter as essential for an effective global zoonotic disease surveillance and response system. Through that comparison, the committee identified some gaps and challenges that need to be addressed, which are summarized in Table 4-2.

Global Coverage of Emerging Zoonotic Disease Surveillance Systems

Irrespective of resource availability, the committee was unable to identify a single example of a well-functioning, integrated zoonotic disease surveillance system across human and animal health sectors. The committee was alarmed by the large gaps in existing disease surveillance networks, including coverage across species and across geographic space. Of particular concern is that in 90 percent of human infectious disease cases the causative pathogens have not actually been identified, even in developed countries (Farrar, 2008).

The current capacity for disease surveillance is strongest in developed countries, particularly when it is linked to response. The United States and Europe are greatly overrepresented in their reports of emerging disease outbreaks, which is directly related to disease surveillance and laboratory capacity (see Figure 2-10). Countries with weak disease surveillance capacity may not capture emerging disease outbreaks, and the level of outbreaks may be grossly underreported. For HIV/AIDS, a disease where substantial global resources have been made available, recent survey results on public health-related infrastructure and capacities revealed that only 21 of 30 country respondents had substantial activities in place for surveillance, meaning that the 9 low- or lower middle-income countries surveyed had insufficient capacity to respond to a known disease (Binder et al., 2008).

Developing countries have often focused their scarce resources on HIV/AIDS, tuberculosis, malaria, and other diseases that cause high morbidity and mortality in human populations. They lack the fundamental resources to conduct zoonotic disease data collection, collation, analysis, interpretation, and dissemination, and to conduct outbreak investigations and implementation of disease prevention and control efforts that should follow. And whatever national or regional data on zoonotic diseases from human and animal health systems might be collected, they often are not accessible or networked for effective global disease surveillance and response. The committee highlighted challenges for a few resource-constrained regions: sub-Saharan Africa, South Asia, and Southeast Asia. Although other regions of the world were not addressed in the same fashion, the lessons learned for Africa and Asia would also apply to other regions of the world.

In Africa

Despite significant advances in disease prevention and control in other parts of the world, communicable and zoonotic diseases still constitute major health problems in Africa (WHO, 2009c). HIV/AIDS is a pandemic of zoonotic origin, though it is now a chronic endemic disease transmitted from human to human. Together with the associated resurgence of tuberculosis in the region, HIV has gained considerable attention and resources commensurate with the threat to human health and development that it represents. Its relevance to zoonotic disease surveillance now is its origins in animals and the impact it has on the use of national and international resources for disease surveillance and response in poor countries. The resurgence of malaria and yellow fever in Africa in the past 15 years is a stark testimony to the serious breakdown of disease surveillance and control efforts (Roberts, 2007). More recently, the emergence and spread of HPAI H5N1 has seriously compromised the poultry industry in Nigeria and a few other West African countries (Xinshen and Liangzhi, 2006). Most public attention has been on the 421 human infections and 262 human deaths in 15 countries recorded through the end of June 2, 2009, with only 1 case and 1 death reported from Africa (both occurring in Nigeria in 2007) (WHO, 2009d).

In many African countries, disease surveillance systems in human and different animal populations function vertically, as they represent specific global initiatives set up to monitor specific diseases. In human health,

TABLE 4-2 Gaps and Challenges in Achieving an Effective, Global, Integrated Surveillance System for Emerging Zoonotic Diseases

Essential Components of an Effective Zoonotic Disease Surveillance System	Identified Gaps and Challenges
Global coverage	• Many countries do not have surveillance for zoonotic infectious diseases in human or animal populations • At the global level, collaboration among WHO, OIE, and FAO is nascent but improving
Multisectoral collaboration for planning, implementation, and evaluation	• In most countries, there are nonexistent or weak channels of communications, or platforms between sectors and multiple disciplines • There is a divide/gap in information sharing between the public and private sectors
Information gathering, dissemination • Disease surveillance in humans • Disease surveillance in livestock and poultry • Disease surveillance in wildlife • Disease surveillance in companion animals • Risk behavior surveillance • Surveillance of risk communication, messaging, public perceptions	• Surveillance is nonexistent or severely limited in human populations at greatest risk of emerging threats, making early detection nearly impossible • Surveillance in livestock and poultry is weak in developing countries; CAFOs may surveil their stock assiduously, but relevant information is not shared with animal or human health public-sector authorities • Disease surveillance in wildlife is nonexistent, inconsistent, or weak in all countries • Integrated disease surveillance in companion animals is mostly nonexistent in all countries • Surveillance of risk behaviors putting people at risk of exposure to zoonotic disease agents is mostly nonexistent • Surveillance of risk communication, messages, public perceptions of danger, threat, cause, and interventions is nonexistent or weak

TABLE 4-2 Continued

Essential Components of an Effective Zoonotic Disease Surveillance System	Identified Gaps and Challenges
Information Technology • Field-based data collection technology with emphasis on the availability of mobile phones • Open source, user friendly, bi-directional information communication tools • Signal detection algorithms and software packages • Improved web-based visualization of outbreaks and hotspots resulting from modeling efforts	• An automated, real-time, and integrated process of data collection, analysis, and interpretation across the multiple sectors concerned is absent • Standard protocols are absent to harmonize epidemiological and laboratory aspects of detection, confirmation, outbreak investigation, and design and implementation of disease control efforts
Laboratory Capacity • Laboratory infrastructure with appropriate biocontainment in resource-constrained regions • Protocols and procedures for sample collection and diagnosis • Adequately trained laboratorians and field staff for sample acquisition in resource constrained areas	• Human, domestic animal, and wildlife sector laboratories are currently not integrated, or operating seamlessly • Assessment of current global capabilities is inadequate and limits ability to develop an integrated laboratory network system nationally, regionally, or globally • Resource constraints limit the ability to further develop the network when plan is in place
Response Capacity	• Due to the committee's limited charge and major gaps and challenges in early detection, a full analysis of response was not addressed by this report. However, the lack of collaboration and communication across sectors was identified as a major gap in planning, implementing, and evaluating an effective response following detection of an emerging zoonotic disease
Human Capacity	• Limited numbers of field-oriented, multidisciplinary training programs and graduates • Expertise lacking in clinic-pathological diagnosis, field epidemiology, laboratory science, social science and communications • Leadership programs are essential but not widely available

NOTES: CAFO = concentrated animal feeding operation; FAO = Food and Agriculture Organization of the United Nations; OIE = World Organization for Animal Health; WHO = World Health Organization.

examples include poliomyelitis, meningitis, cholera, and other vaccine-preventable diseases (WHO-AFRO, 2001). In animal health, examples include rinderpest and foot-and-mouth disease (FMD) in cattle, classical swine fever and African Swine Fever in pigs, and Newcastle disease in poultry—diseases causing loss of production, which threatens local food availability and impedes international trade. These vertical programs have often succeeded in the development and use of disease-specific data collection tools, reporting formats, and surveillance guidelines for diseases of major interest to external donors, but these facilities have been minimally used for the surveillance or control of a country's endemic diseases (WHO-AFRO, 2001). Many specific interventions for vertical disease surveillance and response have not been sustained because of lack of appropriately trained local staff to maintain and sustain the programs when the threat level diminishes and donor support lags.

The many vertical programs have resulted in a landscape consisting of islands of high-quality disease surveillance and laboratory structures brought in and supported by the vertical programs, surrounded by substandard national disease surveillance efforts and laboratory facilities, with dismal working conditions and poorly trained, valued, and paid laboratory scientists. Given this history, it is no surprise that integration of disease surveillance and response is almost nonexistent, there being little to no interest in building local capacity for an integrated approach. Recent trends of importing technologies in lieu of hiring competent laboratory scientists locally have been appealing in the shorter term because fewer resources are required to support personnel costs. Nonetheless, this approach fails to deliver a sustainable system and in the end usually requires recruiting and retaining well-trained, competent, local laboratory science expertise.

The major obstacles to an effective disease surveillance and control system in Africa are insufficient funding, inadequate staffing, inappropriately trained personnel, and a failure to appreciate the cost effectiveness of a reliable disease surveillance system in healthcare delivery (Nigerian Federal Ministry of Health, 2007). Countries often do not have clear guidelines, procedures, and tools for disease detection, analysis, and interpretation of disease surveillance data, and/or the means for timely and complete reporting (WHO-AFRO, 2001). Another major challenge to effective zoonotic disease surveillance in Africa is poor support given to human and animal health laboratories, which are inadequately staffed and often lack basic equipment and reagents. Moreover, few countries have functioning systems for timely transmission of epidemiological information and transport of laboratory specimens to better equipped reference laboratories (WHO-AFRO, 2001). These weaknesses affect disease detection, analysis, and interpretation of data, as well as timeliness and completeness of disease reporting.

Even if zoonotic disease surveillance were sufficient to result in early detection and reporting, most African countries are currently ill-prepared to respond to emerging zoonotic disease outbreaks in humans and animals. Currently, a direct consequence of not having valid and timely surveillance information is that efforts to investigate and target disease prevention and response efforts are typically not guided by surveillance-based information, but rather are based on other political priorities, or just as problematic, a lack of understanding. This can further weaken the perceived role of disease surveillance in the prevention and control of communicable diseases. Thus priority diseases are sometimes not properly monitored, there is often a lack of response, and decisions on disease control can often be made under public and political pressure rather than on the basis of evidence.

In Asia

The Asian region has repeatedly been the source of new zoonotic disease agents. Although Asia was the source for the first alerts of HPAI H5N1 infections in humans (Hong Kong in 1997) and of Nipah virus encephalitis in pigs and swine workers (Malaysia and Singapore in 1997–1998), SARS served as the "wake-up call" for the Asian region. Even though SARS responded to simple containment strategies and even though human-to-human transmission has not been detected since 2003, the SARS outbreak highlighted the need to first strengthen surveillance for emerging zoonotic infectious diseases in both human and animal populations, and then to integrate disease surveillance across these multiple sectors. Because Asia spreads across multiple WHO regions—the Regional Office for Southeast Asia and the Regional Office for the Western Pacific—communication across these regions has been a source of friction within the regional and international network, and has therefore made the sustained early detection of unusual events even more challenging. Fortunately, the joint Asia-Pacific Strategy on Emerging Diseases has moved this critical collaboration forward (WHO, 2006).

At a national level, the Chinese Center for Disease Control and Prevention initiated a web-based disease surveillance system after the SARS outbreak that involves a network of 1,500 centers in China at district, provincial, and central levels. This network is largely involved in human disease surveillance, but a great deal of attention is on the threat of zoonoses. Meanwhile, countries profoundly affected by HPAI H5N1 (e.g., Indonesia, Vietnam, China) have been working to bring the agricultural and human sectors of disease surveillance together. Notable is the "One Health" initiative, which Google among many others is promoting in the Mekong region. The Thailand FETP has trained veterinarians in their epidemiological investigation program, and there are initiatives to make this linkage stronger.

Avian influenza has spurred discussions and information sharing across sectors in many countries, with Thailand being especially engaged. However, early detection will require veterinarians and agricultural workers to closely integrate and communicate with human health authorities. As recently as 2007, a committee member observed that ministry of health workers in one Mekong country still learned of avian influenza outbreaks primarily through the media. In addition, officials from several international health agencies have noted that one Mekong country lacked poultry veterinarians with competent clinical skills to diagnose and differentiate avian influenza from other avian diseases affecting poultry. There was a serious call for assistance to strengthen local and national capacity in the clinical, basic animal husbandry, field epidemiology, and laboratory competencies.

Indonesia, the country most highly impacted by HPAI H5N1, has also experienced many obstacles in carrying out an integrative approach to disease surveillance and response. Politics, economic concerns, trust, and failure to ensure that the benefits of sharing specimens are equitably applied to the populations at risk and the international community have seriously impaired global collaboration on virus sharing (Padmawati and Nichter, 2008).

Multisectoral Collaboration for Planning, Implementation, and Evaluation

Coordination Across Human and Animal Populations and Multiple Governmental Sectors

In addition to geographic coverage gaps, other gaps include the limited coverage of disease surveillance in human populations and in different species of animal hosts, including food-animals, companion animals, and particularly wildlife. Another challenge is the lack of collaboration across the human and animal sectors charged with overseeing health.

Surveillance in Humans As previously mentioned, there are large-scale surveillance programs for major infectious diseases in humans that cause high morbidity and mortality (e.g., HIV/AIDS, tuberculosis, malaria, vaccine preventable diseases). However, with regard to surveillance for emerging zoonotic diseases in human populations, occupational surveillance is nonexistent or is sporadic and weak to detect new infections in people having greatest contact with animals.

Surveillance in Livestock and Poultry Populations Disease surveillance systems for livestock, poultry, and captured or farmed wildlife animal species often monitor the production parameters (such as milk production, egg production, and weight gain), use diagnostic tests to detect specific diseases, and have response plans in place to reduce the spread of diseases when they occur. Several countries, including the United States, have successfully reduced and even eradicated livestock and poultry diseases through effective surveillance and response. Also, brucellosis has been eradicated from countries such as New Zealand, Iceland, and Denmark through field epidemiological diagnosis and confirmatory testing, followed by culling of infected herds. In the United States, eradicating bovine brucellosis from domestic cattle is nearing the final stage, with only 1–3 infected herds reported for 2008–2009. In commercially raised turkeys in Minnesota, occurrences of low pathogenic avian influenza in domestically raised turkeys was dramatically reduced when surveillance data were used to change and evaluate the impact of housing conditions and disease occurrence (Halvorson, 2002).

Despite increased investments in surveillance for HPAI H5N1 and other influenza viruses relevant to U.S. poultry producers, very few influenza viruses that are detected in poultry, swine, and other animals have been characterized. These viruses have potential to genetically reassort and infect across species. If the goal of influenza surveillance is to monitor influenza virus types with potential importance to humans, it can be argued that surveillance should first strategically target animal species most likely to transmit viruses to other animal populations, and then analyze a subset of those viruses for features that may contribute to cross-species transmission and virulence in humans. For resource-constrained regions, it might be most cost beneficial to continuously monitor influenza virus populations where there are a mix of animal species: for instance, when both swine and poultry are in close proximity to each other. Genetic characterization tools are now available to produce a much broader and more robust influenza database, and the NIH-funded Centers of Excellence for Influenza Research and Surveillance tracks and shares influenza data (NIAID, 2008).

Surveillance in Wildlife Even though wildlife populations are known reservoirs for high-impact zoonoses, surveillance for zoonotic pathogens in wildlife populations around the world is fragmented and incomplete at best, and nonexistent at worst. This is due to a variety of factors, including the expense and feasibility involved in reaching and capturing hard-to-reach populations, often leading to small sample sizes. In addition, even if a wild animal is reached, size limitations of birds, bats, and small rodents provide only small volumes of tissues or body fluids for testing.

A few developed countries routinely conduct wildlife disease surveillance for certain zoonotic diseases, such as rabies in bats, foxes, and raccoons in the United States (Blanton et al., 2008). Surveillance for WNV has been ongoing since 2003. With more than 2,000 human cases of WNV reported annually in the United States, it has become the most prevalent vector-borne human pathogen reported in the country (Petersen and Hayes,

2004). Accordingly, state and county departments of health routinely conduct disease surveillance in dead birds and mosquitoes early in the transmission season so that they can follow the spread of emergence and institute preventive interventions (i.e., mosquito control, and health education advising the population to avoid exposure to mosquitoes). The United States has devoted greater investments in wildlife surveillance since the early 1990s, when it was found that tuberculosis-infected wildlife play an important role in bovine tuberculosis emergence in cattle and that cross-infection of U.S. cattle and bison resulted in brucellosis-infected cattle herds.

Even though the human monkeypox virus was introduced into the United States via the wildlife trade, mandatory testing of the 500 million wild animals imported into the United States annually remains confined to a handful of agriculturally significant diseases such as Newcastle disease, FMD, brucellosis, psittacosis, and more recently avian influenza. The remaining potential threats are handled through ad hoc research studies to detect and explore relationships among emerging zoonotic disease agents in wildlife, domesticated animals, and humans. SARS is a good example of such ad hoc efforts in wildlife: There has been significant research interest in the wildlife origins of SARS in China, yet to date there are no coordinated integrated disease surveillance programs for SARS or other pathogens in wildlife. The lack of human SARS cases since 2003 is one factor for waning interest and loss of commitment to conduct ongoing SARS coronavirus surveillance in wildlife reservoirs.

Investigators have monitored the virologic evolution of the HPAI H5N1 virus in southern China markets (Smith et al., 2006), although it appears that a nationally led disease surveillance effort was not created. Australia, with a long history of leadership in zoonotic disease surveillance and control programs, has set up the Australian Biosecurity Cooperative Research Center for Emerging Infectious Disease, which conducts preborder disease surveillance of wildlife and food-animals in countries across Southeast Asia that trade with Australia. *Despite these several positive examples, the committee concludes that the use of ongoing, sustained disease surveillance to detect potential new zoonoses in wildlife remains limited, even in wealthy, developed countries.*

Detecting Subclinical Infections in Livestock, Poultry, and Wildlife Animal and environmental reservoirs are capable of transmitting zoonotic pathogens to others with limited or no impact on individual animal or population health. Disease surveillance is challenging in animal reservoirs because the reservoirs either do not display clinical signs of infection, or if they do, the infection is mild. However, existing knowledge about these reservoir species allows strategic surveillance programs to be designed to continuously assess the agent population for its prevalence in the reservoir and to assess changes that can lead to pathogen emergence in humans or other animal species. Where these characteristic features are known, it is possible to predict emerging events in advance of significant mortality and morbidity.

Many pathogens affecting livestock and poultry cause no clinical illness in individual animals but result in human cases and outbreaks. Cattle infected with brucellosis may not show clinical signs, and the most likely cases are identified by conducting active surveillance through routine specimen collection from herds, then testing with serological assays and bacterial culture.

Asymptomatic infections in food animals can cause significant human food-borne illness and mortality. The H5N1 virus circulates in ducks, an important food source for humans, but the ducks do not show clinical signs of disease. Cattle infected with *Escherichia coli* O157:H7 and *Salmonella enterica* often do not show clinical signs of illness (Dewell et al., 2005), but the bacteria cause illness in humans (Mead et al., 1999; CDC, 2006). Infection can be detected through culture or antigen detection assays including molecular techniques in several organs including lymph nodes, digestive tracts, and even mucosal membrane (Dargatz et al., 2005). In the United States and other selected countries, testing and monitoring of *E. coli* O157:H7 and *S. enterica* in both live and slaughtered animals are conducted as part of a surveillance system for meat safety. Molecular pheno- and genotyping (e.g., PulseNet in the United States) during the past few years has led to the condemnation of infected meat products (CDC, 2008c). However, despite ongoing surveillance efforts, human foodborne outbreaks of *E. coli* O157:H7 persist and underscore the imperfection of existing surveillance methods for these types of infections.

The pandemic influenza A(H1N1) 2009 virus has molecular signatures of swine, avian, and human influenza viruses. Influenza is a mild respiratory disease in pigs and difficult to differentiate from other diseases. Because classical swine fever is a common illness in pigs, strategic sampling of swine and characterization of triple reassortment influenza viruses would be needed to provide a much better base to predict and prevent disease emergence. This could also guide continuous diagnostic assessment modifications in animal and human health laboratories.

Communicating and Collaborating Across Sectors, Professions, and Health Systems

The negative impact of emerging zoonotic disease outbreaks these past few decades has brought increased attention to the need for disease surveillance that links and provides information across human and animal health sectors for early detection and response. Despite considerable understanding of the need for such an integrated approach, one of the biggest challenges has been effectively communicating across these sectors before, during, and after disease outbreaks.

Failed communication among sectors can lead to delays in detecting and confirming emerging zoonotic disease outbreaks. Examples include the failure of human health authorities in 1999 to follow up on a veterinary pathologist's alert that disease outbreaks in birds and humans could be related and caused by the same agent, WNV (GAO, 2001); the failure of the animal health sector to alert human health authorities of sick rodents imported from Africa and housed with prairie dogs, ultimately leading to a human outbreak of monkeypox (CDC, 2003b,c); the failure by animal health officials in Southeast Asia in 2003 to alert WHO about HPAI H5N1 outbreaks in poultry in the region, and leading to delays in confirmation of human HPAI H5N1 cases in Vietnam (WHO, 2004); and the failure of human health authorities in Africa, at least twice, to take action to prevent human exposure to Ebola during 2001–2003 when they were alerted to wild animal outbreaks weeks before human cases occurred (Rouquet et al., 2005).

Other factors can influence the success of cross-sectoral communication. These include protocols for appropriate communication, technological capacity and resources, level of active outreach and persistence by individuals, and political will. It furthermore depends on the extent that professionals in different disciplines understand and respect the expertise of their counterparts in other professions. The busy schedules of professionals and their dispersed office locations also limit opportunities for casual contacts. Therefore intentional meetings are necessary for developing integrated disease surveillance and response strategies, reviewing disease surveillance findings, and reaching joint decisions on prevention and control strategies.

Risk communications by human and animal health authorities and the media can affect how the public understands the disease and affect its actions during an outbreak. Yet relatively few surveillance programs have brought social science professionals on board with their efforts to effectively communicate with the public about ongoing disease risks and behaviors.

Lessons learned to date suggest that easy channels of communication between departments of human and animal health and the public and private sectors are mostly nonexistent. Some efforts to improve communication across medical and veterinary health sectors were launched when HPAI H5N1 began spreading, but those were weak at best. Given the disappointing experiences previously noted—followed by similar communication difficulties encountered with SARS and Nipah virus outbreaks in Hong Kong and Malaysia—several countries have organized special multisectoral coordinating committees and task forces to oversee HPAI H5N1 disease surveillance and response and to formulate appropriate disease control policies (Tanzania, Kenya, Asia). These initiatives will need to be assessed for their success and to determine how best to overcome communication barriers between human, animal, and environmental health officials that seem to exist independently of the resources available to a country.

Employing Strategic Approaches for Effective Surveillance and Response

Emerging zoonotic diseases can emerge at any time in any part of the world, therefore it is difficult to predict which pathogens may emerge, which human and or animal populations it may impact, or how these pathogens may spread. From a growing number of experiences, the world has learned that it is critically important to detect and report emerging zoonotic disease outbreaks that occur in a single country or region. Early detection and reporting at the local level give the international community an opportunity to assist national authorities and implement effective response measures. Because local outbreaks in today's world can quickly spread beyond national borders and have significant global health and economic impacts, it is crucial to invest in disease surveillance capacity for countries that cannot afford it.

No matter the wealth or capacity of any country, resources are often not available and are at best limited at all levels for detecting and responding to zoonotic diseases. National expertise and current levels of disease surveillance system development vary by country and region. Resource-challenged countries are often those geographically located in areas that place them at an increased risk of pathogen emergence and cross-species transmission, due to factors such as climate change, ecosystem degradation, biodiversity, and population density. Thus, strategic approaches tailored to different settings and different resources are needed for surveillance in different animal species—including terrestrial and aquatic wildlife, livestock, and poultry—to detect disease early, improve animal health, and minimize the likelihood of human exposure.

Securing and Providing Information Technology

New disease surveillance data sources hold tremendous potential to initiate epidemiological follow-up studies and provide complementary epidemic intelligence context to conventional sources, yet are subject to a number of potential hazards that need to be studied in depth, including false reports (mis- or disinformation) and reporting bias. An open and accessible IT system assists users in overcoming existing geographical, organizational, and societal barriers to information, a process that can lead to greater empowerment, involvement, and democratization. Because regions with the least advanced communication infrastructure also tend to bear the greatest infectious disease burden and risk, system development needs to be aimed at closing the gaps in these critical areas. Global coverage requires attention to creating and capturing locally feasible channels of communication and making sure that system outputs are more accessible to users in vulnerable regions. Low-bandwidth options, including mobile phone alerts, could be considered for helping to transmit information.

Much work needs to be done to integrate the processes of data collection, analysis, and interpretation across the multiple sectors concerned. Standard protocols are lacking and necessary to harmonize epidemiological and laboratory aspects of detection, confirmation, outbreak investigation, and design and implementation of disease control efforts. The committee could not identify any system that had incorporated routine disease surveillance for human risk behaviors into disease surveillance capacity, a capacity needed for an optimally effective disease surveillance and response system.

The principal concern with any web-based, early alert system is the reliability and verification of information in a similarly rapid and transparent manner. A second high-priority concern is determining how best to use the voluminous amount of information, especially when there is conflicting information. In addition, users need to determine how to filter information to accurately separate actual events from "noise," how to support the diverse set of data reported and who can provide the service, whether it is possible to integrate various data sources into a user-friendly format, and what the cost is for accomplishing all the above. Timeliness and the reliability in confirmation of the diagnosis of the etiologic agent of human or animal outbreaks are two characteristics that can be in direct opposition to one another. For example, the automated media scanning program GPHIN provides a great volume of alerts to WHO's GOARN, which expends significant efforts in confirming these alerts. Given the frequent absence of laboratory confirmation at the time of informal outbreak alerts, an increased rate of false-positive reports is expected. Thus validating the information quickly is essential to minimize false-positive alerts. Moreover, because regions with the least advanced communication infrastructure also tend to carry the greatest infectious disease burden and risk, system development would need to be aimed at closing the gaps in these critical areas.

Another challenge is keeping identifiable data about patients and animal owners secure and confidential. Identifiers can be important for disease tracing, but if divulged can result in unintended harms (e.g., sanctions or stigma). Such outcomes have serious adverse impacts because the willingness to report is based in large part on trust that punitive actions will be avoided. When disease tracing is not essential, data that identify individuals do not need to be collected. When disease trace-backs and trace-forwards are indicated for disease control purposes, however, information on individual people and animals (including the population source) needs to be available. For a disease surveillance system to function effectively, identifying information needs to be kept confidential to engender trust but be available if necessary for an appropriate response to the health risk.

Laboratory Capability and Capacity

Previous National Research Council and Institute of Medicine reports provide a summary of laboratories operated by developed countries, universities, and donor agencies (NRC, 2005; IOM and NRC, 2008). Although the committee was aware of specific laboratories in resource-constrained countries that have outstanding capabilities for zoonotic disease diagnosis (e.g., the CDC–Kenya Medical Research Institute [KEMRI] Emerging Infections Laboratory in Nairobi, the Uganda Virus Research Institute in Entebbe, the Dhaka-based International Center for Diarrheal Disease Research in Bangladesh), it was beyond the committee's scope and task to thoroughly identify existing laboratory capacity and capability for zoonotic diseases on a global basis. Such a database is sorely needed but not available.

The committee was able to request information from WHO, FAO, OIE, and DoD-GEIS about their laboratory locations and capacity. That information provided the basis for the committee's analysis of existing laboratory locations compared to where they are most needed. Figure 4-3 shows DoD-GEIS laboratories and WHO, FAO, and OIE reference laboratories and collaborating centers superimposed on a map depicting the predicted global hotspots for disease emergence, as previously seen in Figure 1-1. There are many additional private and government laboratories with disease diagnostic capabilities that are not designated on this map. It is apparent that there is at a minimum a striking geographic mismatch between intergovernmental organization-designated reference laboratory locations and capacity and hotspot regions suspected to be ideal for pathogen adaptation, selection, and emergence.

It is important to recognize that these reference laboratories typically do not have broad capabilities in disease diagnosis, that they are often research laboratories with agent-specific expertise, or that their mandates are not directed specifically at zoonotic disease surveillance. Thus there is no resource that provides current data on existing zoonotic disease diagnostic laboratory capability and capacity worldwide. In addition, among reference laboratories, there are no common operational protocols as one might expect in a laboratory network, whereas research laboratories often do not seek or maintain accredited laboratory status and are oriented to the particular research program of the institution in which they reside.

There is a critical global shortage of qualified laboratories for zoonotic

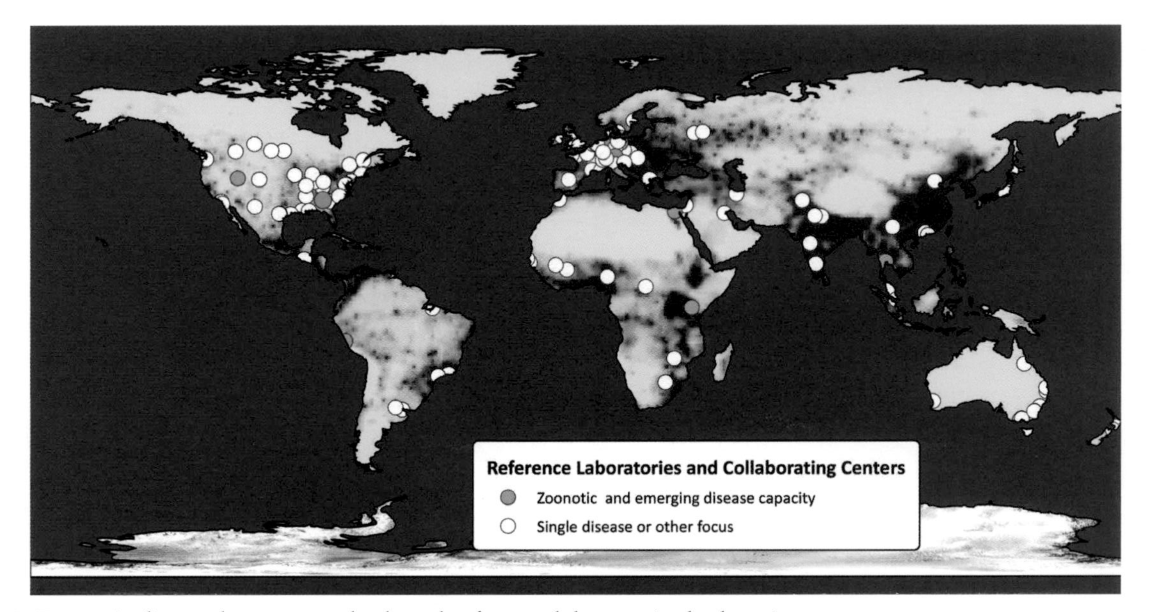

FIGURE 4-3 Zoonotic disease hotspots and selected reference laboratories by location.
NOTE: The white dots signify the location of identified World Health Organization, Food and Agriculture Organization of the United Nations, World Organization for Animal Health, and U.S. Department of Defense reference laboratories and collaborating centers, many of which have a single disease or other focus mandate. Green dots are laboratories that have a broader function in zoonotic and emerging diseases. Locations shaded in red and orange represent hotspot regions. The map does not include university-based research and other laboratories working in the area of emerging disease detection and characterization. Numerous other private-sector and national laboratories may be able to provide laboratory support capability (e.g., those of the Institute Pasteur and Meriux Alliance), but were not included on this map.
SOURCE: Hotspot location data derived from Jones et al., 2008. Reference laboratory data received from committee's communication with Stephane de La Rocque, Tracy DuVernoy, Cassel Nutter, Alejandro Thiermann, and Chris Thorns (2008).

disease surveillance in animals. Investments in laboratory facility renovation and new construction in developed countries can be justified, particularly for animal disease testing laboratories. Furthermore, there is a shortage of trained professionals in diagnostic medicine. These infrastructure and personnel shortages can be met in developed countries if the political will and commitment are made to meet these challenges. In contrast, the laboratory infrastructure status in areas of the globe struggling to develop is distressing. When viewed broadly, all parts of the infrastructure necessary to meet the challenge of surveillance for zoonotic diseases (facilities, trained personnel, equipment, reagents, operational support, informatics) are substandard. For every example of success attributable to laboratory investments made in preparation for pandemic influenza, FMD, or another high-profile targeted disease, many more laboratories have only a single refrigerator available to store reagents requiring cold storage, and it may even operate sporadically. The future of those few laboratories that have been improved with donor funding is questionable at best without further sustained national or international commitment.

Donor agencies and international partners have funded significant laboratory upgrades and new facilities in resource-constrained countries for disease surveillance and research projects on specific infectious agents such as HIV/AIDS and influenza (including but not limited to HPAI H5N1). Some of these are modern research laboratories with trained personnel and the latest equipment that meet all biosafety and biocontainment requirements, and these laboratories are a resource for agent-specific zoonotic disease testing. However, they were neither built for nor do they have a mandate or operating funds to support general zoonotic disease testing. Given the expense of establishing and sustaining laboratory infrastructure in developing nations, duplicating these facilities for broad disease surveillance testing is unlikely. Thus an investigation is warranted to determine the extent that these laboratories can be shared by nations and include capacity for both human and animal laboratory disease surveillance in a single facility—unless the joint use of the laboratory is biologically risky.

The committee applauds the recent efforts by WHO and 13 African countries to strengthen African medical laboratories by developing an accreditation program (Kaiser Family Foundation, 2009). The committee encourages similar efforts with animal health laboratories to build capacity for zoonotic diseases and additional investments in strengthening laboratory capacity.

Gaps in the Global Laboratory Network

Only a limited number of diagnostic laboratory networks exist that embrace and meet the outlined guiding principles. Examples include the U.S.

National Animal Health Laboratory Network and its Canadian equivalent, the U.S. Laboratory Response Network for Bioterrorism, and the European Centre for Disease Control and Prevention's Food- and Water-Borne Disease Surveillance Network. These networks were established to focus on a specific group of agents (e.g., those thought to be the greatest threat from terrorist activities or food- and water-borne agents), some or all of which may be zoonotic, and each operates on the guiding principles of laboratory and network operation.

One of the gains of the vertically organized polio eradication initiative in Africa is the establishment of a reliable acute flaccid paralysis disease surveillance system, backed by a regionwide polio laboratory network in Africa. The 16-member polio laboratory network has been technically upgraded, is accessible to the 46 countries in the WHO African region, and have provided timely and accurate results to national (polio) disease control programs. The success of the polio laboratory network has led to the establishment of other disease-specific laboratory networks, with their associated disease surveillance systems.

Currently in the African region, there are five laboratory networks that cover polio, integrated measles, yellow fever, rubella, HIV, pediatric bacterial meningitis, rotavirus, and human papillomavirus. These networks are functioning despite minimal collaboration among them, either as individual laboratories or networks at large. Figure 4-4 shows the location of WHO laboratories in these networks. Initial efforts to integrate activities of the polio, measles, yellow fever, and rubella laboratories in their respective networks have resulted in some sharing of equipment and facilities, as well as human and financial resources. The similarities in standardized sample collection and testing strategy has led to a higher level of integration of the measles, rubella, and yellow fever laboratories; in training of laboratory staff; use of equipment and reagents; and quality assessment and assurance. Measles labs are routinely required to test samples for rubella when the measles IgM is negative. For resource-constrained countries, the integration of a disease surveillance system, including laboratory services, is required to reduce avoidable duplication of efforts and waste of scarce resources, including trained, skilled, competent laboratory personnel. In addition, joint planning of activities at the laboratory level and joint conduct of internal and external accreditation exercises has taken place. These are certainly positive developments that need to be fully supported.

Laboratory networks in the African region that focus on animal diseases are in their infancy. FAO recently completed an effort to catalog the existing influenza testing laboratory infrastructure for avian samples in sub-Saharan Africa, and it is organizing these animal disease testing laboratories into four African regional networks: Eastern, West/Central, North, and Southern African regions (FAO, 2006). The goal is for each of these regions

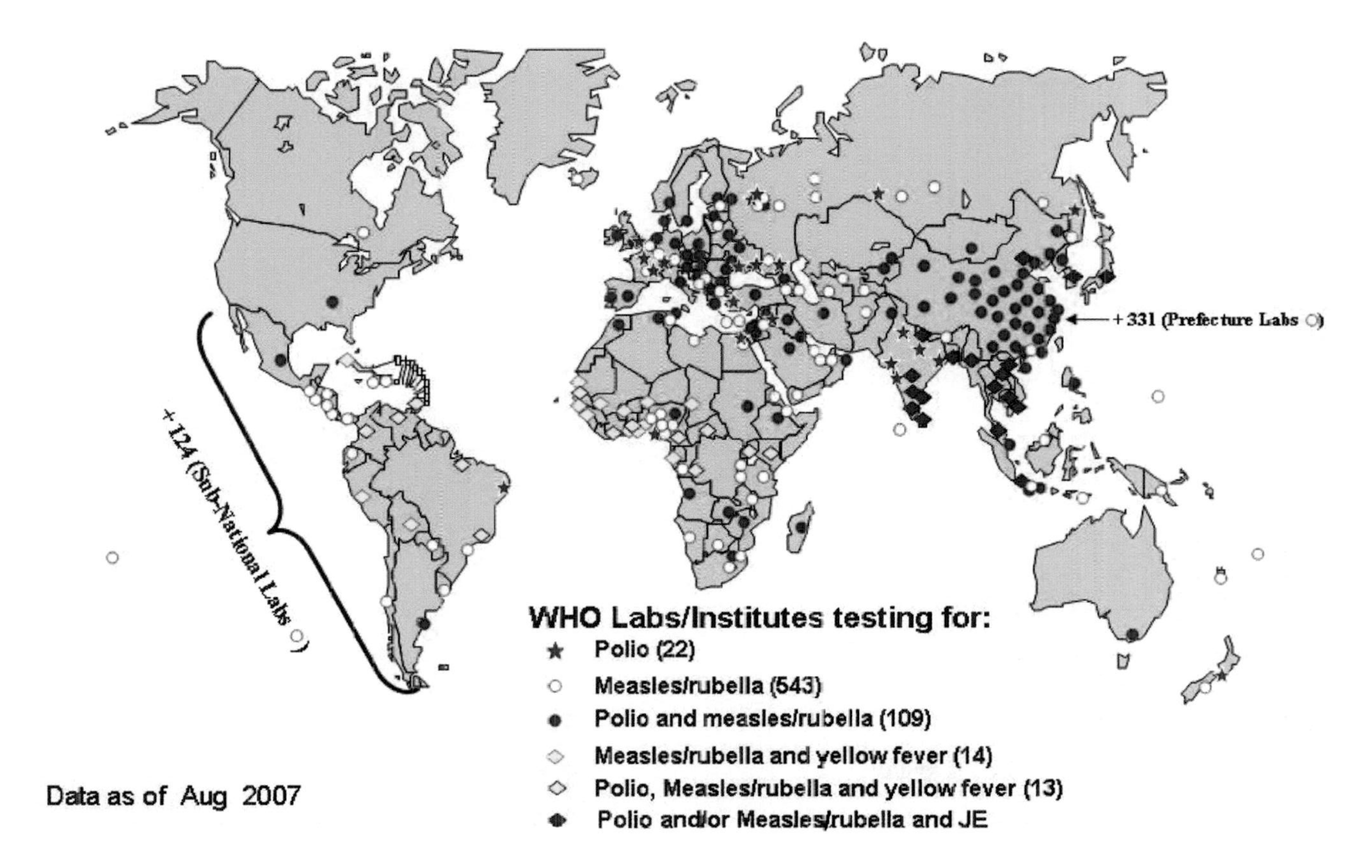

FIGURE 4-4 Global World Health Organization Vaccine Preventable Disease Laboratory Network.
SOURCE: WHO (2007b).

to have one regional reference laboratory, and for the networks to operate on the principles of good laboratory and network practices. Although there are some examples of laboratory infrastructure and disease surveillance systems in developing regions, the overall disease surveillance infrastructure is fragmented at best, and a foundation of laboratory capability and capacity is nearly nonexistent in nearly every country. Worse still, there is virtually no effort to integrate the human and animal laboratory disease surveillance systems for emerging zoonotic diseases. Laboratories need to work together as an effective network to cover testing of specimens from different species and for a broad spectrum of emerging zoonotic disease agents of high priority to human and animal populations.

Field-Oriented Multidisciplinary Capacity-Building Programs and Retaining National Expertise in Resource-Challenged Countries

There is a critical shortage of trained field epidemiology and paraprofessional personnel in both human and animal disease surveillance. Over the past 40 years, training for thousands of health personnel in the human health sector has occurred primarily for vertically funded infectious disease control or disease elimination programs. Yet many such training efforts have not led to a sustained cadre of professionals from different sectors and disciplines, with the expertise and experience needed to implement and manage an effective global, integrated emerging zoonotic disease surveillance system. More promising results have been observed with the more mature FETP programs—in places such as Thailand, Mexico, and the Philippines—where FETP graduates have moved into higher level positions in ministries of health.

In the animal health sector, vertical training programs have also occurred with similar results, but on a smaller scale to those seen in the human health sector. Moreover, funding for training animal health professionals has been insufficient to provide the needed number of trained leaders and experts in this area. Disease recording systems, if they exist, are not coordinated between human and veterinary medical professionals, so the capacity to integrate and synthesize findings and approaches is limited.

Joint human and animal health field epidemiology training programs are absent and needed to improve multisectoral field training, coordination, and communication, and to produce a workforce capable of carrying out zoonotic disease surveillance, outbreak investigation, and response. Existing educational and field-based training programs need further improvements to provide cross-disciplinary training, and new programs are needed in areas where field training programs have not yet been established. The CDC KEMRI FELTP program is a model that could be used for field epidemiology and laboratory training programs.

Finally, in both sectors, trained medical and veterinary health officials from developing countries frequently seek and obtain employment in international agencies, other countries and or regions that offer higher salaries and benefits, places where resources are available for experts to apply their training in the conduct of their work, and where there is greater potential for professional advancement. Taken together, these factors have resulted in many countries having neither human nor animal health personnel available in sufficient numbers. The few that are available are not adequately trained to recognize zoonotic diseases clinically, to conduct a quality outbreak investigation, to design and implement an effective zoonotic disease surveillance system (including risk factor and risk perception surveillance), to provide timely and accurate laboratory confirmation of the etiologic agent causing the outbreak, or to work and communicate effectively as part of a multi-sectoral team.

Countries have the responsibility to train, employ, and retain professionals in their areas of expertise. These training programs therefore need to be implemented through collaborations among relevant ministries, local universities, and extension programs. Individuals could be preferentially targeted to train in geographic areas at higher risk for zoonotic disease emergence so they can properly detect disease.

CONCLUSION

An effective global, integrated zoonotic disease surveillance and response system currently does not exist. National and international commitment to the purpose and goal of such a system are essential. True leadership and collaboration by leaders and professionals in both public and private sectors, and across countries and regions of the world in all relevant health, agricultural, natural resource, education, and other sectors, with financial support and commitment will be critical to building an effective system that meets the purpose and goals of this system. As the system is built, continual assessment and evaluation of surveillance in human, animal, and linked surveillance systems will be needed regarding their comprehensiveness, quality, multisectoral collaborative aspects, and other aspects of the systems. WHO and OIE have begun this assessment and evaluation process, but further effort and support from the international community at large is critically needed to support their efforts.

REFERENCES

AFENET (African Field Epidemiology Network). 2009. *AFENET: Background.* http://www.afenet.net/english/background.php (accessed June 15, 2009).

Beck, L. R., B. M. Lobitz, and B. L. Wood. 2000. Remote sensing and human health: New sensors and new opportunities. *Emerg Infect Dis* 6(3):217–227.

Binder, S., L. Adigun, C. Dusenbury, A. Greenspan, and P. Tanhuanpaa. 2008. National public health institutes: Contributing to the public good. *J Public Health Policy* 29(3):3–21.

Blanton, J. D., D. Palmer, K. A. Christian, and C. E. Rupprecht. 2008. Rabies surveillance in the United States during 2007. *J Am Vet Med Assoc* 233(6):884–897.

Briggs, C., and M. Nichter. 2009 (in press). Biocommunicability and the biopolitics of pandemic threats. *Med Anthropol Q* 23(3).

Brownstein, J. S., C. C. Freifeld, B. Y. Reis, and K. D. Mandl. 2008. Surveillance sans frontières: Internet-based emerging infectious disease intelligence and the HealthMap project. *PLoS Med* 5(7):e151.

Buehler, J. W., R. S. Hopkins, J. M. Overhage, D. M. Sosin, and V. Tong. 2004. Framework for evaluating public health surveillance systems for early detection of outbreaks: Recommendations from the CDC working group. *MMWR Recomm Rep* 53(RR-5):1–11.

CDC (U.S. Centers for Disease Control and Prevention). 1983. Diarrheal diseases control program: Global activities, 1981–1982. *MMWR* 32(25):330–332.

CDC. 2001. Updated guidelines for evaluating public health surveillance systems: Recommendations from the guidelines working group. *MMWR Recomm Rep* 50(RR-13):1–36. http://www.cdc.gov/mmwr/preview/mmwrhtml/mm5030a5.htm (accessed October 17, 2008).

CDC. 2003a. *Bioterrorism agents/diseases: A to z by category.* http://www.bt.cdc.gov/agent/agentlist-category.asp (accessed October 7, 2008).

CDC. 2003b. Multistate outbreak of monkeypox—Illinois, Indiana, and Wisconsin, 2003. *MMWR* 52(23):537–540.

CDC. 2003c. Update: Multistate outbreak of monkeypox—Illinois, Indiana, Kansas, Missouri, Ohio, and Wisconsin, 2003. *MMWR* 52(27):642–646.

CDC. 2006. Multistate outbreak of *Salmonella typhimurium* infections associated with eating ground beef—United States, 2004. *MMWR* 55(7):180–182.

CDC. 2008a. Preliminary FoodNet data on the incidence of infection with pathogens transmitted commonly through food—10 states, 2007. *MMWR* 57(14):366–370.

CDC. 2008b. *Global health e-brief.* http://www.cdc.gov/washington/EGlobalHealthEditions/eGlobalHealth0408.htm (accessed July 1, 2009).

CDC. 2008c. Multistate outbreak of salmonella infections associated with frozen pot pies—United States, 2007. *MMWR* 57(47):1277–1280.

CDC. 2009a. *Behavioral risk factor surveillance system (BRFSS).* http://www.cdc.gov/BRFSS (accessed April 15, 2009).

CDC. 2009b. *Integrated disease surveillance and response.* http://www.cdc.gov/idsr/about.htm (accessed April 15, 2009).

CDC. 2009c. *STOP transmission of polio: Program overview.* http://www.cdc.gov/vaccines/programs/stop/default.htm (accessed June 15, 2009).

CDC. 2009d. *Coordinating office for global health, division of global public health capacity development.* http://www.cdc.gov/cogh/dgphcd/default.htm (accessed April 16, 2009).

Cowen, P., T. Garland, M. E. Hugh-Jones, A. Shimshony, S. Handysides, D. Kaye, L. C. Madoff, M. P. Pollack, and J. Woodall. 2006. Evaluation of ProMED-mail as an electronic early warning system for emerging animal diseases: 1996 to 2004. *J Am Vet Med Assoc* 229(7):1090–1099.

Dargatz, D. A., R. A. Strohmeyer, P. S. Morley, D. R. Hyatt, and M. D. Salman. 2005. Characterization of *Escherichia coli* and *Salmonella enterica* from cattle feed ingredients. *Foodborne Pathog Dis* 2(4):341–347.

Dewell, G. A., J. R. Ransom, R. D. Dewell, K. McCurdy, I. A. Gardner, A. E. Hill, J. N. Sofos, K. E. Belk, G. C. Smith, and M. D. Salman. 2005. Prevalence of and risk for *Escherichia coli* O157 in market-ready beef cattle from 12 U.S. feedlots. *Foodborne Pathog Dis* 2(1):70–76.

DHS (U.S. Department of Homeland Security). 2009a. *The office of weapons of mass destruction and biodefense.* http://www.dhs.gov/xabout/structure/gc_1205180907841.shtm (accessed July 1, 2009).

DHS. 2009b. *Homeland security presidential directive 21: Public health and medical preparedness.* http://www.dhs.gov/xabout/laws/gc_1219263961449.shtm (accessed July 1, 2009).

Doherr, M. G., and L. Audige. 2001. Monitoring and surveillance for rare health-related events: A review from the veterinary perspective. *Philos Trans R Soc Lond B Biol Sci* 356(1411):1097–1106.

Doherr, M. G., B. Oesch, B. Moser, M. Vandevelde and D. Heim. 1999. Targeted surveillance for bovine spongiform encephalopathy (BSE). *Vet Rec* 145(23):672.

EPIET (The European Programme for Intervention Epidemiology Training). 2009. *About EPIET.* http://www.epiet.org/description/index.html (accessed June 15, 2009).

FAO (Food and Agriculture Organization of the United Nations). 2006. *Highly pathogenic avian influenza in Africa: A strategy and proposed programme to limit spread and build capacity for epizootic disease control.* Rome, Italy: FAO. http://www.fao.org/docs/eims/upload//217651/ hpai_strategy_africa_en.pdf (accessed June 24, 2009).

Farrar, J. 2008. A threat but also an opportunity for collaborative international science. Presentation, First Committee Meeting on Achieving Sustainable Global Capacity for Surveillance and Response to Emerging Diseases of Zoonotic Origin, Washington, DC, June 25–26.

Freifeld, C. C., K. D. Mandl, B. Y. Reis, and J. S. Brownstein. 2008. HealthMap: Global infectious disease monitoring through automated classification and visualization of internet media reports. *J Am Med Inform Assoc* 15(2):150–157.

GAINS (Global Avian Influenza Network for Surveillance). 2009. *About GAINS.* http://www.gains.org/AboutGAINS/tabid/57/language/en-US/Default.aspx

GAO (U.S. Government Accountability Office). 2001. *Global health: Challenges in improving infectious disease surveillance systems.* Washington, DC: GAO. http://www.gao.gov/new.items/d01722.pdf (accessed March 9, 2009).

Gilson, L. 2003. Trust and the development of health care as a social institution. *Soc Sci Med* 56(7):1453–1468.

Gilson, L. 2006. Trust in health care: Theoretical perspectives and research needs. *J Health Organ Manag* 20(5):359–375.

Ginsberg, J., M. H. Mohebbi, R. S. Patel, L. Brammer, M. S. Smolinski, and L. Brilliant. 2009. Detecting influenza epidemics using search engine query data. *Nature* 457(7232): 1012–1014.

Glickman, L. T., G. E. Moore, N. W. Glickman, R. J. Caldanaro, D. Aucoin, and H. B. Lewis. 2006. Purdue University-Banfield national companion animal surveillance program for emerging and zoonotic diseases. *Vector Borne Zoonotic Dis* 6(1):14–23.

Google.org. 2008. *Predict and prevent: An initiative to help prevent local outbreaks of emerging disease from becoming pandemics.* http://www.google.org/Predict_Prevent_Brief.pdf (accessed June 24, 2009).

Grein, T. W., K. B. Kamara, G. Rodier, A. J. Plant, P. Bovier, M. J. Ryan, T. Ohyama, and D. L. Heymann. 2000. Rumors of disease in the global village: Outbreak verification. *Emerg Infect Dis* 6(2):97–102.

Halvorson, D. A. 2002. Twenty-five years of avian influenza in Minnesota. In *Proceedings of the 53rd North Central Avian Disease Conference*. Minneapolis, MN: NCADC. Pp. 65–69.

Heymann, D. L., and G. R. Rodier. 2001. Hot spots in a wired world: WHO surveillance of emerging and re-emerging infectious diseases. *Lancet Infect Dis* 1(5):345–353.

Hueston, W. 2008. Public health training for veterinarians. *J Vet Med Educ* 35(2):153–159.

ICLN (Integrated Consortium of Laboratory Networks). 2009. *ICLN Portal*. http://www.icln.org/ (accessed July 1, 2009).

International Society for Disease Surveillance. 2009. *Syndromic.org*. http://www.syndromic.org/index.php (accessed June 24, 2009).

IOM (Institute of Medicine). 2007. *Global infectious disease surveillance and detection: Assessing the challenges—Finding solutions, workshop summary.* Washington, DC: The National Academies Press.

IOM and NRC (Institute of Medicine and National Research Council). 2008. *Achieving sustainable global capacity for surveillance and response to emerging diseases of zoonotic origin: Workshop summary.* Washington, DC: The National Academies Press.

Jones, K. E., N. G. Patel, M. A. Levy, A. Storeygard, D. Balk, J. L. Gittleman, and P. Daszak. 2008. Global trends in emerging infectious diseases. *Nature* 451(7181):990–993.

Kaiser Family Foundation. 2009. Effort launched to strengthen African medical labs. Kaiser Daily Global Health Policy Report. July 28.

Kaufman, J. A. 2008. China's health care system and avian influenza preparedness. *J Infect Dis* 197(Suppl 1):S7–S13.

Kerwick, C., J. Meers, and J. C. Phillips. 2008. Training veterinary personnel for effective identification and diagnosis of exotic animal diseases. *J Vet Med Educ* 35(2):255–261.

Kimball, A. M., M. Moore, H. M. French, Y. Arima, U. Kumnuan, W. Suwit, T. Taylor, T. Sok, and A. Leventhal. 2008. Regional infectious disease surveillance networks and their potential to facilitate the implementation of the International Health Regulations. *Med Clin North Am* 92(6):1459–1471.

Lau, J. T. F., X. Yang, H. Tsui, and J. H. Kim. 2003. Monitoring community responses to the SARS epidemic in Hong Kong: From day 10 to day 62. *J Epidemiol Community Health* 57:864–870.

Lobitz, B., L. Beck, A. Huq, B. Wood, G. Fuchs, A. S. G. Faruque, and R. Colwell. 2000. Climate and infectious disease: Use of remote sensing for detection of *Vibrio cholerae* by indirect measurement. *Proc Natl Acad Sci U S A* 97(4):1438–1443.

Madoff, L. C. 2004. ProMED-mail: An early warning system for emerging diseases. *Clin Infect Dis* 39(2):227–232.

Madoff, L. C., and J. P. Woodall. 2005. The internet and the global monitoring of emerging diseases: Lessons from the first 10 years of ProMED-mail. *Arch Med Res* 36(6):724–730.

Mandl, K. D., J. M. Overhage, M. M. Wagner, W. B. Lober, P. Sebastiani, F. Mostashari, J. A. Pavlin, P. H. Gesteland, T. Treadwell, E. Koski, L. Hutwagner, D. L. Buckeridge, R. D. Aller, and S. Grannis. 2004a. Implementing syndromic surveillance: A practical guide informed by the early experience. *J Am Med Inform Assoc* 11(2):141–150.

Mandl, K. D., B. Reis, and C. Cassa. 2004b. Measuring outbreak-detection performance by using controlled feature set simulations. *MMWR* 53(Suppl):130–136.

Mead, P. S., L. Slutsker, V. Dietz, L. F. McCraig, J. S. Bresee, C. Shapiro, P. M. Griffin, and R. V. Tauxe. 1999. Food-related illness and death in the United States. *Emerg Infect Dis* 5(5):607–625.

Menon, K. U. 2008. Risk communications: In search of a pandemic. *Ann Acad Med Singapore* 37(6):525–534.

Moore, D., B. Gamage, E. Bryce, R. Copes, A. Yassi, and the BC Interdisciplinary Respiratory Protection Study Group. 2005. Protecting health care workers from SARS and other respiratory pathogens: Organizational and individual factors that affect adherence to infection control guidelines. *Am J Infect Control* 33(2):88–96.

NIAID (National Institute of Allergy and Infectious Diseases). 2008. *Centers of excellence for influenza research and surveillance (CEIRS)*. http://www3.niaid.nih.gov/LabsAndResources/resources/ceirs/ (accessed July 1, 2009).

Nichter, M. 2008. *Global health: Why cultural perceptions, social representation, and biopolitics matter.* Tucson, AZ: University of Arizona Press.

Nigerian Federal Ministry of Health. 2007. *Nigeria national medical laboratory services policy, 2007.* Nigeria: ENHANSE-CDC.

NRC (National Research Council). 2005. *Animal health at the crossroads: Preventing, detecting, and diagnosing animal diseases.* Washington, DC: The National Academies Press.

Nsubuga, P., N. Eseko, W. Tadesse, N. Ndayimirije, C. Stella, and S. McNabb. 2002. Structure and performance of infectious disease surveillance and response, United Republic of Tanzania, 1998. *Bull World Health Organ* 80(3):196–203.

OIE (World Organization for Animal Health). 2008a. Animal health surveillance. In *Terrestrial animal health code*, Chapter 1.4. Rome, Italy: OIE.

OIE. 2008b. The new tool for the evaluation of performance of veterinary services (PVS Tool) using OIE international standards of quality and evaluation. Rome, Italy: OIE. http://www.oie.int/eng/oie/organisation/en_vet_eval_tool.htm (accessed April 17, 2009).

Padmawati, S., and M. Nichter. 2008. Community response to avian flu in central Java, Indonesia. *Anthropology & Medicine* 15(1):31–51.

Palmer, S., M. Sully, and F. Fozdar. 2009. Farmers, animal diseases reporting and the effect of trust: A study of West Australian sheep and cattle farmers. *Rural Society J* 19(1):32. http://rsj.e-contentmanagement.com/archives/vol/19/issue/1/article/2686 (accessed July 1, 2009).

Pappaioanou, M., M. Malison, K. Wilkins, B. Otto, R. Goodman, R. E. Chruchill, M. White, and S. B. Thacker. 2003. Strengthening capacity to use data for public health decision making: The data for decision making project. *Soc Sci Med* 57(10):1925–1937.

Perry, H. N., S. M. McDonnell, W. Alemu, P. Nsubuga, S. Chungong, M. W. Otten, Jr., P. S. Lusamba-dikassa, and S. B. Thacker. 2007. Planning an integrated disease surveillance and response system: A matrix of skills and activities. *BMC Med* 5:24.

Petersen, L. R., and E. B. Hayes. 2004. Westward ho?—The spread of West Nile virus. *N Engl J Med* 351(22):2257–2259.

PHAC (Public Health Agency of Canada). 2004. *The global public health intelligence network (GPHIN)*. http://www.phac-aspc.gc.ca/media/nr-rp/2004/2004_gphin-rmispbk-eng.php#6 (accessed April 15, 2009).

Polgreen, P. M., Y. Chen, D. M. Pennock, and F. D. Nelson. 2008. Using internet searches for influenza surveillance. *Clin Infect Dis* 47(11):1443–1448.

Roberts, L. 2007. Infectious disease: Resurgence of yellow fever in Africa prompts a counterattack. *Science* 316(5828):1109.

Rodier, G., A. L. Greenspan, J. M. Hughes, and D. L. Heymann. 2007. Global public health security. *Emerg Infect Dis* 13(10):1447–1452.

Rouquet, P., J. M. Froment, M. Bermejo, A. Kilbourn, W. Karesh, P. Reed, B. Kumulungui, P. Yaba, A. Delicat, P. E. Rollin, and E. M. Leroy. 2005. Wild animal mortality monitoring and human Ebola outbreaks, Gabon and Republic of Congo, 2001–2003. *Emerg Infect Dis* 11(2):283–290.

Salman, M. D. 2008. *Animal disease surveillance and survey systems.* New York: Wiley-Blackwell.

Salman, M. D., K. D. Stark, and C. Zepeda. 2003. Quality assurance applied to animal disease surveillance systems. *Rev Sci Tech* 22(2):689–96

Scoones, I., and P. Forster. 2008. *The international response to highly pathogenic avian influenza: Science, policy and politics.* STEPS Working Paper 10. Brighton, UK: STEPS Centre.

Seto, W. H., D. Tsang, R. W. Yung, T. Y. Ching, T. K. Ng, M. Ho, L. M. Ho, and J. S. Peiris. 2003. Effectiveness of precautions against droplets and contact in prevention of nosocomial transmission of severe acute respiratory syndrome (SARS). *Lancet* 361(9368):1519–1520.

Smith, G. J. D., X. H. Fan, J. Wang, K. S. Li, K. Qin, J. X. Zhang, D. Vijaykrishna, C. L. Cheung, K. Huang, J. M. Rayner, J. S. M. Peiris, H. Chen, R. G. Webster, and Y. Guan. 2006. Emergence and predominance of an H5N1 influenza variant in China. *Proc Natl Acad Sci U S A* 103(45):16936–16941.

Swaminathan, B., T. J. Barrett, S. B. Hunter, and R. V. Tauxe. 2001. PulseNet: The molecular subtyping network for foodborne bacterial disease surveillance, United States. *Emerg Infect Dis* 7(3):382–389.

Teutsch, S. M., and R. E. Churchill. 2000. *Principles and practice of public health surveillance*, 2nd ed. New York: Oxford University Press.

Thacker, S. B., and R. Berkelman. 1988. Public health surveillance in the United States. 1988. *Epidemiol Rev* 10(1):164–190.

Thacker, S. B., A. L. Dannenberg, and D. H. Hamilton. 2001. Epidemiologic intelligence service at the Centers for Disease Control and Prevention. *Am J Epidemiol* 154(11):985–992.

The Center for Food Security and Public Health. 2009a. *USDA highly pathogenic avian influenza diagnostic training course.* http://www.cfsph.iastate.edu/HPAI/hpai_lab_video_list.htm (accessed April 16, 2009).

The Center for Food Security and Public Health. 2009b. *Zoonoses education resources.* http://www.cfsph.iastate.edu/Zoonoses/educationresources.htm (accessed April 17, 2009).

Traub-Dargatz, J. L., L. P. Garber, P. J. Fedorka-Cray, S. Ladely, and K. E. Ferris. 2000. Fecal shedding of Salmonella spp by horses in the United States during 1998 and 1999 and detection of Salmonella spp in grain and concentrate sources on equine operations. *J Am Vet Med Assoc* 217(2):226–230.

UC Davis (University of California, Davis) School of Veterinary Medicine. 2009. *Avian flu school (international).* http://www.vetmed.ucdavis.edu/whc/flu_school/index.html (accessed April 17, 2009).

USAID (U.S. Agency for International Development). 1998. *Reducing the threat of infectious diseases of major public health importance: USAID's initiative to prevent and control infectious diseases.* Washington, DC: USAID.

USAID. 2009. *Building capacity to improve disease surveillance in Ghana.* http://www.usaid.gov/our_work/global_health/id/surveillance/ghanacap.html (accessed April 16, 2009).

USDA (U.S. Department of Agriculture). 2000. *Lameness and laminitis in U.S. horses.* APHIS, National Animal Health Monitoring System. Fort Collins, CO: USDA.

USDA-APHIS (U.S. Department of Agriculture Animal and Plant Health Inspection Service). 2009a. *Bovine spongiform encephalopathy (BSE).* http://www.aphis.usda.gov/newsroom/hot_issues/bse/index.shtml (accessed April 15, 2009).

USDA-APHIS. 2009b. *Foreign animal disease diagnostic laboratory.* http://www.aphis.usda.gov/animal_health/lab_info_services/about_faddl.shtml (accessed April 16, 2009).

USGS (U.S. Geological Survey). 2006. *Surveillance for Asian H5N1 avian influenza in the United States.* http://www.nwhc.usgs.gov/publications/fact_sheets/pdfs/ai/AIFEB06.pdf (accessed April 15, 2009).

Victora, C. G., T. Adam, J. Bryce, and D. B. Evans. 2006. Integrated management of the sick child. In *Disease control priorities in developing countries*, 2nd ed. New York: Oxford University Press.

Wagner, M. M., F. C. Tsui, J. U. Espino, V. M. Dato, D. F. Sittig, R. A. Caruana, L. F. McGinnis, D. W. Deerfield, M. J. Druzdzel, and D. B. Fridsma. 2001. The emerging science of very early detection of disease outbreaks. *J Public Health Manag Pract* 7(6):51–59.

WDIN (Wildlife Disease Information Node). 2009. *Highly pathogenic avian influenza early detection data system (HEDDS)*. http://wildlifedisease.nbii.gov/ai/ (accessed July 1, 2009).

WHO (World Health Organization). 1977. *Evaluation of epidemiologic surveillance systems*. WHO/EMC/DIS/07.2. Geneva, Switzerland: WHO.

WHO. 1980. Expanded programme on immunization training activities. *Wkly Epidemiol Rec* 55(41):316.

WHO. 2004. *H5N1 avian influenza: A chronology of key events*. http://www.who.int/csr/disease/avian_influenza/chronology/en/ (accessed April 17, 2009).

WHO. 2006. *Asia Pacific strategy for emerging diseases, including the International Health Regulations (2005) and avian influenza*. Regional Committee for the Western Pacific, WPR/RC57.R2. Manila, Philippines: WHO.

WHO. 2007a. *International Health Regulations (2005): Areas of work for implementation*. Geneva, Switzerland: WHO.

WHO. 2007b. *Global WHO vaccine preventable disease laboratory network*. http://www.who.int/immunization_monitoring/big_integrated_labs_June06.jpg (accessed July 1, 2009).

WHO. 2009a. *WHO Global Salm-Surv*. http://www.who.int/salmsurv/en/ (accessed June 15, 2009).

WHO. 2009b. *WHO TB epidemiology and surveillance virtual workshop*. http://apps.who.int/tb/surveillanceworkshop/ (accessed June 15, 2009).

WHO. 2009c. Integrated control of neglected zoonotic diseases in Africa. *Wkly Epidemiol Rec* 84(17):147–148.

WHO. 2009d. *Cumulative number of confirmed human cases of avian influenza A/(H5N1) reported to WHO, 2 June 2009 update*. http://www.who.int/csr/disease/avian_influenza/country/cases_table_2009_06_02/en/index.html (accessed June 24, 2009).

WHO-AFRO (World Health Organization for the African Region). 2001. *Integrated disease surveillance and response in the African Region—A regional strategy for communicable diseases: 1999–2003*. Brazzaville, Congo: WHO-AFRO.

WHO and UNAIDS (World Health Organization and Joint United Nations Programme on HIV/AIDS). 2000. *Second generation surveillance for HIV: The next decade*. WHO/CDS/CSR/EDC/2000.5, UNAIDS/00.03E. Geneva, Switzerland. http://data.unaids.org/Publications/IRC-pub01/JC370-2ndGeneration_en.pdf (accessed April 15, 2009).

Wilkins, K., P. Nsubuga, J. Mendlein, D. Mercer, and M. Pappaioanou. 2008. The data for decision making project: Assessment of surveillance systems in developing countries to improve access to public health information. *Public Health* 122(9):914–922.

Wilson, K., B. von Tigerstrom, and C. McDougall. 2008. Protecting global health security through the International Health Regulations: Requirements and challenges. *CMAJ* 179(1):44–48.

Xinshen, D., and Y. Liangzhi. 2006. *Assessing potential impact of avian influenza on poultry in West Africa: A spatial equilibrium model analysis*. International Food Policy Research Institute (IFPRI) Working papers. Washington, DC: IFPRI.

Zeldenrust, M. E., J. C. Rahamat-Langendoen, M. J. Postma, and J. A. Vliet. 2008. The value of ProMED-mail for the early warning committee in the Netherlands: More specific approach recommended. *Euro Surveill* 13(6):8033. http://www.eurosurveillance.org/images/dynamic/EE/V13N06/art8033.pdf (accessed July 1, 2009).

5

Incentives for Disease Surveillance, Reporting, and Response

"It is essential to provide additional incentives. Different incentives will likely apply at different levels. At the local or district level, training, feedback, and epidemiological or clinical assistance to the reporting clinicians and local public health are possible incentives. At the country level, financial incentives and resources are needed to encourage reporting as well as to expand the reach of the primary health care and communications infrastructures. Other psychological incentives, such as increased national prestige for recognizing an unusual disease, should also be considered. Encouragement from the international community, to overcome a country's fear of adverse consequences and help leverage resources, is also necessary."

—Stephen S. Morse
"Global Infectious Disease Surveillance and Health Intelligence"
Health Affairs, 2007

An important lesson from disease outbreaks such as severe acute respiratory syndrome (SARS) is that the ability of the global human and animal health system to respond is only as good as the ability and willingness of local and national systems to detect and report outbreaks. Delays in reporting SARS by China could have resulted in catastrophic consequences worldwide if the pathogen had been more transmissible (Heymann and Rodier, 2004).

Data on an outbreak have to be recognized before they can be reported. Current strategies to contain an avian influenza pandemic are contingent on recognizing human-to-human transmission within approximately 3 weeks of the initial case (Ferguson et al., 2005; Longini et al., 2005). One might argue that suppressing information about an outbreak is difficult in today's world of the Internet, cell phones, and other communication and information technologies. Countries or regions that are less keen on reporting outbreaks—either because they fear trade or travel sanctions or because they have little capacity to control the outbreak once it has been detected—are likely to spend relatively less on disease surveillance (Laxminarayan et al., 2008). All else being equal, incentives to invest in disease surveillance are related to incentives for reporting as part of the entire surveillance system.

BEHAVIORAL AND CULTURAL DETERMINANTS OF INFORMATION SHARING

Socioeconomic and Political Consequences of Reporting

Reporting outbreaks of zoonotic and vector-borne diseases may serve the greater good of the global health community, but the publicity associated with such outbreaks can result in huge national and private-sector costs (Zacher, 1999). Promptly reporting the disease outbreak may not be in the government or ruling political parties' immediate best interests if it will negatively affect trade, tourism, or public confidence in agricultural products (Cash and Narasimhan, 2000). At the national level, government officials may suppress reports of illness among humans as well as animals if they perceive a threat to their careers (Waltner-Toews, 2004; Kaufman, 2008). Government officials may also try to downplay human and animal health system shortfalls and disease outbreaks if those issues lead critics to question an official's performance and the ability of the government to provide basic services to its citizens (Farmer, 1992).

The threat of an epidemic disease may be hidden through the government's use of "rhetorical strategies," such as employing nonspecific terms to describe disease outbreaks to the public. An outbreak of cholera in India was reported instead as gastroenteritis and other nondescript illness categories (Ghosh and Coutinho, 2000). An outbreak of highly pathogenic avian influenza (HPAI) H5N1 in Thailand was first reported as avian cholera (Chuengsatiansup, 2008). Such strategies may be employed as a way to minimize fear and public panic of an epidemic, as well as a means to conceal crises that could have economic repercussions including trade impacts. The strategy of regionally isolating the disease has also been employed to deflect attention away from government responsibility for disease epidemics. This strategy influences both when and how disease outbreaks are reported and perceived. In countries such as Venezuela, it has involved calling attention to the unhygienic behavior of indigenous groups or impoverished sectors of the population, conveying the idea that the rest of the population is relatively safe from disease transmission. This false sense of security is achieved at the cost of impoverished victims of diseases, who are blamed for their own misery (Briggs, 2004).

At the Local and Producer Level

Decisions by officials to proactively engage in the surveillance of zoonotic diseases involve social risk to existing or potential social relations

(Nichter, 2008). Consequently, some producers who discover sick animals may try to sell or dispose of them without reporting infection. Nipah virus outbreak in Malaysia in 1998–1999 is another example of the movement of infected animals without reporting (see Box 5-1). Therefore, local authorities need effective disease surveillance to identify local outbreaks and to rapidly contain them to reduce the risks of zoonotic disease spread to human and animal populations. The information needed to accomplish this exists: Local communities are well aware of infection patterns, but there are barriers to reporting processes because of inefficiency and lack of incentives (PPLPI, 2007).

The 2006–2007 HPAI H5N1 outbreaks in Indonesia show how these socioeconomic and political factors influence disease surveillance activities in the context of decentralized governance (Padmawati and Nichter, 2008). In central Java, Indonesia, local official support for avian influenza surveillance initiatives waned when human cases of the disease did not reach the impending epidemic proportions initially reported in the press. At the time, officials were preoccupied with other pressing needs, such as the occurrence of major earthquakes and the effects of oil price hikes on the costs of basic commodities. HPAI H5N1 was recognized to be an emerging threat to human and animal health, but it was not considered the largest risk facing officials strapped with diminishing funds and growing public demands for assistance. Moreover, as the poultry industry exerted considerable local influence, it was not in the best interests of local politicians to support aggressive HPAI H5N1 control measures that would displease this powerful lobby unless public opinion demanded such actions.

BOX 5-1
Nipah Virus Outbreak in Malaysia

An outbreak of Nipah virus in Bukit Pelandok, Negri Sembilan, lasted from December 1998 to April 1999. Two cases occurred in the state of Selangor, between Perak and Negri Sembilan (CDC, 1999). The transport of infected pigs was accelerated by a "fire sale" that moved grower pigs from Perak to Negri Sembilan, Selangor, Penang, Malacca, and Johore.

In response, the Malaysian government banned the movement of pigs within the country, and neighboring countries stopped pig imports from Malaysia. In 1999, the Nipah virus spread to the state of Sarawak, which is on the island of Borneo (Ahmad, 2000). In response, the state government offered a RM 20,000 reward in return for information on people responsible for smuggling Nipah-infected pigs into Sarawak.

SOURCES: CDC (1999); Ahmad (2000).

Although aggressive measures, such as banning smallholder poultry keeping, were eventually taken to prevent the spread of HPAI H5N1, other municipalities in Indonesia's decentralized state weakly complied with disease surveillance initiatives. Those smallholder poultry keepers questioned the severity of the avian influenza threat to their birds. Many did not view the disease as new, but rather as a form of Newcastle disease, a serious threat they had faced for many years (Padmawati and Nichter, 2008). Some continued to consume and sell diseased dead birds. Small to medium-sized contract poultry farmers feared that government officials might cull their birds before definitive laboratory confirmation of the disease, and they were skeptical of compensation schemes or believed compensation was too low. These poultry farmers reported the deaths of chickens to contractors, who in turn sought the services of private veterinarians to determine the causes of bird death, making effective disease surveillance difficult. Smallholder poultry farmers and keepers feared reporting incidents directly to the government. This fear was not limited to a concern about losing their own birds, but also to the social risk of angering nearby neighbors, whose birds would be subject to culling within a 2–5 km radius of an outbreak location (Padmawati and Nichter, 2008).

Trust and technical skills of government health officers proved to be an important variable in determining whether local stakeholders reported bird death in Central Java and elsewhere in Asia (Kleinman et al., 2008). It was crucial that the local population trust local authorities to provide adequate and timely compensation for culled birds, trust in the efficacy of vaccines to prevent disease and vaccinators who themselves have been associated with spreading disease, and trust that health officials would conduct appropriate scientific tests to ascertain the presence of HPAI H5N1 and not just act on the basis of suspicion. Moreover, local stakeholders had to trust that the provincial and national governments were looking after the public's best interests and not just particular stakeholders. Widely circulating rumors suggested that the Indonesian government was benefiting from HPAI H5N1 through well-publicized appeals for development aid, and that agribusinesses were benefiting from decreased competition in the poultry market because local farmers had the least amount of resources to deal with the losses (Padmawati and Nichter, 2008). *The committee concludes that disease surveillance systems need to effectively combine incentives for collective responsibility and self-reporting, and disincentives for not reporting.*

At the National, Regional, and Global Levels

National authorities face conflicting incentives to report disease outbreaks (Malani and Laxminarayan, 2006). Reporting typically brings medical assistance, which is helpful in containing outbreaks, but also brings the

threat of trade and travel sanctions that can be devastating to the economies of smaller countries. When the risk of sanctions is high, countries may delay issuing an outbreak report or downplay the human and animal health risk of the outbreak. They may put both human and animal populations at risk, but the incentives to do so are strong. The force of sanctions in discouraging reporting can be blunted to some extent by medical assistance, and by external support under the International Health Regulations 2005 (IHR 2005), but the value of medical assistance is often several orders of magnitude smaller than the cost of sanctions. However, the early declaration of a disease outbreak, even if it brings sanctions, also opens possibilities for formal early intervention in containing the outbreak, and thus reducing the costs of eradication.[1] The decision to report depends therefore on a trade-off between the costs of sanctions and the benefits of early outside assistance and a reduction in costs of controlling or eradicating the disease.

Countries may report outbreaks in order to maintain a reputation for reliability and good global citizenship. An example of this is the prompt reporting of cholera by Peru during the late 1990s even though it resulted in significant economic costs (Panisset, 2000). Another is of the willingness of the United States to report a single case of bovine spongiform encephalopathy (BSE). But countries may also be interested in protecting their reputations as healthy places. Health and vital statistics are commonly used to judge development and modernity at province, district, or country levels. Reports of emerging illnesses may be taken as a marker of political and infrastructure problems, as well as a symptom of poverty and underdevelopment. Precisely because disease affects international reputation, tourism, and investment, some governments may prefer not to report outbreaks or to minimally report them (Cash and Narasimhan, 2000). Applying existing least restrictive trade mechanisms (e.g., zoning and compartmentalization sanctions where appropriate) could minimize unnecessary costs of trade sanctions when the countries do effectively demonstrate their ability to detect and appropriately control the disease outbreak with routine, evidence-based responses. The same holds true if countries are able to improve the specificity of reporting to reduce false reports. To avoid the problem of outbreak concealment, it is important to incentivize outbreak reporting within countries by designing outbreak control measures and providing adequate compensation schemes.

Economic considerations are not the only reasons why countries do not

[1]Eradication carries geographically distinct meanings in human and animal medicine. Whereas for human diseases, eradication means purging of the disease from the entire world (e.g., smallpox), in animal health eradication of a specific disease is considered on a nation-by-nation status. To date, no animal disease has been eradicated in the human health sense of the word, but most developed counties can claim to have successfully eradicated various agricultural diseases (e.g., foot-and-mouth disease in Uruguay, screwworm in the United States).

report outbreaks. Disease outbreaks call attention to a government's failure to maintain various infrastructures, and failure to control epidemics may threaten state legitimacy (Farmer, 1992; Ghosh and Coutinho, 2000). As a result, attempts may be made to suppress information about disease outbreaks or classify diseases in ways that minimize collective anxiety. In Cuba, outbreaks of dengue fever in the late 1990s were suppressed and seen as a national embarrassment given the country's highly praised *Aedes aegypti* control program implemented in the 1980s following a major dengue fever epidemic (Van Sickle, 1998). In several Caribbean countries, dengue fever has been glossed over as a nonspecific "viral fever" for fear of affecting the tourist industry. This type of obfuscation has made implementing international health disease surveillance systems, though agreed on in principle, quite challenging (Baker and Fidler, 2006).

At the national health authority level, pressures from stakeholders in other economic sectors may play a role in delaying formal reporting, as was the case in East Asia. After the major economic losses resulting from the earlier SARS outbreak, directors of veterinary services in East Asia were under pressure from stakeholders to delay the declaration of the HPAI H5N1 outbreak. In addition, HPAI H5N1 was unofficially detected in Indonesia in August 2003, whereas the official declaration took place on January 25, 2004 (Dolberg et al., 2005). This delay in official notification to the World Organization for Animal Health (OIE) likely occurred in most countries, as shown by the proximity of the dates the outbreak was declared in different countries—Vietnam, January 8, 2004; Lao PDR, January 14; Thailand and Cambodia, January 23; and Indonesia, January 25—which is epidemiologically highly unlikely. The delayed notification may be partly attributed to inadequate diagnostic facilities and the lack of skilled staff, and partly to political pressure on human and animal health services to suppress information because of the economic consequences in lost domestic and export markets for poultry products and tourism (O'Neill, 2004).

Reporting of disease outbreaks is not a binary event (confirmation or denial), and countries differ significantly in the speed of reporting outbreaks. A case in point is with reporting of foot-and-mouth disease (FMD) by South America's main beef-producing countries, including Argentina and Uruguay. A report by the South American Commission for the Fight Against FMD found that of all the countries in the region, Uruguay was the only country that on average quickly reported outbreaks (Comisión Sudamericana para la Lucha Contra la Fiebre Aftosa, 1996). Yet during a 2001 outbreak in Uruguay, uncertainty about the nature of the outbreak and poor communication with Argentine authorities about Argentina's high cross-border movement fueled the spread of disease and resulted in outbreaks in 4 percent of Uruguay's total livestock (Rich, 2004). Additionally, during an FMD outbreak in Argentina in July 2000, the government failed to acknowledge the severity of the disease spread and did not create an eradication program until April 2001 (Rich, 2004). Consequently, nearly 2.8 million cattle, or 5–6 percent of the cattle population, had been exposed to FMD by the end of the Argentine outbreak in January 2002.

The committee gave considerable thought to existing policies and activities that affect disease reporting at various levels, as well as to those affected by each policy. Table 5-1 summarizes the pros and cons of various types of policies and activities. While it is not exhaustive, it exposes weaknesses in some policies and activities. By learning from the past, new policies can improve and incentivize disease surveillance and reporting efforts that would ultimately protect human and animal health and minimize the loss of livelihood.

Reporting by the Food Production Industry

Voluntary reporting by industry could play an important role in detecting zoonotic disease outbreaks. In developed countries, food producers sometimes issue safety warnings or withdraw their products from the market to protect themselves from legal action from affected consumers. In addition, producers might take these actions to maintain their reputation as being ethically and socially responsible, or they may do so in an effort to improve their reputation.

The efforts of East Asian producers' organizations in containing HPAI H5N1 in 2004 illustrate how actions tend to work better when the industry is organized rather than when it is disjointed with many small producers. Thailand is a major exporter in poultry products and had a direct interest in controlling the disease. The Thai Broiler Processing Exporters' Association, consisting mainly of large scale members, promoted its interest aggressively. It succeeded in convincing the Thai government to institute an immediate culling policy when the outbreak began, particularly targeting small farmyard poultry operations (Chanyapate and Delforge, 2004; Davis, 2005). The same action was expected in intensive farming systems in the country. The poultry producer association of Malaysia was able to coordinate production, but the broad-based nature of its membership and its lack of a cohesive structure denied the organization involvement in formulating shared policy positions and influencing the government. Top-down state-sponsored organizations in Vietnam did not represent farmers: When the government—against scientific evidence—advised people not to eat chicken regardless of whether the chickens were sick, these organizations did little to defend their members' interests (Vu, 2009).

Incentives for industry to report outbreaks may be preempted by national or local regulatory authority action, which may or may not be correct about the source of the outbreak. Moreover, concerns about adversely

TABLE 5-1 Policies That Influence Reporting at Various Levels, Who Is Affected, and Pros and Cons of Each Policy

Policy	Actor	Pros	Cons
Quarantine areas affected by disease	Local and state human and animal health agencies plus agricultural agencies	• Isolates individuals affected by disease • Minimal loss of life (except that animals within quarantine zone will probably be killed) • Decreases the spread of disease	• Producers in affected zone cannot move animals or maybe even cannot get feed trucks in, so great loss in income and animals; can be an animal welfare concern • Potential impacts on tourism
Pay or reimburse for animal culling	State and national governments partnering with industry for agricultural diseases	• Incentivizes reporting of sick animals • Uniform policy • Addresses social equity for poor	• Discourages disease control measures by the industry • Loss of animal stocks and life • Unable to sustain long term • Could spread more among animals if reimbursement is as high as market value (purposeful spread) • Environmental impacts of carcass disposal • Requires funding that many poorer states do not have
Sanction regions impacted by disease	National governments, WTO, international countries	• Prevents importation of diseased animal products • Compartmentalized rather than fully national trade sanctions imposed upon the occurrence of an emerging infectious disease	• Discourages disease reporting because of anticipated sanctions • Difficult to sanction human cases • Loss of income for affected areas, even if no disease present on individual farm
Strengthen informal disease surveillance networks	WHO, FAO, OIE, NAHLN	• Obtains more information about disease outbreaks for quicker alerts • Greater awareness of overall health status, locally to globally	• Increases in false positives that have to be investigated • Attribution/retribution concerns for whistleblowers

Improve quality of testing technology	Local, state, and national human and animal health authorities and agricultural agencies, veterinary laboratories, organizations with international oversight on reporting and laboratory practices standard setting (WHO, FAO, OIE)	• Improves both sensitivity and specificity of diagnostic testing: (1) fewer false positives, resulting in lower costs of outbreak control due to improved management and control of zoonotic diseases; (2) fewer false negatives, which decreases missed outbreaks • Earlier detection reduces cost of outbreak control, which is more timely • A simpler test possibly reduces costs, provides more information • Cost effectiveness combined with a safer test • Feasibility of test, leads to better sample quality	• Inefficiencies in disease diagnosis contribute to larger and longer-lasting outbreaks • Increases in cost due to improvement in infrastructure and training, and need for new reagents • Decreases in feasibility discourages local reporting • Increases need for sharing outside of country or region decreases control of the information • Less feedback from national authorities: In resource-constrained countries, local reporting might decrease vis-à-vis failed feedback from the national level • Safety of laboratory personnel due to new safety requirements
Strengthen capacity at country level to respond to outbreaks	National governments and international partners proving technical assistance or funding	• Increases the feasibility of response to an outbreak • More timely response, thus encouraging early reporting • Improved communication across all levels • More control in country of the containment of outbreaks, resulting in less dependency on foreign technical support	• Increases costs associated with strengthening infrastructure and training • Increases bureaucracy at all levels

NOTES: FAO = Food and Agricultural Organization of the United Nations, NAHLN = National Animal Health Laboratory Network, OIE = World Organization for Animal Health, WHO = World Health Organization, WTO = World Trade Organization.

affecting a trade or industry group could make public authorities more cautious about reporting an outbreak. Both of these concerns emerged during the 2008 *Salmonella* outbreak in the United States that resulted in more than 1,300 human cases of salmonellosis. After a preliminary investigation, the U.S. Food and Drug Administration (FDA) and the U.S. Centers for Disease Control and Prevention (CDC) linked the outbreak to raw tomatoes and issued a warning. After 6 weeks, FDA lifted the warning in July 2008 as it discovered that the real cause had been jalapeño and Serrano peppers grown in Mexico. The United Fresh Produce Association estimated that the tomato industry lost more than $100 million while the warning was in effect and called on Congress to compensate the industry for these losses. It also demanded a stricter burden of proof before the FDA could blame a particular food product for any future foodborne outbreak (Venkataraman, 2008).

Animal Culling and Voluntary Reporting

The culling or intentional slaughter of sick (and potentially infected, although likely healthy) animals is an important part of the human and animal health response to disease outbreaks in animal populations. There are several reasons to compensate private stakeholders for losses incurred as a result of public action, such as paying farmers an indemnity for culling diseased or suspected infected animals for an emerging disease. Justifications to support payment of culling are related to justice, social equity, and incentivizing desirable participation in early disease reporting. The destruction of private property by the state is fair or just when affected citizens are compensated. The poor often depend on food animals for their income and daily nutrition, and ensuring their livelihoods is social equity. Incentivizing local participation in timely disease detection and reporting can encourage farmers or other actors to declare early emergence of a disease, which can then in turn reduce the cost of containment or control.

Compensation schemes that were used to contain cattle plague in the mid-19th century continue to be used today in many countries. The level of compensation is a factor that determines the rate of reporting. Compensation levels that are too low induce producers to hide animals from culling, whereas levels too high would encourage the introduction of animals from outside the region. In general, compensation rates have been around 75–90 percent of market value before the disease outbreak for live animals and lower rates for dead animals (World Bank, 2006).

It is important, particularly for smallholders, to address not only the rate of compensation, but also the timeliness and reliability of payment. At the farm level, farmers may delay reporting because of fears of economic sanctions or inadequate or delayed compensation. Thus, a delay in payment of no more than 24 hours is suggested (World Bank, 2006). Important conditions for early payment include pre-outbreak registration of animals per household, and the current availability and operation in a variety of systems (e.g., levies, insurance) of funding for the compensation payments. Assuming that early identification is a global public good, international funding for poor countries will be required and justified.

ECONOMIC AND TRADE SANCTIONS

SPS and TBT Agreements

Several World Trade Organization (WTO) agreements are relevant to health policy, including the Agreements on the Application of Sanitary and Phytosanitary Measures (SPS) and Technical Barriers to Trade (TBT). Both state that health is a legitimate objective for WTO members to take into account when necessary to protect the health of humans, animals, and plants. A major emphasis in WTO rules is to ensure that trade measures are pursued for recognized reasons and avoid discrimination or unnecessary restrictions on trade (WTO and WHO, 2002). Recognizing that technological developments in recent years have created sensitive early warning disease surveillance systems, rapid and reliable verification procedures, preparedness plans including medication stockpiles, and international response networks, WTO suggests that restrictions should be time-limited and minimally disruptive to international trade and travel (WTO and WHO, 2002).

Under the SPS Agreement, measures may be imposed only to the extent necessary to protect life or health (see Box 5-2) and done so on the basis of scientific information to minimize negative trade effects. Under the harmonization requirement, members are required to use international guidelines, standards, and recommendations (including those for food safety established by the Food and Agriculture Organization of the United Nations/OIE Codex Alimentarius Commission) when available except as otherwise identified in the SPS Agreement.

Trade measures that protect animal and plant life or health usually fall within the scope of the SPS Agreement, meaning that the TBT Agreement would not apply. Under the TBT Agreement, WTO members can also apply technical regulations and standards they consider appropriate—for example: for human, animal, or plant life or health; for the protection of the environment; or to meet other consumer interests. Departures from international standards do require justification if requested by another member state (WTO and WHO, 2002). Despite the differences between the SPS and TBT Agreements, their common aim is to prevent unnecessary trade barriers.

BOX 5-2
Definition of a Sanitary and Phytosanitary Measure at a Glance

Measures Taken to Protect:	From:
Human or animal life	Additives, contaminants, toxins, or disease-causing organisms in their food, beverages, feedstuffs
Human life	Plant- or animal-carried disease (zoonoses)
Animal or plant life	Pests, disease, or disease-causing organisms
A country	Damage caused by the entry, establishment, or spread of pests (including invasive species)

SOURCE: (WTO and WHO, 2002). Reprinted with permission from WHO.

Economic Losses from Trade and Travel Sanctions

Outbreaks of zoonotic diseases impose significant effects on human and animal health and lead to economic consequences on affected countries. Disease outbreak reporting often leads unaffected countries to enact travel and trade restrictions on the affected country that far exceed the actual disease threat (Merianos and Peiris, 2005). This can cripple demand for a country's exports and ripple through its tourism industry, and thus acts as a strong disincentive for outbreak reporting. Thus it is crucial to minimize the spread of diseases across borders, while minimizing trade and travel losses. The 1994 plague outbreak in India provides a seminal example of excessive international sanctions due to panic over disease spread before the creation of WTO or IHR 2005 (see Box 5-3).

Numerous other zoonotic disease outbreaks have led to the enactment of trade restrictions on affected countries and impacted meat and poultry imports and exports. In the United States, a small outbreak of Newcastle disease among Texas poultry in 2003 prompted a number of countries—including Mexico, Russia, Japan, Cuba, and those in the European Union—to place an embargo on all poultry imports from Texas (Romero, 2003). As mentioned in Chapter 2, major beef importing countries temporarily banned imports of beef and beef products from the United States within a week after announcing a BSE outbreak in American cattle in 2003 (USITC, 2008). An earlier U.S. ban on Canadian beef due to a BSE outbreak in May 2003 led to $1 billion in losses for the Canadian beef industry (Grady et al., 2003). In 2007, CDC had imposed trade embargoes on birds and processed bird products from all countries affected by avian influenza (CDC, 2007). Thirty-two countries

BOX 5-3
International Sanctions After a Plague Outbreak in India

In September 1994, seven cases of a highly fatal disease were reported in a hospital in Surat, India. Though no accurate diagnostic tests were available and a number of infectious agents were suspected, the local government reported that the outbreak was pneumonic plague. The outbreak created mass panic throughout India and drew media attention throughout the world. Within 2 weeks of reporting the outbreak, an estimated 500,000 people fled Surat for other large cities in India, and thousands more fled to other countries around the world.

News of the outbreak also led countries to impose massive trade and travel restrictions on India. Bangladesh, Oman, and the United Arab Emirates stopped importing any food from India, and Italy placed an embargo on all Indian products entering Italian ports. Additionally, the United States, Canada, United Kingdom, France, Germany, and Italy issued travel warnings to their citizens. By early October 2004, the outbreak was declared over, and the World Health Organization determined that the outbreak was limited with no cases in any major Indian city. Yet the damage to India's economy was already done, with losses due to trade and tourism restrictions estimated at more than $2 billion.

SOURCES: Burns (1994); Cash and Narasimhan (2000); Gubler (2001).

throughout Asia, Europe, Africa, and the Middle East were included in this embargo, as well as restricted areas within Denmark, France, Germany, Hungary, Sweden, and the United Kingdom.

In 2000, an outbreak of classical swine fever among 35 pig farms in the United Kingdom resulted in import bans of all food animals to the United States, Belgium, the Netherlands, and Spain (Waugh, 2000). Furthermore, after a 2007 FMD outbreak in approximately 60 cattle in the United Kingdom, Britain banned all exports of food animals, meat, and milk in hopes of preventing a larger outbreak, such as the one that occurred in 2001 resulting in $16 billion in losses for that country (CBS/AP, 2007).

Fear of disease spread has also resulted in travel sanctions. During a 2001 Ebola outbreak in Uganda, the government of Saudi Arabia imposed a travel restriction on all Ugandan Muslims who planned to make a pilgrimage to Mecca or Medina that year (Borzello, 2001). The SARS epidemic of 2003 also took a huge economic toll through travel sanctions. During the outbreak in Hong Kong, a number of Southeast Asian countries, including Singapore, Malaysia, and Thailand, issued travel warnings, which resulted in an 80 percent reduction in visitors to Hong Kong from these countries as compared to the previous year (Bradsher, 2003). It also resulted in the Chinese government suspending all international adoption of Chinese babies (Eckholm, 2003). The influenza A(H1N1) outbreaks in Mexico and the United States in April 2009 also saw almost immediate travel advisories imposed by the European Union against Mexico and travelers through the United States, although they were later lifted (McNeil and O'Connor, 2009).

INCENTIVES TO IMPROVE DISEASE SURVEILLANCE AND REPORTING

Disease Outbreak Control Assistance as an Incentive

Countries unable to contain outbreaks are far less likely to report them, and providing assistance for outbreak control is perhaps the most important form of external motivation for disease surveillance and prompt reporting (Laxminarayan et al., 2008). In the case of meningitis in sub-Saharan Africa, an incentive to report the disease was created when the World Health Organization (WHO) made meningitis vaccine available for countries in the meningitis belt (see Box 5-4). By contrast, the Indonesian government has been unwilling to share HPAI H5N1 samples with WHO due to concerns that these samples would be used to create vaccines for developed countries, but not Indonesians (Sedyaningsih et al., 2008). Because Indonesia's own ability to control an outbreak was not enhanced in the process of sharing, the government saw no benefit to cooperating on disease surveillance.

BOX 5-4
Making Vaccines Available to Incentivize Disease Reporting

In the 1990s, a number of countries in the meningitis belt of sub-Saharan Africa were not reporting outbreaks of meningococcal meningitis because they feared their citizens would be barred from the Muslim Hajj. The "meningitis belt" countries include Benin, Burkina Faso, Cameroon, the Central African Republic, Chad, Côte d'Ivoire, the Democratic Republic of the Congo, Ethiopia, Ghana, Mali, Niger, Nigeria, and Togo. Because the $55 meningitis vaccine was not affordable to most citizens of these countries, countries were unable to enforce the Saudi requirement that all Hajj pilgrims be vaccinated against the disease.

Following an outbreak of meningococcal disease that resulted in 250,000 cases and 25,000 deaths, the World Health Organization established the International Consultative Group (ICG) in 1997 to provide meningococcal vaccines to all African countries in the meningitis belt. Countries are now required to provide epidemiological information on their meningitis cases before they can access the vaccine stockpile. By exchanging the tools for outbreak control for information about disease, ICG was able to incentivize reporting by countries.

Incentives to Improve the Quality of Disease Surveillance Information

Investments in surveillance depend on the likelihood that the detected outbreak is a novel disease-causing pathogen that will produce a significant epidemic. The more country officials believe a disease will arise and spread, the more significant the investment in disease surveillance will be. However, this investment can be tempered by the likelihood of false positives: the declaration or reporting a disease outbreak when none exists (Malani and Laxminarayan, 2006). Thus, a trade-off exists between a more sensitive surveillance system and one that is able to have relatively few false positives.

IHR 2005 now calls on national governments to report a wide range of unusual human and animal health events and allows WHO to announce an outbreak, even if it has not first been reported by the government of country of origin (Nicoll et al., 2005). These changes recognize WHO's enormous power in providing information to the world that would allow other countries to protect their citizens and economies from outbreaks in a single country. Moreover, they also alter incentives for disease surveillance and reporting within countries in two ways. First, by preempting a national report from the country, they alert the international community to the possibility of an event from within a country. The onus is then on the country to show there is no outbreak; failure for which could result in trade and tourism bans. Second, countries may be more forthcoming with information if they believe that WHO's report is based on faulty information or false positives. In these instances, the country would benefit from having a strong disease surveillance system that can produce evidence to provide counter information from informal networks. For countries with strong disease surveillance and a reputation for reporting promptly, the credibility of contradictory data from rumor disease surveillance is much weaker.

Improving the quality of information from informal disease surveillance systems—such as the Program for Monitoring Emerging Diseases and the WHO Global Outbreak Alert and Response Network—can also be useful to encourage reporting. If countries recognize that information about an outbreak is no longer theirs alone to provide and that other transnational networks are able to perform this function, they are less likely to suppress information and may derive greater benefit from ensuring that the information reported from these systems is accurate. In instances where the country is able to control an outbreak because it was promptly reported, it may also be in the country's interests to ensure that informal information networks work well to minimize false positives and false negatives in reporting.

The committee was mindful of the global economic crisis at the writing of this report. Nevertheless, pathogens will continue to evolve and emerge. *The committee concluded that despite economic adversity and the potential response of still wealthier nations to reduce international aid for health, poverty alleviation, and other important issues, incentives are needed for optimal disease surveillance and reporting. These incentives are important at the national level, enabling countries to take necessary action for containing zoonotic disease outbreaks and maintaining their access to markets, and at the local and regional levels to encourage early reporting and prevent disease outbreak concealment.*

AUDIT AND RATING FRAMEWORK FOR DISEASE SURVEILLANCE AND RESPONSE SYSTEMS

Trading partners and neighbors frequently restrict the movement of goods and travel contacts based on unreported outbreaks or on outbreaks that have not been officially reported or confirmed (Malani and Laxminarayan, 2006). These sanctions are not necessarily formal or even under the control of partner governments. For instance, concerns about highly pathogenic avian influenza discouraged tourism to Southeast Asia even before any government imposed legal travel restrictions to that region (Tan, 2006). Preemptive sanctions occur when demand for a country's products and services responds to both perceived and actual risks to consumer health. Post-outbreak sanctions discourage reporting by penalizing source countries. Nonreporting or preemptive sanctions (which displace post-outbreak sanctions) are less likely to disincentivize surveillance and reporting by source countries, but these actions could be based on unreliable information. Unlike post-outbreak sanctions, nonreporting sanctions encourage investigation and disclosure because trading partners reply relatively less on post-reporting sanctions to protect themselves from disease outbreaks (Malani and Laxminarayan, 2006). Moreover, these sanctions complement the various policy levers available to the global community.

Of course, not all countries are equally well-informed about the risk of a disease outbreak in any given country. Information about the risk of an unreported outbreak is a global public good in the same way as disease surveillance information. A global emerging disease audit and risk-rating framework would monitor two components: (1) the risk of a novel disease emerging from a given country, and (2) the likelihood that the disease would be undetected (and therefore unreported) by the country's disease surveillance system. Such a framework would also give all countries an incentive to improve their disease surveillance system because a demonstration of prompt disease outbreak reporting would help reduce their rate of risk and alleviate trade and tourism concerns in the event of an unconfirmed outbreak. Any risk identified by the risk-rating framework would alone be insufficient to support a restrictive trade measure for health reasons. However, countries could use this framework to signal their willingness to be transparent about their risk of outbreaks and the likelihood of detection. If countries recognize, apply, and accept existing sanction mechanisms like zoning and compartmentalization—allowing for continued trade of safe products from countries or zones that have reported a disease—they can minimize the unnecessary cost of more restrictive sanctions.

The framework would operate in a manner similar to other mechanisms that rate the risk of sovereign debt default or the risk of a national unreported nuclear weapons program. External assistance for improving a country's disease surveillance infrastructure would be tied to their demonstrated improvements on the framework. Ideally, the framework would not require a new institution, but rather could be housed within an existing global institution that has the scientific and technical expertise to assess the country's risk of disease emergence and nonreporting. Intergovernmental organizations, however, would be excluded to ensure that risk assessments are not affected by political considerations.

To implement IHR 2005 by the 2012 deadline and for WHO member states to comply with the requirement for core disease surveillance capacities, greater efforts will be needed by the member states, WHO, and the international community. WHO is also preparing country guidance for developing core capacities for disease surveillance. Although state parties to IHR 2005 are required to assess the ability of their existing national structures and resources to meet the minimum requirements, it is unclear whether these assessments will be made publicly available.

On the animal health side, OIE uses the Performance of Veterinary Services (PVS) tool to assess the major components of effective veterinary services in a country (see Table 4-1). However, there is no ratified deadline by which countries must report their competency information if they use the PVS tool to assess their capacities. Questions remain for the committee about whether this recommended tool is available as an open-source tool that can be used freely or if countries are formally assessed by OIE. It should be noted that even though there is no deadline for assessment, OIE is developing guidance to help countries systematically use the tool to assess their country's veterinary services and infrastructure.

Capacity assessment information for both human and animal health is essential. It is useful in devising national and local incentives, establishing a disease surveillance system, and in timely disease reporting by local and national participants to protect human and animal health and livelihoods.

ENGAGING MULTI-LEVEL STAKEHOLDERS FOR TIMELY DISEASE DETECTION AND REPORTING

As previously discussed, information from all levels is critical for effective disease surveillance, and therefore data collection will need to include information gathered from the grass-roots level (e.g., Roll Back Malaria).

Disease surveillance will be effective if it is informed by the local knowledge base, particularly for places identified as hotspots. Public and private partners at the international level are also key stakeholders in both collecting and using data to protect human and animal health. Social networking and mobile technologies offer flexible and dynamic new possibilities for community health options. Science journalism, citizen journalism platforms, mobile video, and "sousveillance" (for example, monitoring that is captured by individuals on cell phones and shared through YouTube) are all examples of emerging tools that may lead to greater access and transmission of health-related information by individuals.

Different actors are becoming engaged in disease surveillance because of changes in information and communication technologies. Google.org (the philanthropic arm of Google.com) launched the *Predict and Prevent* initiative, using information and technology to empower communities to know where to look for disease threats, find them earlier, and know how to respond (see Box 5-5). Select multinationals are also engaged as corporate citizens. The Safe Supply of Affordable Food Everywhere (SSAFE) initiative is a public-private partnership that includes multinationals such as Cargill, Nestle-Purina, McDonald's, Pfizer Animal Health, and Coca-Cola. One of its stated goals is to help advance science-based global standards for food, animal feed, environmental health, safety and sustainability, and disease prevention and control (Ades, 2008). SSAFE also serves on an OIE advisory

BOX 5-5
Google.org Predict and Prevent Initiative

Google.org represents a unique model whereby support for information technology to increase collection and reporting surveillance data comes from both internal volunteer efforts and external funding streams. Google.org's Predict and Prevent Initiative (PPI) is a good example of both these roles. As a philanthropic organization, PPI supports efforts in disease detection, ranging from supporting web-based surveillance and data collection efforts (e.g., HealthMap, the Program for Monitoring Emerging Diseases or ProMED-mail, InSTEDD) to collecting specimens for detecting novel zoonotic diseases in the field (Global Viral Forecasting Initiative). Recognizing the potential value of the volunteer efforts of Google software engineers, PPI has deployed novel surveillance systems such as Google Flu Trends and other efforts devoted to mobile communication for disease reporting and health resource mapping. These openly available technologies require the internal capacity of engineers that can properly leverage products such as Google Search, Maps, and Android. Although these efforts play an important role in furthering global technological capacity, singular dependence on any particular company's tools, even if freely available, should be avoided, especially when considering issues of compatibility, scalability, and future availability.

committee of the World Animal Health and Welfare Fund (OIE, 2007). In addition, companies such as Cargill have worked closely with health authorities and provided their own field staff and resources to educate and work with communities on HPAI H5N1 in Thailand (Ades, 2008).

International businesses[2] are increasingly part of the global landscape, as one-third of all global trade takes place directly within international businesses (Moore, 1998). In August 2008, the Kellogg Foundation, ConAgra, Cargill, Kellogg Company, and McDonald's convened at Michigan State University to address how foundations and businesses can create new models of cooperation (Food and Sustainability Conference, 2008). These corporations will increasingly have a greater role in the global food system and will need to be included in discussions about new approaches to improve reporting and sharing of critical information. With the multiplicity of actors, there is a need for improved intersectoral and international coordination, communication, and community of practice to enable environments that facilitate working toward cross-disciplinary collaboration for disease surveillance and response, practice, and research. *The committee concludes that participation by partners at various levels in disease surveillance, monitoring changes in community perception and response to the presence and threat of zoonotic disease, and media coverage of such diseases are essential and should be included for comprehensive disease surveillance systems in human and animal health.*

CONCLUSION

There have been significant improvements in how global health legal frameworks harness new technology to require countries to report disease outbreaks, yet the decision on whether to report and how much effort to expend on disease surveillance remains the province of countries. Even within countries, there may be conflicting economic, cultural, or political incentives to report an outbreak up the chain and those incentives affect whether an outbreak is officially recognized. Yet without prompt reporting of outbreaks, including in resource-poor settings that have the least ability to detect them, the ability of global efforts to prevent the rapid spread of virulent pathogens is limited. Quality disease surveillance information reporting goes beyond just a confirmation or denial: In order to respond effectively, policymakers need a clear assessment of the situation based on reliable scientific information.

The committee concludes that a global zoonotic disease surveillance and response strategy that does not address the fundamental incentives and

[2]The term is used to include multinational, international, transnational, and global companies doing business in other countries.

disincentives of reporting is likely to be unproductive. Therefore, investments need to not only finance a country's disease surveillance activities, but also couple and reinforce incentives needed within each country for disease surveillance and reporting of outbreaks. Additionally, investments for building or upgrading disease surveillance and response capacities would be better spent with an eye on whether these systems will actually be used in the event of an outbreak, or whether these resources are simply crowding out monies that countries would have spent on their own.

REFERENCES

Ades, G. 2008. The role of public-private partnerships in strengthening food systems globally—surveillance to "sousveillance": New models for cooperation. Presentation, Third Committee Meeting on Achieving Sustainable Global Capacity for Surveillance and Response to Emerging Diseases of Zoonotic Origin, Woods Hole, MA, September 30.

Ahmad, K. 2000. Malaysia culls pigs as Nipah virus strikes again. *Lancet* 356(9225):230.

Baker, M. G., and D. P. Fidler. 2006. Global public health surveillance under new international health regulations. *Emer Infec Dis* 12(7):1058–1065.

Borzello, A. 2001. Uganda "almost" Ebola free. BBC News, November 5. http://news.bbc.co.uk/2/hi/africa/1134366.stm (accessed July 20, 2009).

Bradsher, K. 2003. Hong Kong tourism battered by outbreak. *The New York Times*, April 13.

Briggs, C. L. 2004. Theorizing modernity conspiratorially: Science, scale, and the political economy of public discourse in explanations of a cholera epidemic. *Am Ethnol* 31(2):164–187

Burns, J. 1994. Plague in India giving visitors second thoughts. *The New York Times*, October 16.

Cash, R. A., and V. Narasimhan. 2000. Impediments to global surveillance of infectious diseases: Consequences of open reporting in a global economy. *Bull World Health Organ* 78(11):1358–1367.

CBS/AP (CBS News/Associated Press). 2007. U.K. bans exports after disease outbreak. Prime Minister vows to work "night and day" to slow spread of foot-and-mouth. CBS News, August 4. http://www.cbsnews.com/stories/2007/08/04/world/main3133548.shtml (accessed December 11, 2008).

CDC (Centers for Disease Control and Prevention). 1999. Outbreak of Hendra-like virus—Malaysia and Singapore, 1998–1999. *MMWR* 48(13):265–269.

CDC. 2007. *Embargo of birds from specified countries.* http://www.cdc.gov/flu/avian/outbreaks/embargo.htm (accessed November 5, 2008).

Chanyapate, C., and I. Delforge. 2004. The politics of bird flu in Thailand. *Focus on Trade* no. 98, April. http://focusweb.org/publications/FOT%20pdf/fot98.pdf (accessed July 20, 2009).

Chuengsatiansup, K. 2008. Ethnography of epidemiologic transition: Avian flu, global health politics and agro-industrial capitalism in Thailand. *Anthropology & Medicine* 15(1):53–59.

Comisión Sudamericana para la Lucha Contra la Fiebre Aftosa. 1996. *Informe Final.* Rio de Janeiro, Brasil: Comisión Sudamerican para la Lucha Contra la Fiebre Aftosa.

Davis, M. 2005. *The monster at our door: The global threat of avian flu.* New York: New Press.

Dolberg, F., E. Guerne-Bleich, and A. McLeod. 2005. *Emergency regional support for post-avian influenza rehabilitation: Summary of project results and outcomes.* Rome, Italy: FAO.

Eckholm, E. 2003. The SARS epidemic: Avoiding infection—China suspends adoptions and sets edict to fight virus. *The New York Times*, May 16.

Farmer, P. 1992. *AIDS and accusation: Haiti and the geography of blame.* Berkeley, CA: University of California Press.

Ferguson, N. M., D. A. Cummings, S. Cauchemez, C. Fraser, S. Riley, A. Meeyai, S. Iamsirithaworn, and D. S. Burke. 2005. Strategies for containing an emerging influenza pandemic in Southeast Asia. *Nature* 437(7056):209–214.

Food and Sustainability Conference. 2008. Michigan State University, August 2–22.

Ghosh, I., and L. Coutinho. 2000. Normalcy and crisis in time of cholera: An ethnography of cholera in Calcutta. *Econ Polit Wkly* 35(8–9):684–696.

Grady, D., D. McNeil, Jr., A. O'Connor, and C. Campbell. 2003. U.S. issues safety rules to protect food against mad cow disease. *The New York Times*, December 31.

Gubler, D. J. 2001. Silent threat: Infectious diseases and U.S. biosecurity. *Georgetown J of International Affairs* II(2):15–23.

Heymann, D. L., and G. Rodier. 2004. Global surveillance, national surveillance, and SARS. *Emerg Infect Dis* 10(2):173–175.

Kaufman, J. A. 2008. China's health care system and avian influenza preparedness. *J Infect Dis* 197(Suppl 1):S7–S13.

Kleinman, A. M., B. R. Bloom, A. Saich, K. A. Mason, and F. Aulino. 2008. Avian and pandemic influenza: A biosocial approach. *J Infect Dis* 197(Suppl 1):S1–S3.

Laxminarayan, R., E. Klein, A. Malani, and A. Galvani. 2008. *Surveillance and reporting of emerging pathogens.* Washington, DC: Resources for the Future.

Longini, I. M., Jr., A. Nizam, S. Xu, K. Ungchusak, W. Hanshaoworakul, D. A. Cummings, and M. E. Halloran. 2005. Containing pandemic influenza at the source. *Science* 309(5737):1083–1087.

Malani, A., and R. Laxminarayan. 2006. *Surveillance and reporting of disease outbreaks: Private incentives and WHO policy levers.* John M. Olin Program in Law and Economics Working Paper Series. Charlottesville, VA: University of Virginia.

McNeil, D., and A. O'Connor. 2009. Europe urges citizens to avoid U.S. and Mexico travel. *The New York Times*, April 28.

Merianos, A., and M. Peiris. 2005. International Health Regulations. *Lancet* 366(9493): 1249–1251.

Moore, M. 1998. *A brief history of the future.* Christchurch, New Zealand: Shoal Bay Press.

Nichter, M. 2008. *Global health: Why cultural perceptions, social representations, and biopolitics matter.* Tucson, AZ: University of Arizona Press.

Nicoll, A., J. Jones, P. Aavitsland, and J. Giesecke. 2005. Proposed new International Health Regulations. *BMJ* 330(7487):321–322.

OIE (World Organization for Animal Health). 2007. The 75th OIE General Session, May 2007. *Bulletin* no. 2007-3:1–65.

O'Neill, B. 2004. *Disease report and trade responsibilities of OIE member countries.* Speech presented at the 72nd OIE General Session, Paris, France, May 23–28. http://www.oie.int/downld/SG/2004/Speech_oneill.pdf (accessed June 29, 2009).

Padmawati, S., and M. Nichter. 2008. Community response to avian flu in Central Java, Indonesia. *Anthropology & Medicine* 15(1):31–51.

Panisset, U. 2000. *International health statecraft: Foreign policy and public health in Peru's cholera epidemic.* Lanham, MD: University Press of America.

PPLPI (Pro-Poor Livestock Policy Initiative). 2007. *Pro-Poor management of public health risks associated with livestock: The case of HPAI in East and Southeast Asia.* Policy Brief. Rome, Italy: FAO.

Rich, K. 2004. *Animal diseases and the cost of compliance with international standards and export markets: The experience of foot-and-mouth disease in the Southern Cone.* Agriculture and Rural Development Discussion Papers. Agriculture and Rural Development Department. Washington, DC: The World Bank.

Romero, S. 2003. Virus takes a toll on Texas poultry business. *The New York Times*, May 16.

Sedyaningsih, E. R., S. Isfandari, T. Soendoro, and S. F. Supari. 2008. Towards mutual trust, transparency and equity in virus sharing mechanism: The avian influenza case of Indonesia. *Ann Acad Med Singapore* 37(6):482–488.

Tan, P. 2006. Tourism takes a hit. TTG Asia. http://www.ttgasia.com/index.php?option=com_content&task=view&id= 12071&Itemid=48 (accessed April 6, 2009).

USITC (U.S. International Trade Commssion). 2008. *Global beef trade: Effects of animal health, sanitary, food safety, and other measures on U.S. beef exports.* Investigation no. 332–488. Washington, DC: USITC. http://hotdocs.usitc.gov/docs/pubs/332/pub4033.pdf (accessed February 18, 2009).

Van Sickle, D. 1998. Silent treatment: What ails Cuban health care? *The New Republic* 218(25):14–15.

Venkataraman, B. 2008. Amid salmonella case, food industry seems set to back greater regulation. *The New York Times*, July 31.

Vu, T. 2009. *The political economy of avian influenza response in Vietnam.* STEPS Working Paper 19. Brighton, UK: STEPS Centre.

Waltner-Toews, D. 2004. *Ecosystem sustainability and health: A practical approach.* Cambridge, UK: Cambridge University Press.

Waugh, P. 2000. *Crisis for British farming as fever spreads.* http://www.independent.co.uk/environment/crisis-for-british-farming-as-fever-spreads-710257.html (accessed December 11, 2008).

World Bank/The International Bank for Reconstruction and Development (IBRD). 2006. *Enhancing control of highly pathogenic avian influenza in developing countries through compensation: Issues and good practices.* Washington, DC: The World Bank.

WTO and WHO (World Trade Organization and World Health Organization). 2002. *WTO agreements and public health: A joint study by WHO and the WTO secretariat.* Geneva, Switzerland: WTO.

Zacher, M. W. 1999. Global epidemiological surveillance: International cooperation to monitor infectious diseases. In *Global public goods: International cooperation in the 21st century*, edited by I. Kaul, I. Grunberg, and M. Stern. New York: Oxford University Press. Pp. 266–283.

6

Sustainable Financing for Global Disease Surveillance and Response

"At this time of global economic downturn, we face a crossroads. We can cut back on health expenditures and incur massive losses in lives and fundamental capacity for growth. Or we can invest in health and spare both people and economies the high cost of inaction."

—*Ban Ki-moon*
Secretary-General of the United Nations
Remarks at the Forum on Global Health: The Tie That Binds (June 15, 2009)

International funding is necessary to develop and sustain a global zoonotic disease surveillance and response system. Developing an international financing framework especially to assist resource-constrained countries will be challenging. The existing international aid architecture for combating zoonotic diseases is fragmented, and fostering multisectoral cooperation between human and animal health at the local and global levels has proven difficult. Donor funding is unpredictable, especially during a global economic crisis. At the same time, innovative financing tools and mechanisms can provide new ways to modify the existing international aid architecture to create long-term, predictable funding streams. This chapter provides an overview of the current environment for investing in zoonotic disease surveillance and discusses possible financing solutions.

FUNDING ANIMAL DISEASE SURVEILLANCE

As discussed in Chapter 2, the development and implementation of animal disease surveillance—usually as part of national animal health infrastructure, especially in the United States—was a result of the recognition that herds free of selected diseases could translate into broader social and economic benefits. Public investment in animal health helped facilitate export growth by enabling the movement of disease-free animals and related products into new markets and countries. The resulting disease surveillance systems were structured around defined hierarchal relationships addressing governance and geography (local, state, federal), with the government having oversight in the collection, summary, validation, and reporting of disease surveillance information.

Decentralized Disease Surveillance as an Emerging Concept

An important characteristic of animal disease surveillance has been the centralized manner in which data have been captured. This provides for the establishment of defined standards and processes and can help to ensure accuracy and consistency in reporting. It is a clear-cut approach that is most achievable with adequate resources, expertise, legal frameworks, aligned objectives, and sustained commitment; unfortunately this is not the case in many countries. During an international response to a zoonotic disease, a number of different stakeholders may often hold or *withhold* information or resources depending on their particular incentives and objectives. A more centralized approach can mitigate some of this, but the growing reality is that the open economy and globalization are presenting both challenges and opportunities for how future disease surveillance systems will be defined and funded. The issues can be complex and layered, but possibilities are emerging as paradigms shift, relationships change, communities evolve, and technology advances.

Investing in Human and Veterinary Health Infrastructure and Disease Surveillance

The level of global economic interdependence is increasing, and economies are increasingly bound to the overall sanitary status of a country. It would seem only logical that governments and industry would make the necessary and ongoing investments in human and animal health and veterinary infrastructure, but that is not always the case. The public and government officials often have short attention spans and are challenged by competing interests for limited resources. As a result, ongoing investment often takes a back seat to more pressing issues. Government investments may not be commensurate with the significant value of national resources—with proportionally little designated for disease surveillance. This is particularly the case for the veterinary sector. Complicating these issues is the lack of continuity and leadership in ministries of agriculture, where time spent in office can be politically short-lived. This transience filters down through multiple layers in the organization because a change in ministers is usually accompanied by a broader reorganization. Coupled with low salaries in the public sector, this situation requires repetitive cycles of basic training. Furthermore, this hinders the effectiveness of countries as member states within important international organizations. Each year, the World Organization for Animal Health (OIE) convenes its 174 member countries and territories to discuss and approve new international health standards. OIE estimates that a full one-third of the chief veterinary officers who come each year are newly installed. By the time these heads of delegation learn how best to represent their countries, many are replaced and the cycle continues.

The fact that many developing countries request technical support and financial resources from the international community undoubtedly reflects the poor internal state of affairs in their human and animal health infrastructures. In addition, the veterinary workforce is limited even though many infectious diseases are discovered by the veterinary community (e.g., viral cancers, retroviruses, lentiviruses, transmissible spongiform encephalopathies such as bovine spongiform encephalopathy, rotaviruses, papillomaviruses, coronaviruses, ehrlichioses) and later turn out to be of human and animal health importance. Veterinarians and animal health professionals are an essential component in detecting zoonotic diseases earlier and therefore preventing infection in humans. But more fundamental issues may also be inhibiting long-term growth and sustainability. Money is a necessary, but insufficient, condition for sustained change.

The international community jolted into action when it became clear that the highly pathogenic avian influenza (HPAI) H5N1 strain was on Europe's doorsteps and could pose a human health risk at a global level. The United Nations World Health and Food and Agriculture Organizations, the World Bank, and OIE sponsored international conferences. Banks, aid agencies, and nations themselves pledged nearly $3 billion to combat the disease and address the alarming human and animal health concerns (Government of Egypt, 2008). In many ways, the response was a positive example of the international community coming together to attack a global threat. But putting out a fire and addressing the underlying causes of the fire are two different goals.

A recent World Bank report (2006a) on the extent of the needs and gaps in the development of a financing framework for the next pandemic emphasizes the need for medium- and long-term planning. The main concern for this strategy is the endemic nature of HPAI H5N1 in the Southeast Asia region. So far, most of the economic impact of HPAI H5N1 in the East Asia region is occurring in the rural economies. The gross domestic product (GDP) impact in East Asia economies has been restricted to a 0.1–0.2 percent loss in countries such as Vietnam. However, the economic impact associated with severe acute respiratory syndrome resulted in a 2 percent loss of the regional East Asian GDP in 2003 (World Bank, 2005).

Short-term emergency actions such as disease control or eradication[1]

[1]In animal health, eradication of a specific disease is considered on a nation-by-nation status.

and improving infrastructure—including ongoing disease surveillance—while not mutually exclusive, are not a single endeavor. Eradicating a disease requires specific skill sets and large financial outlays for direct field activities such as vaccination or herd depopulation (culling). Building an adequate and sustainable human and animal health infrastructure that includes critical components such as disease surveillance, on the other hand, requires raising the overall levels of technical capacity, building human and financial capital, and fostering collaboration between the public and private sectors. The scope goes beyond any one disease to the ability to proactively address multiple potential emerging diseases and to permanently raise the country's sanitary status.

Independent, international financial support can actually hamper long-term sustainability of a recipient country's infrastructure. In many countries, funding for human and animal health infrastructure competes with other government programs for public funds. Most of that money is tied up in salaries for existing personnel, so little remains for program operations. Tight budgets mean that capital improvements are difficult to carry out. Therefore, what the international community may regard as "bridge funds" until the country can put in place a more sustainable approach becomes the primary source of funding, underwriting a variety of program activities, ranging from field supplies and diagnostic equipment to vehicles and fuel for data collection. Ironically, the presence of a disease makes possible a coveted inflow of resources that might not otherwise be available and reduces the need for elected officials to make tough choices regarding the priority of human and animal health within the national agenda. The inflow of external resources also relieves the immediate pressure to establish longer-term national solutions. When external resources are withdrawn, the infrastructure regresses, but the impacts may not be felt until the next disease crisis.

Countries are at varying stages of advancement regarding their national infrastructure, but two factors stand out as making a critical difference: the level of interaction between the public and private sectors, and continuing leadership. Those countries whose veterinary infrastructures are on the upswing benefit from a private sector that invests its time and resources into strengthening the national infrastructure—similar to its efforts to develop new animal feeds, secure better genetics in its breeding stock, or push into new markets. In countries where animal health infrastructures are weak, mutual distrust or apathy may exist between the two sectors. This can be reflected in allegations of corruption, incompetence, and lack of vision and continuity. Simply injecting external resources under these conditions can mask these underlying problems and prolong the difficult tasks of fortifying national infrastructure and improving relations between the two sectors.

The conditions for sustained investment begin by taking a more integrated approach and creating the foundation for addressing critical competencies, such as current and future disease surveillance—starting with a shared vision between the public and private sectors. Essential to this vision is the establishment of trust, transparency, and clearer communication between the two sectors. Although financial resources are also important, they are not a panacea. Equally important are changing attitudes, building bridges for collaboration, and continuing leadership. When these happen, an environment of empowerment prevails that weans countries from the belief that little can happen without external resources; much is within the country's control, and is less reliant on international organizations or funding. If sustained, the process leads to a set of priorities with measurable outcomes that gauge incremental progress.

Any type of centralized disease surveillance requires a sustained commitment over time on the part of both the public and private sectors. The private sector sees investment in veterinary infrastructure as good business practice, and decreased exposure to disease is a way to protect the private sector's investment and to open new market opportunities. The public sector helps ensure that benefits accrue from the greater public good of an elevated sanitary status. However, in practice, financial resources across countries have varied, ranging from those largely reliant on public expenditures (e.g., United States) to those raising resources from the private sector (e.g., Australia and New Zealand) to appropriate and implement public sector activities.

CURRENT FUNDING EFFORTS

Regardless of the proportions of fiscal contributions from any sector or the relationship between them, the funding needs for developing and sustaining a global disease surveillance system for emerging and reemerging zoonotic diseases will be significant. Recent concerns about a potential highly virulent human influenza pandemic have resulted in coordinated international action to help countries improve their ability to detect disease outbreaks. The funding efforts for these actions have been led by the World Bank, which has used two main mechanisms to provide assistance to countries with avian influenza and to prepare for a possible pandemic. These mechanisms are the Global Program on Avian Influenza (GPAI) and the multi-donor trust fund known as the Avian and Human Influenza (AHI) Facility.[2]

[2]The Global Program on Avian Influenza sanctions loans, credits, or grants up to $500 million from the concessional arm called the International Development Association. The program uses an integrated approach developed with the United Nations World Health and Food and Agriculture Organizations and the World Organization for Animal Health. The

GPAI and the AHI Facility were conceived as concerted efforts by the United Nations (UN) and some multilateral and bilateral agencies to respond to the growing threat of HPAI H5N1 and to prepare for the next influenza pandemic. In September 2005, the United Nations Secretary-General appointed a System Influenza Coordinator (UNSIC). UNSIC was a key factor in developing a strong partnership among technical agencies such as the World Health Organization (WHO), the Food and Agriculture Organization of the United Nations (FAO), and OIE, and other bilateral and multilateral partners, including the World Bank. The partnership focused on developing a flexible and responsive framework to provide financial and technical support at country, regional, and global levels. Both urgent and long-term needs were targeted. The 2008 progress report by UNSIC and the World Bank provides an analysis of the response to HPAI H5N1 and the state of pandemic readiness (UNSIC and World Bank, 2008). Although it does not provide detailed information on spending for areas such as laboratory and human capacity building, there is a detailed breakdown by sector of the funds disbursed by October 2008: 24 percent for animal health, biosecurity, sustainable livelihood; 36 percent for human health and pandemic preparedness; 14 percent for information, education and communication; 11 percent for supporting implementation monitoring, evaluation, and internal coordination; and 15 percent designated as other (UNSIC and World Bank, 2008). Although these World Bank operations identify funding to support country, regional, and global levels of HPAI H5N1 pandemic preparedness, they also support the discussion in Chapter 5 about the importance of and challenges to country incentives to report an outbreak.

Additional resources were provided directly from bilateral sources and the European Commission through WHO and FAO, and to a lesser extent through OIE, and to individual countries. They showed a much better disbursement rate than the multilateral development agencies. As Table 6-1 shows, the bilateral donors committed a total of $1.4 billion USD, of which 90 percent was disbursed on April 30, 2008; the European Union

---

funds can be used to improve the health and veterinary services of the countries, prepare and respond to the pandemic influenza, and minimize the threat to its populations. The funds go through the World Bank's emergency procedures and therefore can be quickly prepared and approved. It is a trust fund located at the World Bank under a partnership that is led by the European Commission, and other donors such as Australia, China, Estonia, Iceland, Korea, the Russian Federation, Slovenia, and the United Kingdom. The Avian and Human Influenza (AHI) Facility helps developing countries meet their financing gaps within their integrated national plans to minimize risk and socioeconomic impact of the pandemic on humans and animals. As of January 2009, the total pledged commitment to the AHI Facility was more than $126 million equivalent (World Bank, 2009a).

TABLE 6-1 Avian and Human Influenza Pledges, Commitments, and Disbursements as of April 30, 2008 (US$ million)

Donor	Inter-ministerial Meetings: Beijing	Bamako	New Delhi	Total[c]	Commitment[d]	Disbursement[e]	Percentage Disbursed	Uncommitted
Australia	56	55		111	100	67	67%	11
Canada		87		87	91	40	44%	—
France	31	10	7	48	50	34	69%	—
Germany	29	8	4	41	41	30	73%	—
Japan	155	67	69	291	297	297	100%	—
Netherlands	14	7		21	22	10	44%	—
Russia	24	8		32	32	29	92%	—
United Kingdom	36	18	10	65	61	51	83%	3
United States	334	100	195	629	629	629	100%	—
Other EU Countries[a]	31	11	11	42	53	48	90%	6
Other Countries[b]	33	4	4	41	33	31	94%	10
Subtotal Bilateral Donors	**742**	**376**	**290**	**1,408**	**1,410**	**1,266**	**90%**	**30[f]**
European Commission	**124**	**83**	**111**	**319**	**241**	**140**	**58%**	**79**
Asian Development Bank	468			468	83	13	16%	385
African Development Bank		15		15	7	4	63%	8
World Bank	501			501	313	69	22%	187
Subtotal Multilateral Development Banks	**969**	**15**		**984**	**403**	**87**	**22%**	**580**
Grand Total	**1,835**	**474**	**401**	**2,710**	**2,054**	**1,494**	**73%**	**689**

NOTES: Donors' reports of amounts committed and disbursed from calendar year 2005 and to April 30, 2008. Uncommitted amounts are net of commitments in excess of pledges.

[a]Austria, Belgium, Cyprus, Czech Republic, Estonia, Finland, Greece, Hungary (which has retracted its pledge due to lack of response from recipient country), Ireland, Italy, Luxembourg, Slovenia, Spain, and Sweden.

[b]Iceland, Korea (Republic of), Norway, Saudi Arabia, Singapore, Switzerland, and Thailand.

[c]Total of pledges from the three international conferences on avian and pandemic influenza (Bamako, Beijing, and New Delhi).

[d]Commitment is defined as the result of an agreement between the donor and recipient for designated purposes; a commitment is a firm decision that prevents the use of allocated amount for other purposes.

[e]Disbursement is defined as the actual budget transfer or release of funds to the recipient for an intended purpose.

[f]This number represents the portion of total donor pledges that remain uncommitted. As some donors have committed more than their pledged amounts, this number does not correspond to the difference between the total (1,408) minus the total commitments (1,410).

SOURCE: UNSIC and World Bank (2008).

committed a total of $241 million USD, of which 58 percent was disbursed on that date; and the Multilateral Development Banks had committed on that same date $403 million USD, of which only 22 percent was disbursed (UNSIC and World Bank, 2008).

In countries where outbreak reporting is hampered only by the availability of financial and technical resources, these programs are likely to improve disease surveillance information. However, in countries that could afford their own programs if desired, these resources would either displace national resources for disease surveillance or be misspent if incentives for the country to detect and report an outbreak are not changed.

In addition, the global approach to finance widespread prevention or control of HPAI H5N1 needs to consist of multisectoral coordination and integrated response at the national, subnational, and global levels as detailed in recommendations from the FAO, OIE, WHO, UNSIC, UNICEF, and World Bank Strategic Framework (2008), prepared for the Sharm el-Sheikh Inter-ministerial Conference on Avian and Pandemic Influenza. Consequently, some of the policy issues that can emerge correspond to the fact that most resource-poor nations lack the resources for the development of training and capacity-building programs of this nature (World Bank, 2006a).

Completing HPAI H5N1 Control Activities

The recent joint Strategic Framework (FAO et al., 2008) reports that of the $2.7 billion pledged in the subsequent international Inter-ministerial Meetings, a total of $2 billion has been firmly committed or already expended for the human and animal control cost of HPAI H5N1. The breakdown of the expenditure provides that about 41 percent ($853 million) was directed to national programs, about 25 percent ($510 million) to international organizations, 15 percent ($301 million) to regional programs, and the remaining 19 percent ($386 million) to other programs, including research. This expenditure pattern seems to deviate somewhat from the earlier declaration of the Beijing conference, the first inter-ministerial meeting on control of HPAI H5N1, which declared that "individual countries are central to a coordinated response." The novelty of the threat, which caught the international community and international organizations by surprise, is probably the reason for the bias toward international programs. Future funds could, most likely, be more directed toward national governments. The Strategic Framework also reports a shortfall in the current programs of $836 million, mainly because of unmet pledges. Most ($440 million, or more than 50 percent) of this shortfall concerns the sub-Saharan subcontinent.

Developing Global Capacity

The Strategic Framework (FAO et al., 2008) makes a very approximate assessment of the costs of a permanent global disease surveillance system, acknowledging that in preparing these estimates, provided in Table 6-2, this was "an art, not a science" (FAO et al., 2008, p. 44). These tentative cost estimates are provided for 49 low-income countries and for all 139 low- and middle-income countries. The table reflects the much higher need per country for the low income countries and includes a basic infrastructure establishment and operation, and special investment in 40 "hotspot" countries, as described in Table 6-3 under global responsibility. The estimates also account for previous investments already carried out under the ongoing AHI Facility. It also includes the approximate estimate of the needs to complete the current campaign, which is based on the considerable number of already prepared Integrated National Action Plans (140 by September 2008). They were based on an assessment by joint human and animal health specialists. They were based on early detection and response to HPAI H5N1. The study committee conducted rough estimates regarding the cost to scale up these HPAI H5N1 needs to survey and control of multiple species. This will come close to International Health Regulations 2005 (IHR 2005) surveillance requirements. The annual additional financing needed over the next 3 years would range from $542 million to $735 million. Based on the working paper for this assessment (FAO et al., 2008), more than 50 percent (nearly $6 billion) would be operating costs, and the remaining $5 billion would be investments in hardware (laboratories, equipment) and human skills (training, etc.). The rather high share of operating costs reflects the major infrastructure investments already made under the current HPAI H5N1 campaign and is based on 2008 prices without inflation.

TABLE 6-2 Estimated Cost of Funding the One World One Health Framework to 2020 (millions of US$)

Category of Expenditure	49 Low-Income Countries	139 Eligible Countries
Prevention:		
Human and animal health services	1,264	3,083
Veterinary services	3,286	5,476
Wildlife monitoring	1,495	2,495
Communication	583	1,167
International organizations	3,180	3,475
Research	420	420
Total	10,228	16,116
Average per year	852	1,343

SOURCE: FAO et al. (2008).

FUNDING A GLOBAL PUBLIC GOOD

Policymakers generally agree that the responsibility for funding has to be divided among international, national, and local public and private sources. The most common key criteria being applied to decide which of these sources is responsible for a particular service are (1) the degree of externalities involved in the consumption of those goods (i.e., the extent to which they are a public good); (2) whether these goods are mostly global, national, local, or private; and (3) the capacity to pay. Conceptually, it follows logically that the global community funds the global public goods; the national and local public sectors, respectively, fund the national and local public goods; and the private sector funds private goods. Because of their transboundary nature, protection from highly infectious zoonotic diseases is generally considered a global public good.

Preventing the emergence and cross-border spread of human and animal highly infectious diseases conforms to this definition and is generally considered to be a global public good. Less infectious and hence more local disease risks—such as rabies or bovine tuberculosis—are by their nature national or local public goods, often also with private goods characteristics. Their control benefits the local population, and in the case of animal diseases, the individual private owner.

One important issue in this context, however, is whether to include equity and poverty alleviation as criteria for deciding whether a disease is a global public good. From a global social perspective, this would be appropriate. Poverty alleviation is a key Millennium Development Goal and explicitly defined as a global responsibility. This discussion has specific relevance in the case of zoonotic diseases because they are typically the "diseases of the poor," affecting their health and that of their animal stock proportionally more than other types of diseases. However, this would have major repercussions on the funding requirements, and hence in this report, the zoonotic diseases of a lesser transboundary nature are classified as a national public or private good.

Most public goods, including those of global characteristics, are impure; they have components with various degrees of exclusion or rivalry and thus present a mixture of public and private goods. Disease surveillance systems have such characteristics. They cannot be disease specific, but underpin the prevention of emergence and spread of highly infectious diseases, which is recognized as a global public good. Surveillance for all diseases is therefore considered a global public good. Control is much more disease specific, and hence becomes a national, local, or private good in the case of disease of low human epidemic nature. These considerations are reflected in Table 6-3.

TABLE 6-3 Disease Prevention and Control Activities at the Human–Animal–Ecosystems Interface and Their Status Level as a Public Good

Activity	Disease of Low Human Epidemic Potential	Disease of Moderate to High Human Epidemic Potential
1. Preparedness		
Risk analysis	Global	Global
Preparedness plan	National/regional	Global
Animal vaccine development	Private[a]	Global
2. Disease surveillance		
Human and animal health, veterinary and wildlife	Global	Global
Diagnostic capacity	Global	Global
Managerial and policy arrangements	National	Global
3. Outbreak control		
Rapid response teams	National/regional	National/global
Vaccination	National/regional/private	Regional/global
Cooperation among human, veterinary, and wildlife services	National	Global
Compensation schemes	National/private	Global
4. Eradication plans	National/regional/private	Global
5. Research	National/regional/private	Global

[a]This may also be a global public good depending on diseases and circumstances (context).
SOURCE: Adapted from FAO et al. (2008).

FUNDING MECHANISMS

All industrialized and middle-income nations need to have responsibility for funding their own disease surveillance systems, a sentiment also reflected in the Sharm el-Sheikh Framework Document. Industrialized countries also have the responsibility to fund research to develop surveillance and control technology. Low-income countries, however, have multiple and immediate needs competing for limited resources. Quite understandably, a long-term investment in the operation of a disease surveillance system cannot be feasibly prioritized above other needs for low-income countries. International funding is therefore necessary and fully justified in view of the global public good involved, as well as the human and animal health and economic benefit that the international community derives from early detection of a potential health or economic (trade) risk.

For the international funding of a disease surveillance and early response system in low-income countries, the classical and still current way such investments are funded consists of a time-bound (mostly 3–5 years), project-based investment, with the external investor (mostly bilateral or international donor or financing agencies) funding most of the infrastructure costs (laboratory, transport, etc.) and some initial operating costs. In this model, the recipient country funds part of the operating costs and is expected to continue to fund the activity after the time-bound project ends. Long-term financing by these international agencies is often not possible because of internal budgetary procedures, parliamentary approval cycles, policy changes, and geopolitical considerations. This model has major constraints funding the national part of the operating costs, with even greater problems in low-income countries to maintain the operation after the international financing stops. Pre-project commitments of governments are difficult to enforce in the post-project period, and the general scenario is one of a high activity level when external financing is available. This is followed by a low activity level when the external funding stops. For a system that is expected to provide a *continuing* service to the global community, such a "boom and bust" model is not recommended.

The constraints of many poor countries may prohibit their provision of the necessary funding and consequently their participation in any globally sustained efforts of disease surveillance for preventing and controlling emerging zoonotic diseases. This situation would result in the typical "weakest link" problem, whereby a country with poor capacity and no resources would jeopardize the efforts of all others. *The committee concludes that the global tasks as described in Table 6-2 clearly require new, more innovative ways of fully or partially funding their costs to replace the current boom and bust model.* The committee reviewed several options for funding this global public good:

- Long-term twinning arrangements between human and animal health institutes of high-income and resource-poor countries, funded by specific budget lines in those high-income countries;
- Long-term commitments of governments to fund WHO/IHR 2005 and FAO/OIE in supporting global disease surveillance systems;
- Establishment of special endowments through nonconventional donors;
- Imposition of a levy on internationally traded meat; and
- Other public–private partnerships.

The fund, regardless of its sources, could provide the full costs of global disease surveillance (infrastructure and recurrent costs) or only the recurrent costs because international funding generally is available for the hardware investments.

Long-Term Twinning Arrangements

Under this option, governments from high-income countries, through their human and animal health institutions, would commit to permanent support for their counterpart institutions in resource-poor countries. This would require a long-term engagement of the high-income country, and, in line with its budgetary procedures, be an integral part of the budget of those agencies in the high-income country, and not be seen as a time-bound contract. For the high-income country, this would have the advantage that it could select the group of countries it likes to sponsor, for example, based on the risk that an outbreak of a zoonotic disease poses to its own human or animal health. Another positive aspect would be the long-term capacity-building opportunities for the recipient country. This model has already been used, for example, by Australia in several Sanitary and Phytosanitary Measures Agreement-related aspects in Southeast Asia. Drawbacks of such an approach would be a certain fragmentation of approaches because the agencies of the high-income countries would tend to establish their own standards and procedures. Another disadvantage would be that the dependence on a national budget, even of a high-income country, would still introduce a degree of fickleness. An additional issue to address in the aspect of twinning support is ensuring that reagents and samples can be exchanged easily. Documentation required for shipment and receipt of biological materials has increased dramatically since 9/11 and can prolong diagnosis time.

Long-Term Commitments of Governments to Fund WHO/IHR and FAO/OIE

This would imply that high- and middle-income countries directly commit to permanently support a funding mechanism established by international agencies. This would enable a full implementation of IHR 2005, which is now underfunded, and would enable FAO and OIE to continue their current support to developing countries, when their funding for HPAI H5N1 concludes, which will be 2010 for most funds. Contributions could be channeled in specific funds, such the OIE World Animal Health and Welfare Fund, and the FAO Special Fund for Emergency and Rehabilitation Activities. A key requirement would be that the three international agencies (WHO, FAO, and OIE) and other donor agencies establish close cooperation mechanisms on priority countries and hotspots, on minimum requirements to be funded and other funding criteria, and fully support the principle of an integrated human–animal (and ecosystem) health system. If not, this option could lead to fragmentation, duplication, and an overall inefficient use of resources. As in the case of the twinning arrangements, this option would also depend on the long-term commitment of high-income countries, and whether that commitment is likely to waver.

Establishment of Endowment Funds Through Nonconventional Donors

Nonconventional donors and foundations and a few conventional donors generally have a longer investment and support span than most other conventional donors, and funding could be on a regular basis or toward the establishment of an endowment. The establishment of an adequately resourced endowment would be the most appropriate solution, but, if established, is unlikely to be able to fund the amounts necessary in the near future, and other sources would need to be identified. This would still have some of the disadvantages of the earlier options.

A Levy on Internationally Traded Meat

Although several levy sources would be possible, one option to be considered in more detail would the imposition of a levy on meat trade from middle- and high-income countries. Developed countries are the major exporters of meat (see Table 6-4). A meat levy to control emerging pandemics would be directly related to the global public good of control of zoonotic diseases. As in the twinning arrangement, it would provide an incentive to the middle- and high-income countries because a strong global disease surveillance and early alert system would help to protect their food-animal production sector against the introduction of other contagious animal diseases, such as foot-and-mouth disease. Under the same argument of enlightened self-interest, the levy would have to be collected by the exporting country, and it would be relatively easy to collect by customs offices. Importing countries have more limited incentives to protect their own food-animal population. The meat levy would not apply to the low-income

TABLE 6-4 Value of Meat Export by Country Income Category Group

Country Group	Bovine Meat	Other Meat and Edible Offal	Live Animals
	(in billions of US$, 2005 data)		
Middle- to high-income countries	20.1	36.5	11.6
Resource-poor countries	0.8	1	0.5
Total	20.9	37.5	12.1

SOURCE: Adapted from ITC (2009).

countries it is raising funds to assist. By exonerating resource-poor[3] countries from this levy, this would allow their emerging food-animal production sectors to grow and develop rather than be penalized.

The meat levy would not include live animals because of the difficulty in collecting such a fee and the potential unintended consequence of sending the live-animal trade "underground," leading to more illegal trade. Disadvantages of such a levy would be the political sensitivity to the imposition of fees (although small as shown below) in an era of globalization and promotion of free trade, its legality under current World Trade Organization rules, and other impacts, such as the potential for increased illegal importation of bushmeat. Strengthened disease surveillance systems, which would be the intended result of such a funding mechanism, would also help to reduce such illegal trade of bushmeat. However, more study and consultation with policymakers and other stakeholders is required to fully assess the political and technical implications of such a financing mechanism.

This proposed funding mechanism would have the advantage of well-established systems of levy collection, as many commodity groups or governments funding national public goods have used such dedicated levies (Nugent and Knaul, 2006). Examples are the imposition of transaction levies on food-animal sales by producers to fund marketing and research on meat in Australia (MLA, 2009); slaughter levies to fund food-animal production insurance against major disease outbreaks in the Netherlands (World Bank, 2006b); and an export levy to fund agricultural research in coffee in sub-Saharan Africa (UNU/INTECH, 1995). These levies directly benefit the producers. The same would apply to the meat levy—this would reduce the risk of infection of notifiable disease and the resulting ban of an exporting country, and therefore be in direct interest of the exporting country. "Check-off" dollars are used in the United States to fund marketing programs for food-animal products. The UK government also just proposed a new arrangement to fund an independent body for animal health, which can raise a levy from food-animal keepers in England through mandatory insurance to recover the costs of dealing with exotic disease outbreaks (Defra, 2009).

While these dedicated levies are generally considered an effective way of securing sustainable funding for a public good, less is known about a levy for a global public good. The proposed levy on air traffic to be managed by the Global Environment Facility (GEF, 2009) to finance climate change adaptation is one example. A similar levy is another possible source of funding for global disease surveillance, but there are no ongoing opera

[3]Defined as the countries meeting the International Development Association eligibility criteria of $1,095 USD (World Bank, 2009b).

tions in the fields of human and animal health. One example, which approaches a global fund similar to the type the committee has advocated, is proposed in a recent feasibility study on the creation of a Global Emergency Response Fund for Animal Epizootics and Zoonoses prepared for the OIE and World Bank conference, *Global animal health initiative: The way forward*. This fund would provide developing and transition countries with immediate funding to cover the cost of control measures and food-animal owners' compensation costs (Alleweldt et al., 2007). It is still in its initial phase, but has had, until now, limited success in attracting funds.

With an estimated 24 million tons of meat exported from developed countries (OECD, 2009), the incremental costs for the meat levy could vary between $0.03 per kilograms (kg) if all costs were covered (of the US $836 million shortfall) and $0.015 per kg for operating costs only. The funds needed to be generated by these permanent streams would thus be reduced by about 50 percent, from $800 million to $400 million.

Public–Private Partnerships

Opportunities should always be left open to also involve the private sector directly in the funding of global surveillance and response teams. Global food supermarket and restaurant chains have a direct interest in preserving animal and human health, as emerging zoonotic diseases have serious economic repercussions. Typically, until now, they have protected their own herd/flock health, but with past experiences, for example in East Asia, where HPAI H5N1 outbreaks have seriously affected demand for poultry products, they have become more interested in contributing to a wider national or regional surveillance and response system.

THE INSTITUTIONAL ARCHITECTURE

The institutional architecture would have to be in line with the funding option selected. Under the twinning arrangement, the main issue is the harmonization of standards and areas to be covered to achieve compatible data and complete coverage. Ensuring a coordinating role of the technical agencies for these tasks is important. To directly support funding the international agency, coordination among international agencies is even more important. In this case, an overall coordinating body might be needed to ensure an appropriate distribution of funds. A similar framework would be required for the endowment and levy option. In these cases, the management could be entrusted to a global entity, such as a United Nations agency (e.g., the United Nations Central Emergency Response Fund or UNSIC) or a fund managed by an international funding agency. Activities could be implemented through the technical agencies WHO, FAO, and

OIE and any additional agency, such as in the areas of communication and wildlife management. However, to avoid conflicts of interest, these technical agencies should not be charged with the overall management of such a funding mechanism. Alternatively, not necessarily all funds have to be channeled through a global-level institution or institutions; individual developed countries can decide based on their own geographical and institutional preferences and maintain their own sustainable funds. Coordination could be provided through annual inter-ministerial meetings (Bamako, Beijing, Delhi, Sharm el-Sheikh). This was the model for the HPAI H5N1 campaign after the Beijing conference, and it has worked well. More study is clearly indicated.

REFERENCES

Alleweldt, F., M. Nell, B. Hirsch, J. Syroka, W. Dick, S. Kara, M. Achten, K. Schubert, N. T. K. Cuc, C. Caspari, and M. Christodoulou. 2007. *Prevention and control of animal diseases worldwide: Feasibility study—A global fund for emergency response in developing countries.* Final Report, Part II. Berlin, Germany: Civic Consulting-Agra CEAS Consulting.

Defra (UK Department for Environment, Food and Rural Affairs). 2009. *Consultation on a new independent body for animal health: A modern governance and funding structure for tackling animal diseases.* London, UK: Defra.

FAO, OIE, WHO, UNSIC, UNICEF (Food and Agriculture Organization of the United Nations, World Organization for Animal Health, World Health Organization, United Nations System Influenza Coordinator, United Nations Children's Fund), and World Bank. 2008. *Contributing to one world, one health: A strategic framework for reducing risks of infectious diseases at the animal–human–ecosystems interface.* Consultation document prepared for the Inter-ministerial Meeting on Avian and Pandemic Influenza, Sharm el-Sheikh, Egypt, October 14.

GEF (Global Environment Facility). 2009. *Innovative financing mechanisms for the GEF.* Paper presented at the First Meeting for the Fifth Replenishment of the GEF Trust Fund, Paris, France, March 17–18.

Government of Egypt. 2008. *The Sharm el-Sheikh vision for the future: Universal solidarity, justice and equity.* The 6th International Ministerial Conference on Avian and Pandemic Influenza, Sharm el Sheikh, Egypt, October 25–26.

ITC (International Trade Centre UNCTAD/WTO). 2009. *International trade statistics by product group and country: Exports 2001–2005 dataset.* http://www.intracen.org/tradstat/sitc3-3d/indexpe.htm (accessed March 14, 2009).

MLA (Meat and Livestock Australia). 2009. http://www.mla.com.au/default.htm (accessed June 22, 2009).

Nugent, R., and F. Knaul. 2006. Fiscal policies for health promotion and disease prevention. In *Disease control priorities in developing countries*, 2nd ed, edited by D. T. Jamison, J. G. Breman, A. R. Measham, G. Alleyne, M. Claeson, D. B. Evans, P. Jha, A. Mills, and P. Musgrove. Oxford University Press and The World Bank.

OECD (Organisation for Economic Co-Operation and Development). 2009. Data extracted on June 22, 2009, from OECD.Stat. In *OECD-FAO Agricultural Outlook 2009–2018.* Paris, France: OECD.

UNSIC (UN System Influenza Coordinator) and World Bank. 2008. *Responses to avian influenza and state of pandemic readiness.* Fourth Global Progress Report. New York: United Nations. http://siteresources.worldbank.org/EXTAVIANFLU/Resources/31244401172616490974/Fourth_progress_report_second_printing.pdf (accessed November 5, 2009).

UNU/INTECH (United Nations University Studies in New Technologies and Development). 1995. Ghana's total expenditures on the pursuit of science and technology. In *In pursuit of science and technology in Sub-Saharan Africa: The impact of structural adjustment programmes*, edited by J. L. Enos. Chatham, Kent: Mackays of Chatham PLC. http://www.unu.edu/unupress/unupbooks/uu33pe/uu33pe0b.htm#ghana's%20total%20expenditures%20on%20the%20pursuit%20of%20science%20and%20technology (accessed June 22, 2009).

World Bank. 2005. *East Asia update: Countering global shocks.* November. Washington, DC: The World Bank. http://web.worldbank.org/WBSITE/EXTERNAL/COUNTRIES/EASTASIAPACIFICEXT/EXTEAPHALFYEARLYUPDATE/0,,contentMDK:20708543~menuPK:550232~pagePK:64168445~piPK:64168309~theSitePK:550226,00.html (accessed October 17, 2008).

World Bank. 2006a. *Avian and human influenza: Financing needs and gaps.* Technical note prepared for the International Pledging Conference on Avian and Human Influenza, Beijing, China, January 17–18. Washington, DC: The World Bank.

World Bank. 2006b. *Enhancing control of HPAI in developing countries through compensation.* Washington, DC: The World Bank.

World Bank. 2009a. *AHIF—Trust fund facility.* http://web.worldbank.org/WBSITE/EXTERNAL/TOPICS/EXTAVIANFLU/0,,contentMDK:21355695~isCURL:Y~menuPK:3124611~pagePK:210058~piPK:210062~theSitePK:3124441,00.html (accessed June 24, 2009).

World Bank. 2009b. *International development association: Borrowing countries.* http://go.worldbank.org/83SUQPXD20 (accessed June 9, 2009).

7

Governance Challenges for Zoonotic Disease Surveillance, Reporting, and Response

". . . [I]nnovations in global health governance, rather than merely increasing investments or incremental improvements in the old systems, are needed to meet the deadly crisis of the new age. These innovations will need to come in the realm of ideas, as the prevailing principles and norms that guide global health governance are redefined and reinvented for a comprehensively and instantaneously interconnected, complex world. They will be needed in the realm of institutions, where new rules, decision-making procedures, resources, and participants are required if the expectations and behaviour of the world's countries and citizens are to converge on the reality, rather than just the ideal, of health for all."

—*Andrew F. Cooper, John J. Kirton, and Ted Schrecker*
Governing Global Health
(May 2007)

As previous chapters have demonstrated, there are many challenges in achieving sustainable global capacities for zoonotic disease surveillance and response. One of the more formidable challenges is identifying governance strategies that will result in an effective global, integrated zoonotic disease surveillance and response system. This chapter addresses the challenges in identifying and implementing these strategies: in particular, how societies organize themselves in ways that are effective in preventing, preparing for, and responding to threats to human and animal health. It also discusses some potential options to address these challenges.

Governance tasks arise within each country and through the interactions countries have with one another. The complexity of multiple governance and scientific contexts at a global level is daunting for human and animal health specialists that are unfamiliar with world politics, intergovernmental organizations, and international law, and for policy and legal experts who lack scientific and technical knowledge about zoonotic diseases. Governance challenges can only be effectively met through strong partnerships among the diverse set of experts needed to craft feasible responses to emerging zoonotic diseases.

The drivers underlying the emergence of infectious diseases are in large part ecological, political, economic, and social forces operating at local, national, regional, and global levels. In diverse ways, they physically bring humans and animals closer together. However, these forces also create domestic and diplomatic problems for governments when changes in the disease drivers require new approaches—integrated approaches—across multiple relevant sectors. The integration of scientifically informed strategies to detect, prevent, and control zoonoses within existing governance structures is proving difficult for states, intergovernmental organizations, and nonstate actors.

Several of the committee's recommendations (see Chapter 8) aim to develop integrated capacities for disease surveillance and response that link human and animal health. These capacities operate from the local to the global level, and need the support of political commitment, normative rules and principles, legal frameworks, and material capacities (e.g., human resources, laboratories) that operate at the same levels.

THE RELATIONSHIP BETWEEN HUMAN AND ANIMAL HEALTH CAPABILITIES AND GOVERNANCE

"Governance" refers to the structures, rules, and processes that societies use to organize and exercise political power to identify and achieve objectives. When we examine governance, we want to know *what* political objectives societies pursue, *why* societies select those objectives, and *how* societies attempt to reach those objectives. Governance includes, but is not synonymous with, government. Societies use governance mechanisms that are not part of the government, for example, when using the market to govern economic behavior. Conflating governance and government means that "global governance" would be impossible because no world government exists.

Typology of Governance

Literature on governance often identifies three governance realms: national, international, and global, which are described in Table 7-1.

National Governance

National governance refers to the way in which a country organizes political power within its territory and controls interactions among local, subnational, and central governmental authorities. The allocation of jurisdiction is particularly important for disease surveillance, which is often a state or provincial function rather than the responsibility of the central or federal government. When it is decentralized, achieving harmonized,

TABLE 7-1 A Typology of Governance

Governance Level	Actors Involved	Sources of Rules	Scope of Rule Applicability
National governance	• Government • Nongovernmental entities (e.g., corporations, producer and consumer organizations, medical and veterinary associations) • Individuals (e.g., as voters)	• Constitutions • Legislation • Administrative regulations • Law of subnational units • Common law • Court decisions • "Soft law" norms	Generally limited to the territorial jurisdiction of the state
International governance	• States • IGOs	• Treaties • Customary international law • General principles of law recognized as international law • "Soft law" norms	Rules apply in relations among states either directly or indirectly through IGOs
Global governance	• States • IGOs • MNCs • NGOs • Individuals	• Treaties • Customary international law • General principles of law recognized as international law • "Soft law" norms • Private governance regimes	Rules apply and affect relations among states and the activities and behavior of nonstate actors and individuals

NOTES: IGOs = intergovernmental organizations, MNCs = multinational corporations, NGOs = nongovernmental organizations.
SOURCE: Adapted from Fidler (2002).

coordinated policies can be complex. Privatized human health and veterinary services also change the context in which a country's governance takes place, particularly with respect to managing private economic incentives that may be in competition with the production of public goods. The demands that global disease threats generate make decentralization and privatization in national human and animal health governance even more challenging.

International Governance

The second level of governance is international, typically defined as the regulation of political interactions among countries. Unlike national governance, which is hierarchical, politics among nations is anarchical—meaning that sovereign states interact, but do not recognize any common, supreme authority. It is for this reason that regimes arise: to form the "persistent and connected sets of rules (formal or informal), that prescribe behavioral roles, constrain activity, and shape expectations" of sovereign states (Keohane, 1984, p. 781). Through regimes, such as international law and intergovernmental organizations (IGOs), countries govern their relations. Many are able to negotiate to identify common interests, forgo the hierarchy of power that always exists, and cooperate in international governance.

In human and animal health, international governance is apparent in the functioning of IGOs, especially the World Health Organization (WHO) and the Food and Agriculture Organization (FAO)—both specialized agencies of the United Nations (UN)—and the World Organization for Animal Health (OIE), an independent non-UN organization that is also a reference organization of the World Trade Organization (WTO) (see Box 1-2). International agreements and programs operated by WHO, FAO, and OIE are mechanisms of international governance because they regulate the interactions of states concerning human and animal health, travel by humans, and food-animal trade, among a number of other issues.

Global Governance

The third governance level is global, which refers to efforts by states and nonstate actors to shape the exercise of political power within and among countries. Global governance differs from international governance because it recognizes that nonstate actors are involved in managing and regulating political activities. In other words, sovereign states are not the only governance actors in world politics. For example, nongovernmental organizations (NGOs) that promote human rights might criticize multinational corporations by using international human rights law developed by states and persuade these corporations to adopt codes of conduct to improve safety and health standards for workers. These codes are instruments of global governance arranged through global political activities among nonstate actors that do not emanate from national governments or IGOs. Global governance strategies cut across traditional boundaries developed in national and international governance.

Governance and Hard and Soft Law

These forms of governance produce diverse normative strategies to channel political power and human behavior to work toward identified goals. Legal experts sometimes categorize these strategies as "hard law" or "soft law" approaches. Typically, hard law is formal, binding law (e.g., a statute in national governance or a treaty in international governance). Soft law, which political scientists include in their definition of "institutions," consists of nonbinding rules, principles, guidelines, and norms that guide individual behavior (e.g., moral precepts), corporate entities (e.g., codes of conduct), and sovereign states (e.g., political understandings among countries concerning shared problems). Each governance realm contains hard law and soft law. National governments enforce hard law rules through criminal or civil sanctions. However, most hard law in international governance—the rules of international law—has no centralized enforcement. Consequently, whatever enforcement takes place happens in an ad hoc, decentralized manner, such as when a country takes countermeasures against another country for violating an international law. Soft law rules are sometimes more effective than hard law, even when the threat of enforcement is nonexistent. In addition, soft law allows political actors an option for developing collective responses without the high transaction costs of reaching binding agreements. Thus, calling a rule or principle "hard law" or "soft law" does not necessarily indicate which approach is likely to be more effective in any given governance context.

Zoonotic Disease Surveillance, Response Capacities, and Governance

Conducting zoonotic disease surveillance to detect threats to human and animal health that cross political borders and to intervene against those threats requires governance strategies and mechanisms that encourage countries to share information and collaborate on responses. For disease surveillance and response systems to be effective, countries must implement international and global governance approaches within their territories, from the local to the national level, and beyond to the international community, through both formal legal rules and informal modes of collaboration. What this governance enterprise contains, why political actors pursue and sustain the effort, and how strategies, mechanisms, and capabilities are built and sustained are critical questions to address, for the answers will determine whether or not a system can work effectively.

Trends in Human and Animal Health and in Governance: Convergence Amidst Fragmentation

The drivers of emerging zoonotic diseases are bringing human and animal health ever closer together. This convergence, and the increasingly diverse circumstances in which humans and animals interact across the world, creates the need for more centralized, harmonized, and rationalized governance across the multiple sectors within and among countries. One expression of this new understanding and appreciation is the greater collaboration among WHO, FAO, and OIE—which is to be applauded and supported. There remains, however, considerable uncertainty about the mechanisms for this collaboration, the definition of responsibility among the three agencies, and their connections to WTO. Convergence in and of itself does not automatically lead to collaboration and integration at national, international, and global governance levels—where unfortunately, fragmentation of sectors remains the dominant theme.

Decentralization and privatization in the human and animal health sectors are creating additional difficulties for national governance struggling with rising globalized disease threats at the local, subnational, and central government levels. This was highlighted when the Indonesian government refused to provide isolates of highly pathogenic avian influenza (HPAI) H5N1 to WHO without guarantees of a return to the nation of resources, whether it was a portion of the intellectual property or the particular products developed using the strains (Padmawati and Nichter, 2008). This experience raised new questions, such as: "How does one implement a high-priority global health program in a decentralized political state? How does one encourage decentralized problem solving while supporting national disease control programs that demand centralized activities and infrastructure?" (Padmawati and Nichter, 2008, p. 46).

The situation has become even more confused, as the traditional position of WHO as the sole intergovernmental authority for human health has been undermined by the growing involvement of other bodies (e.g., the World Bank and WTO) and the creation of new initiatives (e.g., the Global Fund to Fight AIDS, Tuberculosis, and Malaria). In addition, major powers in the international system, particularly the United States, have taken more active and often unilateral roles in global health, substantially affecting international health governance. Examples include the U.S. Global Leadership Against HIV/AIDS, Tuberculosis, and Malaria Reauthorization Act of 2008 (the reauthorization of the President's Emergency Plan for AIDS Relief [PEPFAR]), the U.S. President's Malaria Initiative, the U.S. Agency for International Development's initiative on Avian and Pandemic Influenza Preparedness and Response, the Group of 8 global health initiatives, and most recently, the Obama administration's $63 billion Global Health Initiative, which expands support beyond PEPFAR to address maternal and child health, family planning, and neglected tropical diseases (The White House, 2009).

The landscape of global health governance has also changed with the explosion of new actors and programs, particularly involving nonstate actors (Garrett, 2007). This explosion includes the activities of philanthropic NGOs (e.g., Bill & Melinda Gates Foundation, Google.org), advocacy and service-providing groups (e.g., Médecins sans Frontières,[1] Program for Monitoring Emerging Diseases or ProMED-mail), and private, multinational corporations (e.g., development of corporate-driven food safety standards).

Although the increasing involvement of states, IGOs, and nonstate actors has brought new energy, prominence, and funds to global health, the multiplication of actors and initiatives has led to concerns that global health governance is becoming so fragmented that it is dysfunctional, and in need of new "architecture" to align interests, programs, and funding more effectively (Fidler, 2007). The complexity of zoonotic diseases, with the multiple sectors and players involved, exacerbates this fragmentation even more. The extent to which international and global health governance are susceptible to centralization, harmonization, and rationalization remains an open question, especially in light of the world's transition to a multipolar system (in which multiple states have relatively equal military, cultural, or economic influence)—a context in which multilateralism might become more difficult (National Intelligence Council, 2008). *The committee concludes that the convergence apparent in the drivers of emerging zoonotic diseases means that each level of governance has become increasingly important and interdependent. At the same time the fragmentation in governance, at these same levels, has complicated the development and implementation of strategies to advance governance within each level and to integrate initiatives across all three levels.*

GOVERNANCE PROBLEMS FACING INTEGRATED SURVEILLANCE AND RESPONSE SYSTEMS FOR EMERGING ZOONOTIC DISEASES

Traditional Governance "Silos" for Human and Animal Health

Historically, given their different missions and principal concerns, the approach taken by WHO to develop and adopt the International Health Regulations of 1969 (IHR 1969), which focused on reducing human morbidity and mortality, did not affect how OIE, attentive to reducing the impact of animal diseases on trade and economics, functioned in respect to animal diseases and vice versa. By 1969, experts at WHO recognized that the four legacy diseases subject to IHR 1969—cholera, plague, yellow fever, and smallpox—did not reflect the range of microbial threats at that time (Dorolle, 1969). Given the weaknesses of IHR 1969, the best available governance device to address human and animal health threats was international trade law, which allowed countries to restrict trade when animals

[1] Also known as Doctors Without Borders.

or animal products posed a threat to human health (e.g., food safety) or animal health (e.g., contagious diseases threatening food-animal production in the importing country).

Historical obligations to notify WHO or OIE of diseases reflected concerns about the potential of infectious agents to spread through international commerce, causing adverse effects on trade. In fact, the primary purpose of international trade law was not primarily to protect human or animal health but rather to liberalize trade to generate economic growth and development and avoid trade sanctions. It is not surprising that traditional approaches to disease notification by OIE and WHO were neither closely intertwined nor crafted to be sensitive to the drivers of emerging zoonotic infectious diseases. As a consequence, they had little in the way of enforcement provisions. Much of the governance innovation taking place today with respect to human and animal health and emerging zoonotic diseases attempts to break out of traditional patterns and ways of thinking about disease surveillance, prevention, and control and their subsequent effect on human and animal health at the local, national, and international levels, and to provide greater imperatives to action. The International Health Regulations 2005 (IHR 2005) is a binding legal instrument that obligates the 194 countries that have signed the treaty to improve their surveillance and response capacity, and to report promptly to WHO any disease outbreaks representing an international public health emergency. The purpose is explicitly dual, to protect public health and to limit interference in international traffic and trade. Under IHR 2005 outbreaks of smallpox, wildtype polio, new strains of influenza virus and severe acute respiratory syndrome (SARS) are public health emergencies of international concern.

Regime Contributions to Fragmentation and Weakness Concerning Governance of Drivers of Zoonoses

The fragmentation and weaknesses of regimes involving the drivers of zoonoses limit the ability of national governments, individually or through collective action with IGOs and nonstate actors, to prevent zoonotic threats. For example, the legal regime for international trade is also separate from environmental protection law and unconnected to issues regarding population growth. *The committee concludes that, given the prevailing fragmented and weak governance, the goal of an integrated human and animal health approach is critically linked to efforts to develop and sustain governance mechanisms to promote integrated disease surveillance and response capabilities in countries and is dependent on universal implementation of IHR 2005.*

International Environmental Law: Governance and Environmental Degradation

Chapter 3 discussed examples of environmental degradation driving the emergence and spread of zoonoses—deforestation, encroachment on wildlife habitat, and climate change. Each is complex and has problematic governance regimes. Environmental governance at the international level, for example, does not have a central mechanism as international trade law has in WTO. International environmental law and other forms of global environmental governance have been largely ineffective in affecting the drivers of climate change, which, in turn, could exacerbate the impact of environmental degradation on the emergence of zoonotic diseases and become subject to the governance principles for human and animal health.

Part of the challenge, particularly at the international and global levels, is that environmental problems exhibit different features that require custom policy solutions for each setting, and involve huge economic considerations. Although deforestation and climate change are linked ecologically, they are different problems politically, the former being national and internal, while the latter is external and involves all nations. In part, this explains why governance activities regarding the environment are not well connected or integrated for the purpose of protecting human and animal health.

Deforestation and Encroachment on Wildlife Habitat

Deforestation and other activities related to exploitation of natural resources also facilitate the encroachment of humans on wildlife habitat and in this manner can promote the emergence and spread of zoonotic diseases. Under international law, the state in which the forest or the wildlife habitat is located has sovereignty over such resources, and has the right to develop those resources according to its strategies for economic and social development. Without an agreement, other states potentially affected by the exploitation of that resource have no grounds to intervene in the domestic affairs of the "exploiting" state. Although countries have concluded treaties on the type of problems described above (e.g., desertification, protection of endangered animal species), deforestation has only been subject to soft law instruments, such as the Statement of Forest Principles (United Nations, 1992). In this instance, compliance with soft law has not proved compelling for countries interested in exploiting their forest resources. The same dynamic generally holds for efforts to protect wildlife habitat from human encroachment. Typically, countries potentially affected by the exploitation of environmental resources in another nation are unwilling to provide sufficient resources to offset the costs of not exploiting these resources.

Climate Change

Climate change is a fundamentally different issue than deforestation and encroachment on wildlife habitat. A single country can reduce deforestation or encroachment on wildlife habitat within its territory by pursuing and enforcing conservation policies, but any individual attempt to address climate change is bound to be futile because the threat arises from the actions of many countries. Pollution of the atmosphere, a resource subject to the sovereignty or jurisdiction of no state or group of states, also cannot be addressed by one nation. As the travails of the Kyoto Protocol and the impasse concerning post-Kyoto strategies demonstrate, hard and soft law instruments have not persuaded countries to make common sacrifices to reduce threats related to climate change. The same situation exists with respect to efforts to help developing countries handle the effects of climate change on their economies and societies.

The committee concludes that the weakness of regimes concerning environmental degradation, deforestation, and climate change suggests that pursuing governance strategies in these realms is unlikely to contribute much in the near term to preventing and protecting against emerging zoonotic diseases. This reality heightens the importance of having disease surveillance and response capacities in order to detect and intervene in a timely manner against outbreaks.

Other Drivers

Food Security

Chapter 3 also identified food security as a driver contributing to zoonotic diseases. The global food crisis in 2008 highlighted how vulnerable food security is even though, under human rights law, governments are required to take necessary actions to ensure that every individual has access to adequate food (High-Level Task Force on the Global Food Security Crisis, 2008). The impact of food insecurity, of course, is not just its legal implications, or even the specter of starving people. Malnutrition is a determinant of the resistance of an individual to infectious diseases, and starvation forces people to search for alternative food sources, some of which (e.g., bushmeat) increase the risk of emerging zoonotic diseases.

Although counterintuitive, efforts to achieve food security at affordable costs, and thus fulfill the right to health, could lead governments to increase food production at the expense of food safety, animal hygiene, and environmental protection. Ensuring food security at a reasonable cost, in the context of population growth, requires increasing food resources through increased production, harvesting of wildlife, or imports—each of which has implications for emergence and movement of new zoonotic diseases. Without economies of scale in agricultural production, countries are vulnerable to becoming more dependent on imported food and at risk of volatile world food prices, as happened in 2008.

Population Growth and Population Movements

Although states, IGOs, and NGOs have long attempted to craft regimes to control population growth and movement, these efforts have largely failed. The metrics of global population increases, especially in the developing world, illustrate the failure of schemes to limit world population growth. Internal migration from rural to urban areas or populating previously uninhabited areas to exploit natural resources are predominantly, if not exclusively, matters for national governance because of the international legal principles of sovereignty and nonintervention in domestic affairs of other states. Cross-border movements of populations affect international and global governance, but existing principles and mechanisms merely attempt to manage the consequences of population growth (e.g., poverty) rather than addressing the underlying causes of migration.

GOVERNANCE INNOVATIONS SUPPORTING INTEGRATED DISEASE SURVEILLANCE AND RESPONSE IN HUMAN AND ANIMAL HEALTH

Four objectives for foreign policy and diplomacy were described as conceptually important innovations for surveillance and response capacities in Chapter 1. What has not been seen until recently are disease surveillance and response capacities relevant to all four of these objectives of foreign policy simultaneously. Governments, IGOs, and NGOs now talk of threats—such as SARS, HPAI H5N1, pandemic influenza due to any influenza virus (e.g., the new triple reassortment influenza A(H1N1) 2009 virus strain sweeping the world), HIV/AIDS, and bio- and agro-terrorism—to national and international security; national economic welfare and power; development objectives (e.g., as threats to the health-related Millennium Development Goals); and human rights and human dignity. The factors that motivate political interest in these realms connect to objectives that governments pursue in their foreign policies and diplomatic agendas. Without these changes, the potential of convincing governments to expend political, economic, and diplomatic capital to improve these capacities for human and animal health is limited.

Strategic Innovations: Priorities for Emerging Zoonotic Disease Surveillance and Response Systems

A second innovation in the past decade has involved a strategic re-thinking of what government disease surveillance and response capacities should be. The traditional approach has been to focus on reporting specific diseases to IGOs for disease surveillance purposes, without requiring countries to improve their abilities to respond to disease events. For example, IHR 1969 had rules concerning human and animal health capabilities at ports of entry and exit, but did not reach beyond those contexts. Although IHR 1969 tried to ensure that health measures restricting trade to prevent the spread of the listed diseases had a scientific basis, these provisions did not obligate state parties to develop response capabilities.

Recent strategic thinking about what human and animal disease surveillance and response systems ought to monitor and address, through the application of hard and soft law, has pushed WHO and OIE to attempt to broaden national and global disease surveillance systems to capture more than a limited number of diseases. Disease surveillance experts have also realized that a more direct connection between human and animal disease surveillance capabilities would provide a more comprehensive picture of disease trends of known or unknown potential threats. In adopting IHR 2005, which requires notification of some specific diseases (e.g., SARS, novel influenza virus subtypes) and any event that is deemed to constitute a public health emergency of international concern, WHO member states have agreed that the international surveillance system should be able to address known and unknown diseases, including emerging zoonotic human diseases.

IHR 2005 may help close the governance gap to address human and animal health threats. According to Katz (2009) recent evidence of the unfolding accounts of the detection of and response to pandemic (H1N1) 2009 demonstrated that effective multilateral plans and agreements developed in recent years in response to provisions of IHR 2005 contributed to successful and timely communication, determination of an epidemic of international public health concern, and disease mitigation as the agreement had been designed to achieve. This success also exemplified the need for "sound international health agreements" and served as a clarion "call to action for all nations to implement these agreements to the best of their abilities" (Katz, 2009, p. 1).

OIE members have amended the OIE Terrestrial Animal Health Code to require notifications of not only listed diseases or infections, but also any "emerging disease with significant morbidity or mortality, or zoonotic potential" (OIE Terrestrial Animal Health Code, Art. 1.1.3, §1e). The OIE Terrestrial Animal Health Code defines "emerging disease" as "a new infection resulting from the evolution or change of an existing pathogenic agent, a known infection spreading to a new geographic area or population, or a previously unrecognized pathogenic agent or disease diagnosed for the first time and which has a significant impact on animal or public health" (OIE Terrestrial Animal Health Code, glossary, p. 4).

Many barriers to transparency and transmission of information to effectively address disease threats require solutions. WHO and OIE, together with FAO, are establishing new collaborative disease surveillance strategies. The Global Early Warning System (GLEWS) is a major step for the three organizations to combine information from their respective individual efforts into a broader, more robust capability to identify emerging and reemerging zoonoses more rapidly. GLEWS is a hybrid governance mechanism in that it combines disease surveillance supported by hard law (e.g., IHR 2005, OIE Terrestrial Animal Health Code) and soft law.

IHR 2005 also exemplifies strategic rethinking for response capabilities. For the first time in international law on infectious diseases, it contains requirements for state parties to develop and maintain minimum response capabilities in addition to disease surveillance activities (IHR 2005, Articles 5 [disease surveillance] and 13 [response] and Annex 1 [core capacities]). WHO has obligations to provide assistance to state parties to IHR 2005 on request (e.g., IHR 2005, Articles 10.3, 13.6, and 44.2) to address needs from the local to the national levels, providing a comprehensive foundation for improved disease surveillance and response capabilities within countries. Despite these obligations, progress to achieve the "minimum" capacities has been spotty and slow. Without their achievement, the global effort will fall, necessarily, short of the desired goals. This is not so much a limitation of intent, but rather of resources to build the necessary capacity. OIE's new tool for the Evaluation of Performance of Veterinary Services (PVS tool) is intended to assist OIE members in assessing the quality of their national veterinary capabilities, including the ability to respond to animal disease threats and zoonoses. Use of the PVS tool is voluntary, in contrast to IHR 2005. Nevertheless, the PVS tool represents an OIE initiative that captures new thinking about what veterinary services can do to improve animal health and is a starting point to requiring systematic assessment of the veterinary services of countries.

Operational Innovations: How Can Implementation of Disease Surveillance and Response Efforts Be Strengthened?

Three operational innovations have emerged to change the landscape of disease surveillance and response efforts nationally and globally: the exploitation of information technologies developed for other purposes, the involvement of nonstate actors, and the enhancement of IGO authority in disease surveillance and response governance.

Exploitation of Information Technologies Developed for Other Purposes

The emergence and global spread of new information technologies, especially the Internet, electronic mail, and mobile devices (e.g., cell phones), has transformed disease surveillance in human and animal health. These technologies did not develop to service health needs, but leaders in both health areas have moved to exploit them to increase the speed and flow of epidemiological information to permit earlier detection and verification of diseases and facilitate more rapid interventions to control identified threats.

The capabilities created by new technologies have supported a proliferation of early-warning and disease surveillance networks, described more fully in Chapter 4, including "networks of networks," such as WHO's Global Outbreak Alert and Response Network. OIE's World Animal Health Information System (WAHIS) provides members with a faster, more reliable way to submit disease notifications and other information. Using the World Animal Health Information Database Interface, WAHIS allows OIE to organize and make available a wider range of information on animal diseases, except in wildlife and companion animals. Information technology has the potential to help overcome the "silo" effect by enabling the collection and sharing of information outside traditional approaches and to increase the effectiveness of early-warning and disease surveillance activities for human and animal health.

As discussed in Chapter 5, the exploitation of information technologies has had another governance impact—changing the calculations of how states handle information about outbreaks. The failure of states to report outbreaks undermines international governance targeted at promoting and protecting human and animal health and trade. Given the potential imposition of harmful trade sanctions, states often have not reported outbreaks in order to avoid the adverse economic consequences that often followed. Until the global governance can address economic losses associated with transparency and rapid reporting, nations reporting outbreaks promptly, as they are bound to do, are at risk of trade and tourism losses, as Mexico experienced in its rapid sharing of information and samples after the influenza A(H1N1) outbreak in April 2009.

Today's global networking ability has reduced, but not yet eliminated, a state's ability to hide or ignore outbreaks. Both WHO and OIE search the Internet and other nongovernmental sources for news or indications of disease events and seek verification from countries about potential events. This reality has changed state incentives regarding transparency about disease problems and engagement in international cooperation—precisely the directions in which human and animal health experts want governments to follow.

Information technologies have had less impact on response capabilities because response interventions involve more than collecting and sharing information. Although information on the outbreak of HPAI H5N1 in Southeast Asia was known by January 2004, it took WHO, FAO, and OIE until November 2005 to develop a coordinated or joint strategy against this threat. In contrast, the manner in which surveillance, information technologies, and response planning for SARS was coordinated, including the day-to-day public reporting of the progress of the outbreak and the pandemic status updates by WHO, demonstrates how health officials can use new technologies to improve responses to disease threats.

Involvement of Nonstate Actors

A second operational innovation has been the involvement of nonstate actors in governance functions for human and animal health, especially early warning and disease surveillance activities. As described in Chapter 4, nonprofit entities such as ProMED-mail, pioneered the use of information technologies and networks of stakeholders, and transformed early warning and disease surveillance strategies. Bigger and better funded NGOs are following these pioneers and taking nonstate actor involvement to new levels of activity and importance in governance (see Boxes 8-1 and 8-2).

In human health, WHO has integrated nonstate actors into its governance system. In developing IHR 2005, WHO acknowledged that its duty to rely only on government-provided information was a handicap to an effective regime. The 2001 World Health Assembly first authorized WHO to use nongovernmental information (WHO, 2001) to gather and assess information generated by nongovernmental sources. Eventually WHO members raised this authority to the level of international law in IHR 2005, Article 9. Statistics show the importance of WHO access to nongovernmental sources of information. Of the 1,704 substantiated events tracked by WHO from January 1, 2001, until December 31, 2008, 54.8 percent were initially reported by nongovernmental sources (WHO, 2009; see Table 7-2).

IHR 2005 also permits WHO to verify information it receives from governments, giving WHO leverage to approach governments with "unofficial" information to determine whether reports are rumors or true hazards, and, if they are hazards, how to address them (IHR 2005, Art. 10). With the majority of initial event reports to WHO coming from nongovernmental sources, WHO has many opportunities to seek official verification of information. Less than 9 percent of all disease events tracked by WHO from January 1, 2001, to December 31, 2008 (n = 2,503) were determined

TABLE 7-2 Substantiated Events by Initial Source of Official Information

Year	Substantiated	Initial Information from Official Source	% Official
2001	133	68	51.1
2002	152	61	40.1
2003	249	95	38.2
2004	282	133	47.2
2005	220	108	49.1
2006	198	88	44.4
2007	217	90	41.5
2008	253	127	50.2
Grand total	1,704	770	45.2

SOURCE: WHO (2009).

TABLE 7-3 Events and Final Designation and Year of Reporting

Year	Discarded	No Outbreak	Substantiated	Unverifiable	Grand Total
2001	7	12	133	17	169
2002	7	34	152	38	231
2003	15	173	249	41	478
2004	15	36	282	18	351
2005	21	29	220	39	309
2006	14	35	198	35	282
2007	30	61	217	11	319
2008	35	52	253	24	364
Grand total	144	432	1704	223	2,503
% of total	5.8%	17.3%	68.1%	8.9%	

SOURCE: WHO (2009).

to be "unverifiable," meaning that WHO received no information from responsible national authorities and was unable to assess the events properly (WHO, 2009; see Table 7-3).

Looked at another way, the 91 percent response rate demonstrates that the monopoly states once possessed on what information WHO could access and use has been broken. As one WHO official stated publicly, "Changes to the International Health Regulations in terms of being able to respond to rumours, as opposed in the past to official notifications, have made a huge difference. We are now able to go to a country . . . to ask specifically what is going on, and that country realises that the world knows a particular country has a problem" (House of Lords, 2008, p. 45). IHR 2005's inclusion of nongovernmental sources of information and WHO's verification authority represents an excellent example of global governance for human health.

The circumstances for animal health are different. OIE has experienced greater difficulty in getting its members to comply with their legal duties to notify OIE of specified animal disease incidents. As an OIE official stated, "There are too many instances where member countries haven't fulfilled their obligations and some serious disease events have not been reported to the OIE, or reporting has been very slow. The situation results in the very real risks of diseases being spread and loss of credibility of countries involved" (O'Neill, 2004, p. 1). In recognition of this in May 2009 through Resolution 17, OIE member states reminded each other of their obligation to make information about relevant animal diseases available to OIE. Since that obligation is already specified in the Terrestrial Animal Health Code and the Aquatic Animal Health Code, concern remains for how, and by whom, this new resolution can be enforced given the limited staff at OIE and its limited presence in the member countries.

OIE has also introduced nonstate actors in their governance with new communication technologies. The International Committee—OIE's highest policymaking organ—authorized the Central Bureau in May 2001 "to question any delegate of a member country regarding animal health incidents reported in the media (newspapers, scientific journals, ProMED-mail, etc.)" (Jebara and Shimshony, 2006, pp. 435–436). Using this authority, OIE collects and analyzes information from nongovernmental sources, as illustrated by OIE's participation in networks such as GLEWS that receive and assess unofficial information for early warning and disease surveillance purposes. OIE then seeks verification of unofficial information from OIE members. The number of requests has been increasing steadily since 2002, but OIE members still fail to respond to requests more than 25 percent of the time. In fact, in more than 50 percent of the cases of verification requests, OIE members had not provided official notifications until after a verification request (Table 7-4).

TABLE 7-4 OIE Verification Requests and Responses to Them

Year	Number of OIE Verification Requests	Replies (% of Requests)	Nonresponses (% of Requests)	Official Notifications (% of Requests)	Invalidated Nonofficial Information
2002	32	18 (56%)	14 (43%)	18 (56%)	0%
2003	29	24 (79.2%)	5 (20.8%)	14 (48.27%)	30.93%
2004	85	67 (78.8%)	18 (21.2%)	39 (48.75%)	30.05%
2005	97	74 (76.28%)	23 (23.71%)	36 (37.11%)	39.17%
2006	113	80 (70.79%)	33 (29.2%)	66 (58.4%)	12.38%
2007	140	103 (73.57%)	37 (26.42%)	71 (50.71%)	31.06%

SOURCE: Thiermann (2008).

Enhancement of Intergovernmental Authority

IHR 2005 also authorizes the WHO Director-General to declare a public health emergency of international concern, even over the opposition of the affected country (IHR 2005, Art. 12). If the Director-General makes such a declaration, then his or her temporary, albeit nonbinding recommendations, must be issued to state parties on how best to address the emergency (IHR 2005, Art. 15). With these authorities, WHO can exercise power and play a leading role in responding to and shaping country-level responses to declared public health emergencies of international concern.

The OIE regime does not grant similar powers to the OIE Director-General, which may be a major reason for the difference in verification of reporting rates. However, all OIE member states are also member states of WHO, except for Liechtenstein, New Caledonia, and Taiwan. Except for these three entities, all OIE member states under IHR 2005 have already accepted the binding legal obligation to respond to WHO requests for verification of information that WHO receives from nongovernmental sources about disease events in their territories. With this precedent, creating the same binding requirements on reporting to OIE about animal diseases in general and zoonotic threats in particular should be possible. However, the authorities granted to OIE by member states have several shortcomings. In particular:

1. OIE members states do not have a binding obligation under international law to respond to OIE requests for verification of information received from nongovernmental sources of information;
2. OIE does not have the policy or legal authority to publicly disseminate information received from nongovernmental sources without confirmation by the affected OIE member state;
3. OIE does not have the policy or legal authority to declare an animal health emergency of international concern and to issue recommendations about how OIE member states should respond to such emergencies; and
4. OIE members do not have legally binding obligations to develop and maintain minimum core disease surveillance and response capabilities for risks to animal health, including zoonotic diseases.

Substantive Harmonization: The Same Rules for Human and Animal Health Governance Systems

One potential goal is to ensure that human and animal health governance systems operate under a harmonized set of substantive rules. This would require revising WHO or OIE rules (or both) and obtaining agreement by FAO so that disease surveillance and response in human and animal health contexts operate seamlessly, eliminating gaps that might undermine governance efforts. A high-profile substantive harmonization proposal has in fact been recently made by a British House of Lords Select Committee (HLSC) on Intergovernmental Organisations (House of Lords, 2008). The HLSC recommended that the United Kingdom (UK) urgently pursue through relevant IGOs "the creation of an event-reporting system for animal diseases along the same lines as the new IHRs relating to human health" (House of Lords, 2008, p. 46). The committee explored this proposal to assess its persuasiveness.

The House of Lords Committee's Recommendation

In comparing the regimes governing human and animal health, the HLSC observed that (1) the two areas "operate separately rather than in an integrated manner"; (2) there is a lack of local animal health capacity (e.g., veterinary services, human and material resources) that impairs animal health surveillance and response efforts; and (3) OIE's regime does not contain rules that have made IHR 2005 such an innovation in global health governance (House of Lords, 2008, pp. 44–45).

The HLSC concluded that "given that some three quarters of emerging infections in humans originate from animals, this asymmetry between the new IHRs . . . and the regulations governing the declaration of diseases in animals is worrying" and that "the present disjunction between the management of animal and human diseases are too great for it to be allowed to continue" (House of Lords, 2008, pp. 45–46).

The HLSC was most concerned about OIE's inability to use nongovernmental information and to seek its verification from OIE members, but it did not mention WHO's authority in IHR 2005 to declare human and animal health emergencies of international concern and to issue temporary response recommendations. In the end, the HLSC recommended that the UK government pursue the creation of a disease reporting system for animal diseases in the image of IHR 2005. The approach would create substantive harmonization of the rules in the human and animal health governance realms.

In terms of how to accomplish this recommendation, the HLSC "considered whether the new IHRs should be amended so as to cover explicitly threats to human health from diseases which are detected in animals" (House of Lords, 2008, p. 45). The HLSC report contained differing views on whether amending IHR 2005 in this manner was feasible.

Committee Analysis of the House of Lords Select Committee's Recommendation

IHR 2005 already permits WHO to investigate potential zoonotic threats by authorizing WHO to collect information from any source on disease events of whatever origin that might threaten human and animal health. This strategy captures emerging or reemerging zoonotic disease threats. During the committee's background research, WHO officials described two examples—one involving Rift Valley fever and one concerning Crimean-Congo hemorrhagic fever—in which WHO used IHR 2005 to alert health officials of potential zoonotic threats.

In fact, human health officials promoted the governance approach in IHR 2005 by arguing that it would allow WHO to catch new zoonotic diseases earlier and more often than previously. WHO's response to SARS functioned as a "roll out" of the approach, which was formally adopted in IHR 2005. As a result it requires notifications of any case of SARS or human influenza caused by a new subtype, which reflects concern with zoonotic diseases (IHR 2005, Annex 2). The decision instrument that guides state parties to IHR 2005 in determining whether a disease event may constitute a public health emergency of international concern is also sufficiently robust to catch zoonotic disease emergence or reemergence (IHR 2005, Annex 2).

Given the implicit mandates within IHR 2005 regarding zoonotic diseases, the treaty may not need to be amended to explicitly include zoonotic diseases within its disease surveillance approach. Rather, the more salient concern appears to be fine-tuning disease surveillance to detect animal diseases earlier so experts have more time to prevent diseases from threatening other animals and, potentially, humans. Under this reasoning, the OIE treaties, not IHR 2005, would require amendment.

However, the highest policymaking body in OIE has authorized the organization to collect, analyze, and seek verification of nongovernmental sources of information within WAHIS and as part of its participation in GLEWS. The OIE data in Table 7-4 indicate that when verification of unofficial information was sought between 2002 and 2007, OIE members provided official or unofficial responses to verification requests more often than they failed to respond (providing responses to nearly 74 percent of verification requests made by OIE to its member states). Thus, the HLSC belief that OIE does not or cannot use nongovernmental sources of information is incorrect.

The committee concludes that the performance of OIE member states to provide official notification of disease events and to respond to OIE verification requests is suboptimal for achieving an effective global system of emerging infectious disease surveillance and response. OIE efforts to increase incentives for better compliance and responses need stronger policy or legal measures.

The HLSC report never mentions IHR 2005's obligations on state parties to develop and maintain minimum core disease surveillance capabilities. IHR 2005 recognizes that disease surveillance capacities are fundamental public goods for which national governments are, and always will be, primarily responsible. A strategy for animal diseases comprehensively informed by IHR 2005 would have to include requirements on OIE members to develop and maintain minimum core disease surveillance capabilities—requirements that OIE agreements do not impose. This change would further underscore how radical a strategy of substantive harmonization on the basis of IHR 2005 would be for the OIE governance system.

The HLSC analysis also did not consider whether IHR 2005 offers a template for improving response capacities in animal health. As described above, IHR 2005 moves response governance closer to becoming an integral part of disease surveillance for two reasons: (1) the regulations only empower WHO to declare a public health emergency of international concern and authorize WHO to issue temporary recommendations to guide countries' responses to such emergencies; and (2) it mandates that state parties develop and maintain minimum core response capabilities from the local to the national levels.

OIE itself has not developed strong national animal health response capabilities because it relies on OIE members to develop and maintain them. FAO might possess more response capabilities at the country level, but FAO capabilities are currently inflated because of HPAI H5N1 and might not be sustainable in the long run. Dividing responsibilities between OIE (early warning) and FAO (early response) to animal outbreaks would require amendments and policy changes across two IGOs.

The committee concludes that implications of including minimum core disease surveillance and response capacities for animal health at national levels are enormous. In many countries, and particularly in the poorest ones, the disease surveillance and response capabilities of public and private veterinary services are limited and fragmented. Strengthening these capabilities within and across countries, while integrating them with capacity building for human and animal health, would require major amendments, resources, and attitude changes at the national governance level.

The committee also acknowledged that under a broad-based substantive harmonization strategy, the task of amending OIE treaties to reflect IHR 2005's rules would be an even more significant undertaking than implied by the HLSC analysis. Whether such a transformation of OIE governance would be feasible is difficult to assess, but the committee was aware that the revision process that produced IHR 2005 lasted 10 years (1995–2005), and perhaps would not have been finished by then without the shock of SARS in 2003. Amending the OIE treaties to produce substantive harmonization with IHR 2005 would not be a quick and controversy-free diplomatic endeavor. However, it would be feasible with sufficient pressure and resources.

MOVING TOWARD A GLOBAL, INTEGRATED DISEASE SURVEILLANCE AND RESPONSE SYSTEM: FUTURE GOVERNANCE STRATEGIES

The threat that zoonotic diseases currently pose, and will continue to pose for the foreseeable future, counsels against complacency and in favor of strengthening national, international, and global governance strategies concerning disease surveillance and response capabilities—even though progress has been made in governance contexts in and between human and animal health. The strategies that countries, IGOs, and NGOs should pursue is difficult to determine. The answer must balance what reforms might be feasible with what disease surveillance and response capabilities are required of stakeholders. The committee identified three principal reform options that can be explored: (1) structural centralization, (2) structural coordination, and (3) intensified implementation.

Structural Centralization: One Regime for Integrated Human and Animal Health

Unifying disease surveillance and response efforts for zoonotic threats under a single institution and set of rules would be the most radical reform option. Rather than trying to ensure coordination among WHO, FAO, and OIE, structural centralization would empower one of the existing intergovernmental entities to exercise primary responsibility for disease surveillance and early response. This approach would provide a streamlined architecture for global health governance on zoonoses. If established and supported by all three organizations, it would be the most expedient option.

However, such a structural centralization has several disadvantages. First, WHO, FAO, and OIE are unlikely to support this approach; second, the transaction costs of negotiating a single regime would be significant; and third, the time needed for complicated negotiations among the three organizations would be counterproductive given the pressing needs the zoonotic threat creates. Finally, structural centralization in itself would be insufficient because such centralization does not produce the functional capabilities that are needed for surveillance and response for zoonotic diseases. In addition, streamlined architecture might not produce more effective governance because the new regime would disrupt patterns of collaboration currently developing and could have difficulty with tapping into the major existing bodies of knowledge and experience of WHO, FAO, and OIE.

Structured Coordination: Maintaining Core Competencies of Existing Agencies While Establishing Transparent Coordinating Mechanisms

Under this second option, WHO, FAO, and OIE would maintain their own respective mandates, but a permanent coordination mechanism would be established with the authority and provide a means of bringing the technical agencies together in order to act quickly, develop common standards and, in the case of a potential emerging zoonotic disease outbreak, prepare a joint response strategy. This would make permanent the current arrangement under the United Nations System Influenza Coordinator's (UNSIC's) office, except that it would not be disease specific. The rationale is that it would be based on the very successful intervention of UNSIC, which, the committee believes, has made a major difference in the overall coherence and efficiency of the HPAI campaign. The coordinating mechanism could be entrusted to UNSIC, making its mandate broader to include all zoonotic infectious diseases and making it permanent (it is currently expected to end in December 2010), or entrusted to another high-level UN agency, independent of WHO, FAO, and OIE. The establishment of a permanent coordination mechanism would still carry some transaction costs for the three technical agencies, but these costs would be much less than in the case of the structural centralization integration option. It would also use the available resources of the technical agencies.

Intensified Implementation: Integrated Human and Animal Health Capabilities

A third option, but not mutually exclusive with the second option, would involve intensifying efforts to implement and integrate WHO and FAO/OIE activities that seek to strengthen local, subnational, and national disease surveillance and response capabilities. The inadequacies and weaknesses of human and animal health systems in many countries is recognized to be a serious impediment to effective planning to prevent, protect against, and control zoonotic diseases (Vallat and Mallet, 2006). Rather than focusing on the structural or substantive aspects of international and global governance, reform could concentrate on implementing and coordinating the obligations and initiatives contained in the IHR 2005 and FAO/OIE strategic plans related to national disease surveillance and response capabilities.

In animal health, intensified implementation would require greater commitment, including financial resources, to advance the strategy of improving disease surveillance and response capacities within countries as proposed by FAO and OIE in their joint strategy on *Ensuring Good Governance to Address Emerging and Re-Emerging Animal Disease Threats* (OIE and FAO, 2007). It would also require a further definition of the respective tasks of OIE and FAO, which could entail a clearer focus of OIE on setting standards as well as a commitment from FAO to support its member countries in the implementation of these standards. For human health, IHR 2005 would provide the strategy for intensified implementation, especially helping developing countries comply with their obligations to develop and maintain core disease surveillance and response capabilities.

Nonetheless, the intensified implementation strategy faces daunting challenges. Concerns already exist about whether WHO members will adequately implement IHR 2005. Although the committee considers this implementation essential for an integrated surveillance and response system for zoonotic emerging infectious diseases, and WHO is undertaking implementation activities, there is not yet a clear and adequately funded strategy for achieving IHR 2005 compliance. WHO does not have the necessary internal capacity to fully focus on IHR 2005 implementation challenges. Sufficient, sustainable financing to implement and improve disease surveillance and response capacity in low-resource countries, discussed in Chapter 6, is a particular strategic challenge.

OIE's and FAO's *Ensuring Good Governance* strategy (2007) also confronts implementation problems that arise from, among other things, the scale of the capacity-building task and the lack of sufficient human and financial resources. As noted earlier, OIE has no internal capabilities to task with capacity building within OIE member states, and most of FAO's capabilities are focused on the HPAI H5N1 threat and will rise and fall with the perceived magnitude of the threat. More permanent financing mechanisms, as recommended in Chapter 6, are essential to implement this option.

A strategy of intensified implementation that is executed inadequately could raise other dangers for human and animal health governance. The controversy sparked by Indonesia's decision not to share samples of HPAI H5N1 viruses with WHO for disease surveillance purposes illustrates these dangers. Indonesia questioned the legitimacy of sharing virus samples for global disease surveillance when it, like other developing countries, received little if anything in return. Indonesia has been specifically concerned about its lack of equitable access to influenza vaccines and drugs, which represent an important response against this disease threat. *The committee concludes that the HPAI H5N1 campaign, as discussed in Chapter 2, has shown the strong value and feasibility of enhanced coordination among WHO, FAO, and OIE, as provided by UNSIC. UNSIC's establishment significantly enhanced the coherence and efficiency of the campaign, and its credibility vis-à-vis the donors.*

International Trade Agreements

Supporting Liberalization of Trade in Agricultural Goods and Food Products

Chapter 3 identified trade as a driver influencing zoonotic disease emergence and reemergence. Trade agreements have facilitated the growth of global trade in agricultural and food products. Governance mechanisms that increase trade, such as the General Agreement on Tariffs and Trade (GATT) under WTO, have proved successful.

The current WTO round of trade negotiations—called the Doha Development Round and ostensibly focused on helping developing countries—has been suspended largely because of disagreements on how to liberalize trade in agricultural products and reduce agricultural subsidies, particularly the significant production subsidies that the European Union and the United States provide to their respective domestic agricultural sectors. Despite this setback in multilateral trade talks, global trade in agricultural products is likely to continue to expand for three reasons. First, WTO agreements facilitate the existing levels of agricultural trade and make any retrenchment difficult to accomplish. Second, progress in the Doha Round will have to be based on agreements to liberalize agricultural trade and to reduce the market-distorting impact of agricultural subsidies. Third, regional and bilateral trade agreements are proliferating and often include commitments to liberalize trade in agriculture products. These regional and bilateral agreements might expand trade in agricultural products despite the difficulties presented in the Doha talks.

Rules on Trade-Restricting Measures to Protect Human or Animal Health

Trade agreements have provided one way of handling problems related to human and animal diseases. In 2002, WHO and WTO jointly published "WTO Agreements and Public Health: A Joint Study by the WHO and WTO Secretariat," which describes the increasingly coordinated activities on the technical and policy levels for the organizations with acknowledgement of the common ground between trade and health. There was also acknowledgement that their respective policymakers could benefit from closer cooperation to ensure coherence between their different areas of responsibilities.

The WTO agreements explicitly allow governments to take measures to restrict trade when pursuing national health and other policy objectives in order to protect health. The publication discusses the rationale and benefits of IHR 2005 as well as the risk-based preventive approach in FAO/WHO Codex Alimentarius (Codex) because the WTO Agreement on the Application of Sanitary and Phytosanitary Measures (SPS Agreement) formally recognizes the food safety standards, guidelines, and recommendations established by the Codex Commission. "The link between the standard-setting work of the Codex and the scientific input from the WHO is important in that it lends some dynamics to the trade rules" (WTO and WHO, 2002, p. 143). The report also acknowledges the formal mechanisms and activities used for coordinated communication and mutual participation between and among WTO, WHO, FAO, and OIE—such as WTO's reliance on WHO's scientific expertise to resolve trade disputes arising from health concerns, the mutual observer status and active participation to provide advice on the SPS Agreement, Technical Barriers to Trade Agreement, Trade-Related Aspects of Intellectual Property Rights Agreement, and World Health Assembly meetings; and their mutual participation in regional and national meetings related to capacity building for disease surveillance to detect and control diseases that could pose a threat to health, especially via trade activities.

For example, WTO members can violate GATT if a trade-restricting measure is necessary to protect human, animal, or plant life or health (GATT, Article XX(b)). Under the SPS Agreement, WTO members have the right to restrict trade to protect life or health of humans, animals, or plants under certain conditions (e.g., in the context of food safety for human health) and subject to specific obligations (e.g., the measure must be based on a risk assessment and be supported by adequate scientific evidence).

Concerning food safety, the SPS Agreement provides that WTO members that apply standards established by the Codex Commission are deemed to comply with the SPS Agreement (SPS Agreement, Art. 3.2). The same "safe harbor" applies if WTO members base trade-restricting animal health measures on OIE standards. In this way, the SPS Agreement gives legal significance to Codex and OIE standards that they do not have within WHO, FAO, or OIE. Outside the SPS Agreement, Codex, FAO, and OIE standards are nonbinding recommendations. Pegging compliance with the SPS Agreement on conformity with Codex and OIE standards raises the legal importance of these standards in ways Codex and OIE did not achieve in their own realms.

Thus, GATT and the SPS Agreement increase the legitimacy of trade-restricting health measures when they are harmonized according to international standards. The SPS Agreement allows WTO members to apply standards that are more protective than international standards, as long as the WTO members comply with their other obligations (e.g., conducting a risk assessment, providing sufficient scientific evidence) (SPS Agreement, Art. 3.3). Countries have violated obligations to report human and animal disease outbreaks because they fear trade sanctions or travel restrictions, and their resulting negative economic consequences. Furthermore, countries have often applied trade and travel sanctions in irrational ways, causing unjustified harm to exporting countries. Even though the SPS Agreement requires a risk assessment (Art. 5.1) and a scientific basis for trade-restricting health measures (Art. 2.3), a country experiencing an outbreak can still be damaged by illegitimate trade sanctions and effectively have no recourse. Even the WTO dispute settlement body provides no mechanism for compensating an exporting member for losses caused by unjustified trade-restricting measures because of a perceived health threat (Van den Bossche, 2008).

Problems and Potential: WTO's Recognition of Codex and OIE Standards

Attempts to cover up outbreaks for fear of economic sanctions are increasingly impractical, as discussed earlier in the chapter. WTO members' rights to restrict trade for health purposes at levels more protective than international standards allows developed countries, and even private associations of importers (e.g., GlobalG.A.P.[2]), to impose more stringent requirements for agriculture and food imports. Although higher standards might create incentives to produce higher value products and generate increased employment, these standards impose costs on exporters in developing countries that are increasingly difficult to meet (Bobo, 2007). Compliance with international standards may still not provide increased market access when developed country or private-sector standards become more stringent.

The potential in WTO's recognition of international standards as a "safe harbor" arises because this approach increases the incentives of exporting nations and exporters to upgrade their SPS strategies and capabilities at home and in export sectors. Such upgrades could help the effort to protect against zoonotic diseases by providing incentives to produce food and agricultural products according to the highest internationally accepted standards.

OIE has leveraged the SPS Agreement's use of its standards to work more with its members on improving their ability to meet OIE standards. The WTO Secretariat has increased assistance to developing countries to help their exporting enterprises meet international standards. Without the lure of export markets, efforts to have countries improve their production processes in this manner would be less effective.

---

[2]GlobalG.A.P. (formerly known as EUREPGAP) "is a private sector body that sets voluntary standards for the certification of agricultural products around the globe. The aim is to establish one standard for Good Agricultural Practice (G.A.P.) with different product applications capable of fitting to the whole of agriculture" (GlobalG.A.P., 2009).

WTO Rules and Integrated Surveillance and Response Capacities for Zoonotic Diseases

Controversies with GATT and the SPS Agreement have focused on whether their rules leave WTO members with sufficient policy space to protect human, animal, or plant life or health from import-borne diseases in the context of expanding and intensifying trade in agricultural and food products. Although important, these controversies do not illuminate how countries, multinational corporations, IGOs, and NGOs should strengthen surveillance and response capacities for zoonotic diseases from local to global levels. National and global action against emerging zoonotic threats could be strengthened by engaging in WTO-based activities that seek to balance human and animal trade and health interests. WTO members could operate these WTO rules more effectively *if* national, international, and global surveillance and response capacities for human and animal diseases were integrated and robust. However, achieving that objective is not the function of GATT, the SPS Agreement, any other WTO agreement, or any regional or bilateral trade agreement for that matter.

Other Regulatory or Policy Options to Address Zoonoses

Also discussed in Chapter 3, wildlife trade is too often ignored as a significant driver for zoonotic disease emergence and spread. The cultural food preferences and practices of people, as well as increased interest in exotic pet ownership, reinforce its relevance and often create lucrative incentives for increased trading of wildlife. Thus, legal and illegal wildlife trade activities deserve more concerted attention in disease surveillance, prevention, response, and control. Strengthening governance mechanisms on drivers of zoonotic disease emergence and spread would help reduce threats to human and animal health at all geographic levels. Addressing the issue in existing or new international treaties, or in domestic policies, that may directly address health or wildlife trade, or whose activities may have unintended consequences to them, are options to consider.

CONCLUSION

The current environment to integrate and improve surveillance and response capabilities for diseases of zoonotic origin is fraught with structural problems in the form of governance "silos" for human health and animal health as well as fragmentation and weaknesses in regimes that address the drivers of zoonotic disease emergence and spread. Despite these structural problems, conceptual, strategic, and operational governance innovations have improved disease surveillance and response capabilities nationally and globally. However, the progress enabled by these innovations has not been sufficient to produce the necessary integrated disease surveillance and response capabilities for zoonotic diseases.

The committee's analysis suggested this objective will only be achieved through a set of national, international, and global efforts focused on the threat of zoonotic diseases. In the increasingly complicated, challenging, and dangerous interconnections between human and animal health, we have no single simple intervention to end the threat. The committee believed, however, that stakeholders could craft a set of integrated, coordinated actions and activities that will measurably improve governance of zoonotic disease threats. The beginnings of this effort are discernable, especially in the increasing collaboration the HPAI H5N1 crisis has created among WHO, FAO, OIE, and other UN agencies and organizations, and the impact this collaboration has at national levels of governance. Aspects of this are also apparent in other governance contexts, such as the manner in which WTO recognizes OIE and Codex standards and the ability of IHR 2005's disease surveillance strategy to catch zoonotic disease emergence or reemergence.

Although governance challenges can look foreboding, never before has there been so much policy and diplomatic activity focused on zoonotic disease threats. The opportunity to harness the momentum generated by SARS and HPAI H5N1 to create a more permanent governance structure—that is flexible and robust enough to handle zoonotic disease emergence and spread rapidly, efficiently, and effectively—has never been greater.

REFERENCES

Bobo, J. A. 2007. The role of international agreements in achieving food security: How many lawyers does it take to feed a village? *Vand J Transnat'l L* 40:937–948.

Cooper, A., J. Kirton, and T. Schrecker. 2007. *Governing global health: Challenge, response, innovation*. Aldershot, UK: Ashgate Press.

Dorolle, P. 1969. Old plagues in the jet age: International aspects of present and future control of communicable diseases. *WHO Chron* 23(3):103–111.

Fidler, D. P. 2002. *Global health governance: Overview of the role of international law in protecting and promoting global public health*. Key Issues in Global Health Governance Discussion Paper no. 3.

Fidler, D. P. 2007. Architecture amidst anarchy: Global health's quest for overnance. *Global Health Governance* 1(1):1–17. http://diplomacy.shu.edu/academics/global_health/journal/PDF/Fidler-article.pdf (accessed July 20, 2009).

Garrett, L. 2007. The challenge of global health. *Foreign Aff* (January/February). http://www.foreignaffairs.org/20070101faessay86103/laurie-garrett/the-challenge-of-global-health.html (accessed July 20, 2009).

GlobalG.A.P. 2009. *What is GLOBALGAP?* http://www.globalgap.org/cms/front_content.php?idcat=2 (accessed July 20, 2009).

High-Level Task Force on the Global Food Security Crisis. 2008. *Comprehensive framework for action.* New York: United Nations. http://www.un.org/issues/food/taskforce/Documentation/CFA%20Web.pdf (accessed July 20, 2009).

House of Lords, Select Committee on Intergovernmental Organisations. 2008. *Diseases know no frontiers: How effective are intergovernmental organisations in controlling their spread?* First report of the Session 2007–2008. London, UK: The Stationery Office Limited.

Jebara, K. B., and A. Shimshony. 2006. International monitoring and surveillance of animal diseases using official and unofficial sources. *Vet Ital* 42(4):431–441.

Katz, R. 2009. Use of revised International Health Regulations during influenza A (H1N1) epidemic. *Emerg Infect Dis* 15(8):1165–1170.

Keohane, R. O. 1984. *After hegemony: Cooperation and discord in the world political economy.* Princeton, NJ: Princeton University Press.

National Intelligence Council. 2008. *Global trends 2025: A transformed world.* Washington, DC: National Intelligence Council.

OIE and FAO (World Organization for Animal Health and Food and Agriculture Organization of the United Nations). 2007. *Ensuring good governance to address emerging and re-emerging animal disease threats—Supporting the veterinary services of developing countries to comply with OIE international standards on quality.* Paris, France: OIE. http://www.oie.int/downld/Good_Governance07/Good_vet_governance.pdf (accessed July 20, 2009).

O'Neill, B. 2004. *Disease report and trade responsibilities of OIE member countries.* Speech presented at the 72nd OIE General Session, Paris, France, May 23–28. http://www.oie.int/downld/SG/2004/Speech_oneill.pdf (accessed July 20, 2009).

Padmawati, S., and M. Nichter. 2008. Community responses to avian flu in Central Java, Indonesia. *Anthropology & Medicine* 15(1):31–51.

The White House. 2009. Statement by the President on global health initiative. Office of the Press Secretary, May 5.

Thierman, A. 2008. *International Animal Health Regulations and the World Animal Health Information System (WAHIS).* Presentation, Institute of Medicine Forum on Microbial Threats Workshop on Globalization, Movement of Pathogens (and their hosts) and the Revised International Health Regulations, Washington, DC, December 16–17.

United Nations. 1992. Non-legally binding authoritative statement of principles for a global consensus on the management, conservation, and sustainable development of all types of forests. A/CONF.151/26 (Vol. III). New York: United Nations. http://www.un.org/documents/ga/conf151/aconf15126-3annex3.htm (accessed July 20, 2009).

Vallat, B., and E. Mallet. 2006. Ensuring good governance to address emerging and re-emerging animal disease threats: Supporting the veterinary services of developing countries to meet OIE international standards on quality. *Rev Sci Tech* 25(1):389–401.

Van den Bossche, P. 2008. *The law and policy of the World Trade Organization*, 2nd ed. Cambridge, UK: Cambridge University Press.

WHO (World Health Organization). 2001. Global health security: Epidemic alert and response. WHA54.14. Geneva, Switzerland: WHO.

WHO. 2009. *WHO epidemic threat detection 1 Jan 2001–31 Dec 2008 (n=2503)* (on file with committee).

WTO and WHO (World Trade Organization and World Health Organization). 2002. *WTO agreements and public health: A joint study by WHO and the WTO secretariat.* Geneva, Switzerland: WTO.

8

Recommendations, Challenges, and Looking to the Future

"Sustainability is not just about securing predictable financial resources. It is also about strengthening health systems while fighting disease, and using the extraordinary opportunities provided by disease programmes to deliver other health benefits. It is about training and empowering the health workforce. It is about drawing on the experience of the private sector to help us innovate and measure risk and results."

—*Ban Ki-moon*
Secretary-General of the United Nations
Remarks at the Forum on Global Health: The Tie That Binds (June 15, 2009)

Nations bear the responsibility to provide for the security, education, development, and the health and welfare of their citizens. This includes responsibility for disease surveillance and response. It is now clear that contemporary threats from infectious diseases require a system capable of providing sustainable global coverage, an objective that can only be achieved through more intensive cooperation among all nations, international organizations, and nongovernmental stakeholders.

In studying what will be required for a sustainable global integrated system for surveillance and response to emerging infectious diseases of zoonotic origin, the committee found significant weaknesses in the ability of all nations, but particularly low-income countries, to address their need for a functional, sustainable, and integrated surveillance and response system for emerging human and animal diseases. Limited surveillance and response capacities at the national level represent more than just a national threat; they are, in fact, a serious global threat, especially in countries in which the drivers of zoonotic diseases are most concentrated and where experts predict that zoonotic disease emergence is most likely to occur. The implication of this is clear: that all countries, in partnership with private and public stakeholders, should develop, maintain, and globally coordinate integrated surveillance and response capabilities to prevent, detect, and respond to the emergence of zoonotic diseases in order to limit loss of life and livelihoods.

RECOMMENDATIONS

The National Research Council report, *Animal Health at the Crossroads*, addressed the importance of strengthening collaborations at the national and international levels. The report specified the need for the United States to commit new resources and develop shared leadership roles with other countries and international organizations in order to promote global systems for preventing, protecting against, detecting, and diagnosing emerging animal disease threats (NRC, 2005). The committee concurs with that report and reemphasizes the importance of U.S.-supported collaborations at the international level for the development and promotion of such a global system, including a U.S. commitment to provide technical assistance to other countries and to increase its participation in developing international animal health standards for preventing, detecting, and responding to zoonotic diseases. An effective zoonotic disease surveillance and response system needs to be integrated across sectors and disciplines so that it identifies and responds to human and animal disease threats at the earliest time possible, without regard to national boundaries or professional discipline.

The committee therefore offers the following 12 recommendations for improving zoonotic disease surveillance and response by priority and category areas (see Table 8-1). The recommendations are grouped as technical, economic, and political actions needed to achieve the desired system. Recommendations assigned as high priority are foundational for a global, integrated zoonotic disease surveillance and response system. The remaining recommendations are considered priority and are not listed in rank order, though they are all considered essential to achieving the goal. The committee understands that it may be necessary to implement these recommendations according to different timetables, depending on how the United States and its partners are able to mobilize the necessary resources. Ultimately, an effective, sustainable system will require attention to each of the 12 recommendations.

High-Priority Recommendations

Technical: Strengthen Surveillance and Response Capacity

Establish Surveillance and Response Strategies

Recommendation 1-1: The U.S. Departments of Health and Human Services (HHS), Agriculture (USDA), Homeland Security (DHS), and the Interior (DoI) should collaborate with one another and with the private sector and nongovernmental organizations to achieve an integrated surveillance and response system for emerging zoonotic diseases in the United States. In addition, these government agencies, including

TABLE 8-1 Recommendations for Improved Zoonotic Surveillance and Response by Priority and Category Areas

	Technical	Economic	Political
	Strengthen Surveillance and Response Capacity	*Financing and Incentives for Surveillance and Response*	*Governance of Global Efforts to Improve Surveillance and Response Capabilities*
High priority	Establish surveillance and response strategies (*Recommendation 1-1*)	Establish sustainable funding strategies (*Recommendation 2-1*)	Create a coordinating body for global zoonotic disease surveillance and response (*Recommendation 3-1*)
Priority	Improve use of information technology to support surveillance and response activities (*Recommendation 1-2*)	Create an audit and rating framework for surveillance and response systems (*Recommendation 2-2*)	Deepen the engagement of stakeholders (*Recommendation 3-2*)
	Strengthen the laboratory network to support surveillance and response activities (*Recommendation 1-3*)	Strengthen incentives for country and local reporting (*Recommendation 2-3*)	Revise OIE governance strategies (*Recommendation 3-3*)
	Build human resources capacity to support surveillance and response efforts (*Recommendation 1-4*)		Mitigate disease threats from wildlife and trade (*Recommendation 3-4*)
	Establish a zoonotic disease drivers panel (*Recommendation 1-5*)		

NOTE: OIE = World Organization for Animal Health.

the U.S. Department of State and the U.S. Agency for International Development (USAID), should collaborate with the World Health Organization (WHO), the Food and Agriculture Organization of the United Nations (FAO), and the World Organization for Animal Health (OIE) to spearhead efforts to achieve a more effective global surveillance and response system, learning from and informing the experiences of other nations.

Given finite resources and the complexity of the challenge, an integrated zoonotic disease surveillance and response system can succeed only if the

U.S. government demonstrates its commitment to develop and strengthen the needed capacities at the national level, and to engage others at the global level. The following strategic approaches are necessary to achieve an effective, global zoonotic disease surveillance and response system:

First, departments or ministries of health, agriculture, and natural resources, with external support as needed, should work with researchers to develop and use science-based criteria to determine and measure the distribution and magnitude of the drivers of zoonotic disease emergence. Rapid changes in ecology, environmental degradation, population density, population movements, animal production systems, and close interaction of humans with livestock, poultry, and wildlife are just a few drivers to study. From these studies, targeted surveillance would then be designed to focus on countries and regions within countries where drivers increase the risk for zoonotic disease emergence.

Second, in countries where disease surveillance in animal populations is absent or weak, ministries of health, agriculture, and natural resources should collaborate as broadly as necessary to develop, enhance, and implement disease surveillance and response systems in human populations that are at high risk for zoonotic disease infection. For example, surveillance is needed in the following high-risk human populations:

- Occupational groups that are at high risk for infection with zoonotic diseases. Such workers include livestock, dairy, and poultry workers; live-animal market workers; veterinarians and animal health technicians; hunters of bushmeat and other wildlife; food preparers (and restaurant workers handling food prepared from bushmeat and exotic animals); slaughterhouse workers; and laboratory scientists and technicians working with animals;
- Healthcare workers who could spread zoonotic diseases to the general public;
- Household and village members who keep live animals within their living quarters or come in close contact with animals in village settings; and
- People engaging in high-risk behaviors known to increase risk of exposure to zoonotic diseases. Such high-risk behaviors include close contact with wildlife and exotic animals; preparing and consuming bushmeat; culturally traditional animal husbandry practices and livestock production systems; failure to use personal protection equipment; failure to follow recommended hand-washing practices.

Third, to reverse the trend where human outbreaks serve as sentinels for animal disease, ministries of agriculture and natural resources should develop and strengthen livestock, poultry, and wildlife zoonotic disease surveillance systems, particularly where surveillance in animal populations is currently limited. In partnership with the private sector, ministries of agriculture should conduct active and passive disease surveillance in animal populations that are raised in high-density conditions but lack good biosecurity measures, that are located in areas of dense human populations (e.g., Asia, Latin America, and Eastern Europe), and/or that are interspersed with smallholder livestock farms. Ministries of agriculture and natural resources should also conduct high-priority surveillance in livestock, poultry, companion animals, and wildlife whenever species are clustered, mixed, and inhabit areas in close proximity to human populations (e.g., co-habitation with humans in homes, villages, or are transported to, housed, and sold in live-animal markets). To detect subclinical or unnoticed infections, ministries of agriculture and natural resources should develop capacity to systematically test laboratory specimens from domesticated animals and wildlife that are at high risk of serving as zoonotic disease reservoirs (e.g., bats, wild aquatic birds, and nonhuman primates). This will enable responses to be targeted and can limit pathogen transmission and prevent or minimize their impact on the health of human and domesticated animal populations. Ministries of agriculture, natural resources, and health should build capacity to institute active sentinel surveillance in wildlife—such as bats, wild aquatic birds, great apes, and rodents—and other important reservoir species that are in close contact with humans to continuously assess the "baseline" population with pathogens of concern (e.g., influenza, Ebola, Nipah, hendra, rabies, Rift Valley fever [RVF], coronaviruses, tularemia, plague). Targeted wildlife populations should be those most likely to interact with humans, either directly or indirectly through domesticated animals. The list of pathogens needs to be established by consensus at the global, regional, and local levels (see Recommendation 3-1 on the recommended coordinating body) and resources should be commensurate to the identified need for surveillance.

Fourth, ministries of health, agriculture, and natural resources will need to develop and formalize a system wherein surveillance information from these different human and animal populations will be integrated and synthesized for analysis. These ministries will also need to develop and formalize effective communication and reporting systems to ensure real-time reporting of linked surveillance data from human and animal populations nationally and internationally to those responsible for planning and instituting prevention, protection, and response interventions. The Danish Zoonosis Centre could be a model of an effectively integrated national program for zoonoses (see Box 8-1).

Finally, science-based nongovernmental organizations (NGOs) have a critical role to play in national and global efforts to develop an integrated surveillance system. In many cases these organizations have extremely

BOX 8-1
Model of an Integrated National Program for Zoonoses

The Danish Zoonosis Centre was established in 1994 in response to the major fragmentation of the surveillance systems and increasing incidence of zoonotic diseases at that time. The Centre is part of the National Food Institute of the Technical University of Denmark, and it has special responsibilities for prevention, surveillance, and outbreak tracking of zoonotic infections by compiling surveillance data on food-borne zoonoses and by developing prevention strategies. As such, it is an integral part of the national contingency plan for outbreaks of food-borne diseases. Funding comes partly from the Danish government and partly from income generated through the provision of research and advisory service to the private sector. The Centre has a staff of 13 specialists but relies heavily on industry for data collection. It is an inter-sectoral center, meaning that representatives of the Danish Board of Health Food and Veterinary Administration are a part of its Steering Group, and producer boards and nongovernmental organizations are included in its coordination groups. The integration of public and private sectors has been a critical element of the Centre's success. Its excellent performance continues to make it a reference center for the World Health Organization and the European Food Safety Authority, among others, on zoonotic disease-related issues.

wide geographic reach, with offices and trained staff based in countries with the highest risk for new zoonoses. They have often developed the most effective and closest relationships and collaborations with local communities. NGOs have the capacity to act nimbly to rapidly refocus resources on outbreaks during crises, and they are usually not encumbered by geopolitical constraints. Science-based NGOs—such as Wildlife Conservation Society, Wildlife Trust, The Consortium for Conservation Medicine, and EnviroVet—have launched programs specifically targeted at many of this committee's recommendations and should be actively involved in future efforts to address them. While the focus of this committee is primarily scientific, it recognizes that advocacy groups can also provide an important push for integrated surveillance by urging relevant policy changes involving food production, wildlife conservation, poverty alleviation, and global health.

Economic: Financing and Incentives for Surveillance and Response

Establish Sustainable Funding Strategies

Recommendation 2-1: USAID—in partnership with international finance institutions and other bilateral assistance agencies—should lead an effort to generate sustainable financial resources to adequately support the development, implementation, and operation of integrated zoo-

notic disease surveillance and response systems. An in-depth study of the nature and scope of a funding mechanism should be commissioned by these agencies, and the study should specifically consider a tax on traded meat and meat products as a potential source of revenue.

The committee concluded that an integrated global surveillance and response system should be designated a national and global public good. As observed in recent outbreaks, emerging zoonotic pathogens are rapidly transmitted across borders and from one continent to another. Too often, responses are either slow but evidence-based or quick but inappropriate (e.g., non-evidence-based restrictions on travel, transport of goods, culling of animals). This has resulted in large political, economic, and social impacts on national and global human, animal, and economic health.

Although primary responsibility for creating and maintaining such a system remains at the national level, the needs of low-income countries for assistance and the complexities of building an integrated global system will require both smarter expenditure of existing resources and additional funding. Without such financial support, the global public good that an integrated system could produce will not be achieved.

The current global economic crisis underscores the need to develop sustainable financing strategies to produce this global public good. Countries with greater resources will need to show leadership by supporting low-income countries and international organizations to create a global system. With the continued spread of H1N1 virus to developing countries, United Nations (UN) Secretary-General Ban Ki-moon stated that the UN would need more than $1 billion to combat the pandemic for the remainder of 2009 alone and made a plea for assistance from developed countries (Maugh, 2009). The inadequacy of traditional donor support, the limited duration of commitment, and the competition for resources generated by other global health problems require the U.S. government, other countries, and intergovernmental organizations to design and implement strategies that will provide sustainable resources for zoonotic disease surveillance and response. National government access to realigned and new funding should be made conditional on fulfillment of agreed criteria of participation, including the willingness to conduct national assessments of surveillance and response capacity and have such assessments independently reviewed (see Recommendation 2-2).

While countries need to be encouraged to invest in developing the capacity to detect, investigate, and report suspected disease outbreaks and thus prevent sporadic cases from escalating to epidemics (especially of known diseases), resource-poor countries undoubtedly will need external support and assistance for this purpose. The challenge of maintaining global surveillance capacity calls for identifying sustainable funding sources rather than depending on development aid budgets, which historically have fluctuated with donor priorities or changes in leadership. Although a number of possible suggestions are provided in this report, the committee did not have the mandate or expertise to conduct a thorough investigation of the implications of these options. The committee therefore calls for an in-depth study to further identify innovative funding mechanisms that can continuously support the need for surveillance and response systems.

Revenue sources should be, in principle, tied to levies on activities that increase the risk of emergence and movement of zoonotic pathogens. This has led to the committee's recommendation for further study on a product tax for internationally traded meat and meat products, which represent an important route for the emergence and spread of zoonotic diseases. This levy would be imposed primarily on wealthier exporting countries (see Table 6-4). One of the potential adverse consequences of imposing a levy may be that it increases product smuggling in an attempt to evade taxes. There may well be other unintended consequences of this strategy; therefore the committee concluded that a thorough study of the pros and cons for this, or other sustainable approaches, is a necessary prerequisite before making final decisions on the optimal mechanism to fund the required actions.

The committee considered other funding options. These include long-term commitments from high- and middle-income countries to contribute directly to a global fund established for this purpose; long-term commitments from governments to fund specific WHO, FAO, and OIE programs; establishment of endowment funds; increased contribution from foundations and nonconventional donors; and public-private partnerships. These remain options that could be considered when more intensive and targeted discussions are initiated.

Initial access to global funding for a recipient country could be made dependent on its commitment and participation in an assessment of its national surveillance capabilities (see Recommendation 2-2). Further funding could be conditional on its subsequent performance to integrate human and animal health systems and its contribution to pay for the surveillance and response systems' operating costs. While the committee did not explore these options and the institutional arrangements necessary to manage them, the committee concluded that it would be prudent if the recommended independent global funding mechanism (e.g., the Global Fund) would not be administered by a government entity or international governmental organization.

Political: Governance of Global Efforts to Improve Surveillance and Response Capabilities

Create a Coordinating Body for Global Zoonotic Disease Surveillance and Response

Recommendation 3-1: USAID, in cooperation with the UN and other stakeholders from human and animal health sectors, should promote the establishment of a coordinating body to ensure progress toward development and implementation of harmonized, long-term strategies for integrated surveillance and response for zoonotic diseases.

As discussed earlier in this report, WHO, OIE, and FAO have improved their coordination efforts on zoonotic diseases, especially through the creation and operation of the Global Early Warning System (GLEWS) for major animal diseases including zoonoses. In addition, WHO and OIE have independently revised their central legal agreements—the International Health Regulations 2005 (IHR 2005) and the Terrestrial Animal Health Code, respectively—to facilitate better governance strategies for zoonotic disease threats. The committee concluded that these positive developments can and should be supplemented by the establishment of an overarching global coordinating body. Building on the foundation laid by GLEWS, the adoption of IHR 2005, changes to the OIE's Terrestial Animal Health Code, and better collaboration between OIE and FAO, this coordinating body could raise the profile of zoonotic disease surveillance and response efforts and provide the necessary high-level political support to advance national, regional, and global coordination efforts. The approach developed in the UN System Influenza Coordinator (UNSIC) strategy is widely perceived as an effective effort and could serve as a model for the coordinating body needed for an integrated zoonotic disease surveillance and response. The zoonotic disease coordinating body should work to ensure that all relevant stakeholders are consulted and involved in coordinating activities. The mechanism could also draw attention to problems and challenges faced in implementation of IHR 2005, OIE agreements, OIE/FAO strategies, and GLEWS. The coordinating body could also facilitate improved and additional funding streams for zoonotic disease control (see Recommendation 2-1).

Priority Recommendations

Technical: Strengthen Surveillance and Response Capacity

Improve Use of Information Technology to Support Surveillance and Response Activities

Recommendation 1-2: With the support of USAID, international organizations (such as WHO, FAO, OIE, and the World Bank) and public- and private-sector partners should assist nations in developing, adapting for local conditions, and implementing information and communication technologies for integrated zoonotic disease surveillance.

BOX 8-2
Philanthropic Support for Information Technology Development and Management

The Rockefeller Foundation supports an "eHealth" initiative, along with a portfolio of grants on surveillance networks. This initiative focuses on a number of aspects of design and implementation of eHealth including open-source software development for medical records, laboratory records, and disease reporting. In 2008, a series of Bellagio meetings were held that brought experts together from many parts of the world to discuss architecture, standards, training needs, and other activities. The Rockefeller Foundation is supporting the creation of the Centers of Excellence for Informatics in a number of low-resource settings to facilitate the implementation of eHealth, including the development and implementation of standardized tools for disease reporting and public health response. The Bill & Melinda Gates Foundation recently funded a planning grant for the American Medical Informatics Association to outline plans for a global informatics scholars program. Such philanthropic input and support is critical for developing countries where such expertise will greatly advance efforts at streamlining information systems for surveillance. While these initiatives are directed towards human disease, they represent technical models that can be applied to animal disease.

Effective use of such technologies facilitates acquisition, integration, management, analysis, and visualization of data sources across human and animal health sectors and empowers information sharing across local, national, and international levels. To establish, sustain, and maintain this technologically sophisticated system, both leadership and investment are critically needed.

Leadership and investment should emerge within each country; however, low-income countries will need support to engage in broader training and capacity building. This effort should integrate key nongovernmental actors, including private philanthropies with interests in infectious disease surveillance and management (see Box 8-2 for an example); industry partners in food production, information technology, and data management; and nongovernmental organizations involved in global health. Organizations should follow the lead of actors such as Google.org., which contribute both external funding as well as internal efforts to support the development of open source surveillance technology (see Box 5-5).

Strengthen the Laboratory Network to Support Surveillance and Response Activities

Recommendation 1-3: USAID should promote and initially fund the establishment of an international laboratory working group charged with designing a global laboratory network plan for zoonotic disease surveillance. The working group's objective would be to design a laboratory network that supports more efficient, effective, reliable, and timely diagnosis, reporting, information sharing, disease response capacity, and integration of human and animal health components. In addition, a long-term coordinating body for zoonotic diseases, perhaps modeled after the United Nations System Influenza Coordinator's (UNSIC's) office (see Recommendation 3-1), should implement the global laboratory network plan, manage it, and assess its performance in consultation with the international laboratory working group.

The international working group charged with developing the global laboratory network plan should include representation from several groups. These include international organizations (e.g., WHO, FAO, and OIE); national human and animal health laboratories with experience in laboratory network development and support (e.g., U.S. Centers for Disease Control and Prevention [CDC], Department of Defense [DoD] Global Emerging Infections Surveillance and Response System, USDA Animal and Plant Health Inspection Service [USDA-APHIS] National Veterinary Services Laboratory, Canadian Science Centre for Human and Animal Health, Australian Animal Health Laboratory); professional laboratory organizations, such as the Association of Public Health Laboratories and the American Association of Veterinary Laboratory Diagnosticians in the United States and their counterparts in other nations; wildlife health specialists; and private for-profit and not-for-profit entities with a stake in zoonotic laboratory network development. Integration of animal and public health laboratory infrastructure, operations, and personnel should be a driving factor in development of the global plan.

To develop the plan the working group should take steps that include

1. conducting an inventory and assessing the quality of the current global capacity for laboratory diagnosis and reporting of zoonotic diseases in human and animal health laboratories;
2. based on this inventory, designing the optimal laboratory network structure with emphasis on utilizing existing regional laboratories in high-risk regions as reference labs capable of the work necessary for identifying emerging diseases, and sentinel surveillance laboratories within those regions;
3. identifying where new laboratory infrastructure is necessary;
4. creating the environment (e.g., common space, common platforms,

and quality assurance standards and practices) for integrated zoonotic disease diagnostics at the laboratory and network levels;

5. ensuring that the operational procedures for sample collection and priority secure transport to sentinel and reference laboratories are established; and

6. ensuring optimal information flows from the local to the national, regional, and international levels and back, in order to permit integrated data analysis and provide opportunities for networking.

Once developed, a coordinating expert body (see Recommendation 3-1) should take steps to implement the global laboratory network plan modeled after the U.S. Integrated Consortium of Laboratory Networks. Funding of the laboratory network infrastructure and network operation needs to be a primary consideration in developing a global zoonotic disease surveillance financing plan (see Recommendation 2-1 on strengthening funding for zoonotic surveillance and response). Implementation would include oversight and monitoring of infrastructure development; developing performance standards for network laboratories; reviewing, recommending, and harmonizing diagnostic assays for zoonotic diseases, standardized equipment, and standard operating procedures; overseeing development and validation of new assays when needed; assisting with the provision of reference standards and reagents; ensuring an integrated laboratory network information system (Recommendation 1-2 on information and communication technologies); identifying and providing personnel training opportunities; and conducting proficiency tests for network laboratory personnel.

The coordinating body also needs to establish a mechanism for monitoring performance of the laboratory network, by, for example, sponsoring tabletop exercises and scenario testing, and engaging in continuous monitoring of assay performance. The coordinating body and international laboratory working group would need to work closely with epidemiologists and field personnel in determining how and which samples are collected, preserved, and transported to a local or national laboratory, and with which accompanying clinical and epidemiological information. The coordinating body would also need to work closely with regulatory agencies to review what barriers may exist in transporting and submitting specimens to regional or international reference laboratories, and how these can be addressed so that delays (such as those that occurred with the transport of the influenza A(H1N1) 2009 virus from the Mexican government laboratories to the CDC) are precluded. Although there are many factors to be considered—the nature of the agents and the risk they pose, type of laboratory capacity required, security of the transport mechanism—samples need to be able to reach reference and academic laboratories with the requisite safe facilities to identify new agents, determine pathogenesis, identify targets for diagnostics, drugs and vaccines, and develop and disseminate diagnostic kits and products to prevent or treat infections in animals and/or humans.

Build Human Resources Capacity Building to Support Surveillance and Response Efforts

> **Recommendation 1-4: Given the need for increased human capacity to plan, conduct, and evaluate integrated zoonotic disease surveillance and response, U.S. government agencies should take the lead in developing new interdisciplinary educational and training programs that integrate human and animal health and allied fields. Existing national and regional training programs in field epidemiology, clinical, and laboratory diagnosis supported by HHS, USDA, and DoI should be improved to include a better balance of human and animal health concerns, incorporate contributions from laboratory and social science professionals, and connect with one another where appropriate.**

The National Institutes of Health's Fogarty International Center—collaborating with CDC, USDA Agricultural Research Service, USDA-APHIS, USDA National Institute of Food and Agriculture (the former Cooperative State Research, Education, and Extension Service), and U.S. Geological Survey (USGS)—should be funded to partner with educational institutions and relevant ministries to develop field-based, integrated, interdisciplinary model curricula and training programs for emerging zoonotic disease surveillance and response. Educational institutions should include U.S. medical, veterinary medical, and public health schools; colleges and universities of agriculture, natural resources, and social sciences; nongovernmental organizations, especially those engaged in wildlife disease surveillance and training; and international human and animal health organizations. Model curricula and training programs would need to be interdisciplinary, field-oriented, and address the interfaces of human, animal, and environmental health. Support for open-source curricula would be valuable to ensure that the needed quality can be attained and accessed freely. In addition, countries and partner educational institutions would need to develop a strategy to retain faculty expertise and a trained professional workforce to conduct and support emerging zoonotic disease surveillance and response.

Education and training is needed in all nations so that trained professionals can properly detect and diagnose known diseases in animals and humans at the earliest point possible, and know how to proceed when there is the possibility of an emergence of a new pathogen or disease. Training in the areas listed below is essential for producing a skilled workforce capable of conducting surveillance and initiating proper response actions. Training programs should, to the greatest extent possible, include human and animal health professionals and paraprofessionals, together with community and public health professionals, to maximize the opportunity to improve interdisciplinary communication. Training topics would include

- Leadership, multidisciplinary and multisectoral collaboration, and communication with surveillance and response teams across local, national, and international levels.
- Clinical and pathological diagnosis of emerging zoonotic diseases in humans and animals.
- Specimen collection, storage, and transportation for laboratory-confirmed diagnosis.
- Laboratory procedures and protocols to prepare and test specimens with appropriate assays that would identify and confirm the cause of outbreaks.
- Epidemiology, routes of transmission, and methods for outbreak investigations, and prevention and control of zoonotic diseases, including knowledge of population-based, public health strategies for disease control in human populations, and population-based, herd health approaches for animal disease prevention and control.
- Advanced quantitative methods for analyzing and modeling epidemiological data.
- Monitoring human risk behaviors associated with increasing risk of human exposure to zoonotic diseases.
- Monitoring human practices associated with disease drivers that increase the risk of zoonotic disease emergence.
- Better risk communication methods and skills aimed at informing the media and the public on the extent and cause of disease outbreaks, factors that place humans and animals at risk of exposure, and evidence-based options for response.
- Monitoring the perception of risk and knowledge of risk factors and prevention options by communities in response to media messages.
- Skills in handling policy and legal concerns and challenges that arise with surveillance and response activities.

It is especially important to create and support field-based training programs in low-income countries, because it is valuable for trainees to identify social and cultural factors, incorporate local approaches, and find acceptable solutions. For instance, by identifying, training, and utilizing local trainers in low-income countries (such as surveillance personnel trained by the polio eradication program), programs can more effectively develop the human capacity to collect and preserve samples and carry out preliminary tests. It will also teach trainees to recognize emerging problems and connect with international reference laboratories.

For low- and middle-economic countries, USAID should continue supporting and providing resources for (1) the implementation and expansion of multidisciplinary applied field training programs, such as the Training Programs in Epidemiology and Public Health Interventions Network; and (2) the development, recruitment, and retention of faculty in schools of medicine, veterinary medicine, and public health who understand the importance of specializing in integrated human and animal health surveillance and response. Special attention is needed to incorporate surveillance and response strategies with wildlife populations, human risk behaviors, and risk perception in communities.

USAID, in collaboration with counterparts in other countries, should organize or sponsor workshops to (1) develop an integrated surveillance and response curricula; (2) develop protocols, procedures, and other approaches for multisectoral collaboration and communication regarding disease detection and integrated response; and (3) train human and animal health professionals and paraprofessionals together on these methods and procedures. USAID should also fund and support professional development opportunities, including participation at international conferences and workshops. Over time, a goal is for local staff in developing countries to serve as the organizing groups for such sponsored conferences and workshops. Low-income countries will need to work with the international community to accomplish these activities, gain relevant skills for training their workforce, and develop plans for retaining faculty and a trained workforce in country.

Regardless of the resources available to a country, the committee found gaps and challenges that all countries will need to address in order to ensure a competent workforce. To field a capable workforce that can prevent, detect, and respond effectively to emerging zoonotic infectious diseases, there needs to be political will, priority assigned to surveillance and response, commitment across disciplines, adequate funding, and strong coordination at national and international levels.

Establish a Zoonotic Disease Drivers Panel

> **Recommendation 1-5: The U.S. Department of State, in collaboration with WHO, FAO, OIE, and other international partners, should impanel a multidisciplinary group of technical experts to regularly review state-of-the-science information on the underlying drivers of zoonotic disease emergence and propose policy and governance strategies to modify and curb practices that contribute to zoonotic disease emergence and spread.**

Many drivers for zoonoses and the measures for controlling them are transnational in nature. The U.S. Science and Technology Advisor to the

President and Department of State's Science and Technology Advisor to the Secretary could co-lead the effort and bring the results of the panel's findings to the attention of important stakeholders and diplomatic forums, including the UN, Group of Eight (G8), Group of Twenty (G20), and regional intergovernmental organizations. This international panel would be composed of representatives from national, international, and intergovernmental agencies, nongovernmental entities, and technical experts from academic institutions selected on the basis of demonstrated disciplinary expertise to examine the broad set of zoonotic disease drivers. The coordinating body for zoonotic disease surveillance and response (see Recommendation 3-1) would be a member of this panel. This panel could be modeled after the Intergovernmental Panel on Climate Change,[1] an international group that provides decisionmakers and other interested parties with an objective assessment of scientific, technical, and socioeconomic information about climate change. OIE has recently formed an ad hoc group to address the role of climate and environmental changes on emerging and reemerging animal diseases. It is essential that there is an organizational architecture to ensure that specialized groups such as this do not function in isolation, but are a part of an integrated global system.

Economic: Financing and Incentives for Surveillance and Response

Create an Audit and Rating Framework for Surveillance and Response Systems

> **Recommendation 2-2: USAID should convene a technical working group to design and implement, by the end of 2012, an independent mechanism to audit and rate national surveillance system capacities for detecting and responding to emerging zoonotic disease outbreaks in humans and animals.**

---

[1]The Intergovernmental Panel on Climate Change (IPCC) provides its reports at regular intervals, and they immediately become standard works of reference, widely used by policymakers, experts, and students. The comprehensiveness of the scientific content is achieved through contributions from experts in all regions of the world and all relevant disciplines including, where appropriately documented, industry literature and traditional practices and a two-stage review process by experts and governments. However, the IPCC does not conduct any research nor does it monitor climate-related data or parameters. Governments often participate in plenary sessions of the IPCC where main decisions about the IPCC work program are taken and reports are accepted, adopted, and approved. The IPCC work aims to support the promotion of the United Nations' human development goals. The IPCC Second Assessment Report of 1995 provided key input for the negotiations of the Kyoto Protocol in 1997 and the Third Assessment Report of 2001, and Special and Methodology Reports provided further information relevant for the development of the United Nations Framework Convention on Climate Change and the Kyoto Protocol (IPCC, 2009).

In structuring an integrated global surveillance and response system, countries should be encouraged to develop their national capacities and be encouraged to take steps that enable them to participate in a global system. At present, there is no independent mechanism to review progress toward the needed integrated surveillance and response system capabilities, increasing the likelihood that integration will remain uneven and incomplete. By creating an independent framework to engage in constructive, transparent assessments of national efforts, this would contribute in a major way to more efficient and effective policies targeted at creating integrated zoonotic surveillance and response capabilities.

A technical working group to establish the audit and rating framework for surveillance and response systems would include representatives from WHO, FAO, OIE, and academic experts, nongovernmental organizations, and the private sector. The timing is consistent with the deadline for state parties to develop the core minimum surveillance and response capacities required by 2012 under the International Health Regulations 2005 (IHR 2005).

Countries already participate in national assessments of human and animal health systems under the IHR 2005 and OIE programs, respectively. An independent audit and rating framework can help public and private stakeholders identify problems and develop common strategies in an effort to improve national and global capabilities for integrated human and animal zoonotic disease surveillance and response. To the fullest extent possible, information generated by these reviews should be made publicly available.

Under the audit and rating framework, participating countries would provide a national assessment of the country's risk of a disease outbreak and its reliability in reporting. The framework would then independently review and verify such information and finally rate the countries on their ability to detect and mitigate disease threats. A global emerging disease risk-rating framework that provides specific information on the risk of an outbreak by country and the likely speed of outbreak detection by national authorities would help trading partners, neighbors, and other stakeholders incorporate zoonotic disease risk into their trade and travel decisions. Participating in these audits would benefit countries because it signals a willingness to be transparent about the country's risk of outbreaks and likelihood of detection. This then translates into trading partners and potential tourists having greater confidence in that country's practices.

By demonstrating a commitment to co fund their national efforts and participate in the audit and rating framework, countries would qualify to access the global funds specified in Recommendation 2-1 for improving their national surveillance and response capacity.

In order to maintain independence and credibility, this audit and rating framework would be housed within an independent global technical consortium composed of members from relevant ministries, such as the Global Fund, rather than within intergovernmental organizations that answer politically to their member states. Creating a new institution is not necessary, although that may ultimately be deemed the most feasible option.

Because information on national risk is a global public good, resources to support this activity should be sourced through the global funding mechanism described in Recommendation 2-1. The cost of auditing surveillance and response systems should be borne by this centralized global funding mechanism and not by individual countries to ensure that the process is seen as independent, unbiased, and credible.

Strengthen Incentives for Country and Local Reporting

> **Recommendation 2-3: To reduce incentives to conceal outbreaks and mitigate the negative social and economic repercussions of early disease reporting (e.g., stigma of disease, food safety concerns, culling, and trade and travel disruptions), financial incentives at the following levels are needed through partnerships among bilateral aid agencies, the international community, and national governments:**
> **(a) Country level: USAID—in partnership with international finance institutions and other bilateral assistance agencies—should implement economic incentives to encourage middle- and low-income countries to report human, animal, and zoonotic disease outbreaks.**
> **(b) Local level: National governments, with added support from the international community, should identify and provide the resources needed for financial incentives to promote early disease reporting and to engage in effective responses at the local level.**

Current methods to control outbreaks include culling of livestock and poultry, and they also influence social and economic incentives to report outbreaks. Although there is now increased sensitivity in some countries to the importance of not disincentivizing future reporting when implementing control measures, this is by no means uniform. Economic disincentives to reporting include culling of livestock and poultry without adequate compensation, and food product warnings, recalls, or bans without evidence of reasonable risk. Evidence-based practices need to be designed and implemented to assist outbreak containment. Efforts to control the international spread of zoonotic diseases include trade and travel restrictions that place significant economic hardship on reporting countries.

The international community can also minimize the unnecessary cost of sanctions at both levels by using existing regulatory mechanisms, like zoning and compartmentalization, where appropriate. International community application and acceptance of these initiatives allow for continued trade of safe products from countries or zones that have reported a disease.

Although efforts to date have focused on upgrading surveillance capacity in countries that are less able to report outbreaks in a timely manner, the committee recommends that bilateral aid agencies and international organizations pay closer attention to the economic incentives for reporting disease outbreaks (e.g., vaccination campaigns and reimbursements for livestock and poultry culling). Resources earmarked to upgrade surveillance capacity should consider whether these systems will actually be used in the event of an outbreak or whether these resources are simply crowding out monies that countries would have spent on their own. In addition to funding for upgrading surveillance capacity, guaranteed assistance with outbreak containment needs emphasis, including the availability of vaccines for humans or animals. Without such guarantees, countries have fewer incentives to report disease outbreaks, regardless of international legal obligations.

National governments should explicitly plan to increase incentives for surveillance and reporting by allocating financial resources to pay for adequate reimbursement to those who stand to lose from reporting, while decreasing disincentives by reviewing and reducing the unwarranted use of outbreak control measures such as travel restrictions, quarantines, and culling.

Political: Governance of Global Efforts to Improve Surveillance and Response Capabilities

Deepen the Engagement of Stakeholders

> **Recommendation 3-2: In its work on zoonotic disease surveillance and response, USAID—in collaboration with WHO, FAO, and OIE—should convene representatives from industry, the public sector, academia, nongovernmental organizations (NGOs), as well as smallholder farmers and community representatives to determine how best to build trust and communication pathways among these communities in order to achieve the efficient bi-directional flow of both formal and informal information needed to support effective, evidence-based decisionmaking and coordinated actions.**

The complexity of achieving sustainable, integrated national and global surveillance and response systems for zoonotic diseases requires deliberate and intensified efforts to engage and connect all relevant stakeholders at each governance level—local, national, and global. Moreover, high stakes for trade or industry groups—as illustrated by the detection of bovine spongiform encephalopathy (BSE) in three cows in the United States between 2003 and 2006, causing great economic harm to that industry with a total loss of $11 billion—necessitate their involvement as well.

To achieve better surveillance and response, different players will be challenged to work effectively together, as they often have vested interests and their actions can affect or alter the occurrence, transmission, or spread of the disease. These players include animal producers and related industries that have an economic interest in the trade-off between quality and yield, governments that have a political interest in the trade-off between improving the levels of sanitary health on behalf of citizens and the freedom of international commerce, and the public that desires higher levels of health and less risk of disease. Despite these often mutually beneficial interests, different sectors can still be resistant to working together. To overcome such barriers, it is critical to engage relevant stakeholders from all levels to help build transparency and trust.

Ultimately, stronger systems of surveillance and response lead to improved sanitary environments and higher levels of health for both human and animal populations. Benefits can also extend beyond disease prevention or control, including increased productivity, higher trade competitiveness and market expansion, increased levels of availability and food security, reduction in the risk of bioterrorism targeting the safety of food, and greater options for growth in sectors such as tourism. Articulating the range of possible benefits can improve public acceptance of measures and behaviors designed to reduce the risk of zoonotic disease emergence and spread.

Starting at the local level, steps for cooperative action include a better articulation and quantification of potential benefits and costs. A greater understanding of how cooperative action can be beneficial is needed, and examining community-based models would be helpful to understand nonfinancial incentives that make such initiatives sustainable. Some community-based models have demonstrated their success and sustainability, and those benefits and costs accrue across multiple sectors. The Bangladesh Rural Action Committee serves as one example in the human health sector that is worth further exploration; however, there are likely other working models outside health sector initiatives that engage different actors to achieve common objectives. Finally, there needs to be a willingness to test different approaches in pilot applications so that they systematically build upon one another for success.

Revise OIE Governance Strategies

Recommendation 3-3: To protect animal health and international trade, and to contribute significantly to the reduction of human and animal

health impacts from zoonotic diseases, OIE members states should take the necessary steps to:

(a) Adhere to Resolution 17 (adopted on May 28, 2009), which reminds OIE member states of their obligation to make available to OIE all information on relevant animal diseases, including those that are of zoonotic potential.

(b) Create legally binding obligations for OIE members to develop and maintain minimum core surveillance and response capabilities for animal health risks, including zoonotic diseases.

(c) Authorize OIE to publicly disseminate information received from nongovernmental sources, in the event OIE member states fail to confirm or deny such information in a timely manner, or when denials of such information run counter to persuasive evidence that OIE has obtained from other sources.

(d) Empower the OIE Director-General to declare animal health emergencies of international concern with respect to emerging or reemerging zoonotic diseases that constitute a serious animal or public health risk to other countries and issue recommendations about how countries should address such emergencies.

As discussed in Chapter 7, the committee analyzed existing similarities and differences in the governance strategies and legal obligations embedded within WHO's IHR 2005 and OIE's approaches, rules, and resolutions. Although they have more similarities than some comparative analyses have recognized, the committee concluded that the OIE rules lack important provisions found in IHR 2005 that should be operative to promote animal health. The first three parts of this recommendation identify the key provisions. These call on OIE member states to adhere to the most recent resolution on reporting relevant animal diseases to OIE, to establish binding obligations for members to develop and maintain minimal capability for zoonotic disease surveillance and response, and to provide the Director-General with the authority to make public credible information on zoonotic disease outbreaks, even without concurrence of the member state. The fourth would also provide the Director-General with the authority to declare an animal health emergency of international concern, analogous to the authority the Director-General of WHO now has under IHR 2005.

Adopting these principles will strengthen OIE's ability to ensure that its member nations have the minimal capacity for effective surveillance and response to animal diseases, ultimately improving the potential to control animal diseases before they decimate animal populations and impact human health. In addition, their implementation by OIE would create a more harmonized set of global principles that would apply to both human and animal diseases. The committee is convinced that this would provide a stronger foundation for coordination and collaboration among human and animal health organizations, ministries, and experts.

Mitigate Disease Threats from Wildlife and Trade

Recommendation 3-4: To mitigate and decrease the threat of zoonotic diseases emerging from wildlife, U.S. government entities and their international partners, especially OIE, should proactively take the following initiatives:

(a) Conduct a comprehensive review of federal and state laws on trade in wildlife as a prelude to optimizing the policy and regulatory options to identify gaps and weaknesses in such laws, and to enact new legislation, regulations, or administrative rule changes to strengthen the government's ability to protect human and animal health from diseases carried by wildlife traded through foreign or interstate commerce.

(b) Incorporate efforts and initiatives that support actions to prevent, prepare for, protect against, and respond to threats to human and animal health into current and new international negotiations and cooperative processes that address drivers of zoonotic diseases (e.g., exotic pet trade, food safety and security, environmental degradation, and climate change).

(c) Pursue negotiations for a new international agreement on trade in wildlife species that improves international collaboration on reducing the threat that such trade presents to human and animal health. The objectives of the negotiations and the agreement would be to make wildlife-related zoonotic disease prevention and control a higher priority in the international management and control of legal and illicit trade in wildlife species, the production and distribution of food and animals, and environmental protection.

(d) Incorporate wildlife diseases and zoonoses into the OIE World Animal Health Information System (WAHIS) and integrate reporting on wildlife diseases and zoonoses in GLEWS. OIE should also expand the role and capability of its Working Group on Wildlife Diseases in order to more effectively meet the growing zoonotic threat that wildlife diseases represent.

U.S. government entities that should take the lead for these recommendations include the Department of Commerce, USDA, HHS, DHS, and DoI. Other relevant entities include the U.S. Postal Service and the U.S. Trade and Development Agency. There is growing awareness of the wildlife trade as a conduit for zoonotic pathogens of public health concern, and of others that directly affect livestock or wildlife, at the same time it is apparent that there is extremely limited ability to monitor and control this trade. In the United States, the Fish and Wildlife Service is responsible for inspecting import consignments for conservation and trade requirements, but not for detecting disease. The USDA is responsible for testing imported livestock for disease, but not wildlife. To overcome the current fragmentation of responsibility in the United States, a first step to address this lack of coordination would be establishing an inter-agency working group to recommend a collaborative strategy for improved oversight and action. At the same time, the USGS National Wildlife Health Center should be tasked to conduct a risk assessment of the potential health impact of imported wildlife, the extent of the illegal importation into the United States, and a cost-benefit analysis of control measures.

Internationally, OIE has authority to list a disease as notifiable to protect trade in animals. This is usually applied to diseases that would hinder trade in livestock, but it may also be applied to diseases which can affect the environment, including wild animals. OIE should adopt a broad view of its remit and form an ad hoc committee to assess the most significant disease risks in the international wildlife trade, including those of potential impact to human, livestock, and environmental health. The ad hoc committee should make recommendations on which diseases should be listed as notifiable for these reasons. It is important to remember that many diseases of zoonotic potential are nonpathogenic in traded wildlife. OIE, WHO, FAO, and U.S. government agencies (including USDA and USAID) should fund pathogen discovery programs to identify potentially zoonotic, novel agents in wildlife currently traded between countries.

CHALLENGES TO SUCCESSFULLY INTEGRATING AND COORDINATING INTERNATIONAL DISEASE SURVEILLANCE AND RESPONSE SYSTEMS

The committee acknowledges that achieving a sustainable, integrated global zoonotic disease surveillance and response system is a complex and daunting task. The goal requires unprecedented collaboration on multiple levels: at global, regional, and national levels; among government, industry, academia, and the public; between human and animal health communities, including those working on livestock, poultry, companion animals, and wildlife; and across many disciplines, such as field epidemiology, clinical science, pathology, laboratory science, animal husbandry, social science, communications, economics, and national and international law and governance.

These challenges need to be overcome in order to successfully implement a multisectoral and integrated approach to zoonotic disease surveillance and response. The committee believes its success depends on the following factors:

- Sufficient national and global surveillance and response capacities;
- Enhanced compliance and implementation of relevant international agreements, especially the IHR 2005 and OIE instruments, and global strategies, such as the joint OIE/FAO framework and GLEWS;
- Better utilization of existing financial resources and generation of new funding for zoonotic disease surveillance and response;
- Effective communication and cooperation across sectors, relevant disciplines, and institutions;
- Joint resource use and greater equity in resources for implementing surveillance and for human and animal health prevention and control interventions;
- Improved cross- and interdisciplinary training in medical and veterinary education and allied fields;
- Attention to understanding the nonbiological social, political, and economic drivers and consequences involved with zoonotic disease and human and animal health;
- Generation of political will to commit political, economic, and intellectual capital for zoonotic surveillance and response capabilities;
- Better understanding of zoonotic disease surveillance and response capabilities as priority national and global public goods; and
- Greater mutual respect and trust between human and animal health communities, academic institutions, and practitioners.

Uncoordinated Approaches in Designing and Implementing Zoonotic Disease Surveillance and Response

The committee frequently referred to multiple players involved in designing, implementing, and evaluating disease surveillance and response systems at local, national, and international levels. The result of these multiple players has been many different, often vertical and single-disease oriented systems that generally have incompatible implementation approaches. Multiple guidelines have been developed and recommended (e.g., by USDA, CDC, WHO, and OIE), and different methods for evaluating disease surveillance and response systems have been employed. In general, there is a lack of standard or harmonized laboratory, field epidemiology, and disease prevention and control protocols. There is also considerable variation in protocols for disease surveillance in human, food-animal, and wildlife populations. In addition, aside from rabies, there are no protocols for zoonotic disease surveillance in companion animals for pathogens such as toxoplasmosis or visceral larval migrans (due to *Toxocara canis* or *T. cati*). For disease agents associated with wildlife, laboratory diagnostics that are reliable, sensitive, and specific to wildlife specimens are lacking.

At the same time, the generally adverse trade and tourism impacts of disease outbreak reporting can lead to political interference, thereby precluding the rapid release of important information to the global community for implementing a rapid and effective response. The committee therefore believes it is essential to develop and apply a standard method for conducting and evaluating the effectiveness of zoonotic disease surveillance systems in different countries, independent of political interference (such as suppression of information or corruption).

The Complexity of an Integrated Approach

The training mechanisms and health systems necessary to build human capacity for an integrated zoonotic disease surveillance and response system have developed as separate and unequal systems. In the past, greater resources have been available for training in human health, thus additional investments to train those in animal health are needed while at the same time not reducing existing support to train human health professionals. Opportunities to jointly train human and animal health professionals together are particularly valuable. Joint program initiatives, joint workforce education and training, and joint performance standards for emerging zoonotic diseases will need to be given priority to support the widespread changes essential for implementing a more integrated and effective system. There will likely be resistance to moving forward, funding may be difficult to find, and there will be issues at many levels over control. These problems should be anticipated and will require leadership from both the human and animal health sectors to overcome them.

For many years, various scholars have argued in favor of increased collaboration among professionals in the human and veterinary health communities (Schwabe, 1983; Murphy, 1998; NRC, 2005; Hadorn and Stark, 2008). The committee is deeply concerned to note that despite these appeals for action, progress to increase collaboration between the human and animal health systems has been limited. In response to the fragmented national and international responses to human and animal health emergencies, there is now considerably greater attention with respect to the need to increase and strengthen multisectoral and multilateral collaborations for emerging zoonotic disease surveillance and response. This multidisciplinary approach is being promoted under the banner of "one health," defined as "the collaborative effort of multiple disciplines—working locally, nationally, and globally—to attain optimal health for people, animals and the environment" (AVMA, 2008, p. 4). The committee supports all such efforts.

Zoonotic disease surveillance and response offers an opportunity to realize the vision of integrated human, animal, and environmental health in a practical and visible way. Information sharing, laboratory infrastructure, sample collection, trained workforces, laboratory analysis, and response teams can and need to be integrated. Community health workers and paraprofessionals can be trained to collect samples from both humans and animals, or at a minimum can work side by side to assess and sample human and animal populations where zoonotic agents are residing, evolving, and moving. Shared cold chains can deliver samples from humans and animals to laboratories analyzing all samples using assays that are well-characterized, validated, and equivalent, if not identical. Information from laboratories and regulatory agencies can be distributed back to the community level to all human and animal health workers. When zoonotic disease outbreaks occur, health teams—that at a minimum include physicians, veterinarians, public health professionals, and other disease experts when appropriate, such as medical entomologists and wildlife biologists—can work together to unravel the problem and set in motion the response component from the outset.

Political Will and Elevating Integrated Surveillance and Response to Emerging Zoonotic Diseases as a Priority

Among the many challenges for establishing an integrated surveillance and response system is the lack of political will to address emerging zoonotic health threats. Furthermore, health is often assigned as a low priority by political leadership; there is an accompanying lack of commitment to finance the system; ownership of the disease surveillance system is unclear; and there are often conflicting partner priorities. The sequential emergence of infectious diseases of zoonotic origin in the past few decades—such as HIV/AIDS, *Escherichia coli* O157:H7, severe acute respiratory syndrome (SARS), HPAI H5N1, and most recently influenza A(H1N1) 2009, which contains genes from human, pig, and bird influenza viruses—have captured the public's attention and raised the level of engagement of political leaders. With attention and engagement has come funding; however, it has been disease specific and primarily oriented to address consequences affecting human health. It has also failed to build the broader surveillance and response system that is necessary and described in this report.

In many developing countries where the human health system is inadequate, it is not surprising that the priority placed on the interface of human and animal health is low. These realities for both human and animal health are amply demonstrated by the lack of funding, inadequate staffing, poor quality or inappropriate training for existing personnel, and the failure to appreciate the cost effectiveness of a reliable disease surveillance system in healthcare delivery. Given the low priority and limited expertise, decision-makers often do not understand how to interpret and use available information on emerging zoonotic diseases. Even if they know what they should do, they typically lack the authority and resources needed to rapidly respond (Pappaioanou et al., 2003). Furthermore, the fear of sanctions and economic losses as immediate consequences of reporting trumps any instinct to act quickly. As these countries are also confronted with HIV/AIDS, malaria, and tuberculosis, for which considerable international assistance has been generated, it is no wonder that national policymakers are unable to allocate scarce resources to newly emerging zoonotic diseases. Any support that has been directed towards zoonotic disease control has come mainly through external and vertical targeted programs. The lack of funds for veterinary and environmental agencies is a particularly serious impediment to effective action (GAO, 2001; NRC, 2005). In Kenya, for example, the Ministry of Health could deploy five times more staff to address the recent outbreak of RVF than could the Veterinary Service that is actually charged with controlling the main source of human RVF infection.[2]

Disease surveillance systems function vertically in many African countries, because they were set up to support global initiatives for monitoring and controlling specific diseases. These include poliomyelitis, bacterial meningitis, measles, cholera, yellow fever, and other vaccine-preventable diseases. The ad hoc system of establishing specific disease surveillance systems has in many ways prevented the establishment of a reliable and comprehensive national disease surveillance system. The vertical programs may have succeeded in the use of disease-specific data collection tools, reporting formats, and disease surveillance guidelines for donor-targeted disease. However, the facilities are minimally used for disease surveillance or control of other emerging zoonotic infectious diseases. Where there are facilities, often the same person or team performs all disease surveillance activities, limiting their ability to attend to other problems.

At this time, there is sufficient global concern to mobilize international leadership because of the potential for influenza A(H1N1) 2009 to return later in 2009 with considerably greater virulence, perhaps through reassortments with avian influenza A(H5N1) viruses. This is an opportune time for international organizations—such as WHO, FAO, and OIE—and national governments and local authorities to take ownership of the surveillance and response system. These various actors need to make the commitment and take the first steps towards creating the local to global systems: If there is no local "ownership" of the disease surveillance system, especially at the health district level where most epidemics originate, it is hard to generate and sustain political will at the higher levels to take action. On top of this,

[2]Ester Schelling, International Livestock Research Centre, personal communication, 2008.

inflexible regulatory constraints are commonly imposed by donor agencies for their own administrative and reporting requirements. This hampers the maximum use of facilities and especially human and financial resources for integrating disease surveillance systems.

Implementing the International Health Regulations

The adoption of IHR 2005 by the World Health Assembly represents a giant leap forward because it provides a comprehensive framework for human disease surveillance (Fidler, 2005). The committee recognizes that IHR 2005 took 10 years to develop, and its slow implementation in many countries restricts the ability to reach the full promise of IHR 2005. The committee reemphasizes that full implementation of IHR 2005 is the bedrock for building a new integrated and sustainable human and animal surveillance and response system for emerging zoonotic diseases.

With increasing disease risks related to globalization of trade, in 2007–2008, the OIE refined an evaluation tool originally developed in collaboration with the Inter-American Institute for Cooperation on Agriculture to produce the Performance of Veterinary Services tool. This was developed to assist the country's veterinary services by assessing their existing level of performance, identifying gaps and weaknesses in their capacity to comply with OIE international standards, and developing new strategies and approaches for the public and private sectors to collaborate in addressing the identified gaps and challenges (OIE, 2008). In general, by strengthening veterinary services and infrastructure with enhanced capacity to implement strategic and sensitive surveillance methods, this will allow local and national integrated health systems to better detect the emergence of new zoonoses. However, if IHR 2005 is not fully implemented, there is little chance that OIE efforts can be successful. For this reason, the committee recognizes the critical importance of full implementation of IHR 2005 and registers its concerns about the pace of progress.

Fostering Trust

An effective disease surveillance system is one in which diseases are detected early and reported in a timely fashion. That is fully dependent on achieving a level of trust between (1) the local population and (2) local, national, and international human and/or animal health authorities. Too often, those locally reporting disease in animal populations are confronted with what appears to be arbitrary loss of their food animals without compensation. Countries reporting zoonotic diseases internationally may face unilaterally imposed trade restrictions, often accompanied by the imposition of travel advisories and a subsequent drop in tourism. In order for timely, transparent, and credible information to transfer up the line, and for information and support to come back down to the community, prior trust needs to be established between the community, scientists, and the political system at the local, national, and global levels. Building trust will also involve understanding how individuals assess risk and behave in response, and whether various stakeholders believe their concerns can be fairly addressed in the surveillance and response system.

LOOKING TO THE FUTURE

Since the Institute of Medicine released its 1992 report *Emerging Infections: Microbial Threats to Health in the United States*, there has been a growing awareness of the frequency with which new and reemerging infectious diseases are appearing. As the number and frequency of emerging threats increases, the committee realizes that the old veterinary maxim, "When you hear hoofbeats on the covered bridge, don't think about the zebra," needs re-working in today's environment to "When you hear hoofbeats on the covered bridge, at least think about the possibility of a zebra." Most newly emerging infections are zoonotic in origin, for which a limited but broad-based set of microbiological, ecological, and behavioral drivers have been identified. The United States and other well-resourced nations have increased their research efforts and held conferences, leading to an increased number of scientific publications, reports, and to some extent improved disease surveillance and global response on a disease by disease basis. However, more effort is needed, as demonstrated by the first pandemic of the 21st century caused by influenza A(H1N1) 2009, the recent emergence and rapid spread of SARS (albeit to a limited number of countries), and the discovery of West Nile virus in the United States (which has become endemic across the country within a few years). With the exception of pandemic (H1N1) 2009, these examples show how delayed information reporting can result in the further spread of disease. Although subsequent catch-up efforts in the latter two examples finally enabled human and animal health experts to effectively connect and collaborate with each other, those connections had to first be forged during the outbreak, enabling the disease to spread and making it more difficult to contain. Together with HPAI H5N1 as an emerging threat to both humans and animals, these events have captured public attention as never before, highlighting the ongoing risk these pathogens represent and the deficiencies in our disease surveillance and response mechanisms. They further demonstrate to the public the need for continued investment in disease surveillance, as another veterinary adage best describes how it is cheaper to invest in some good stall doors than to try to collect all the horses after they leave the barn.

Perhaps the most serious concerns identified in this report are the continued separation of human and animal health expertise and infrastructure, the vertically organized responses to the recent threats of SARS and HPAI H5N1 infection, and the lack of coordinated governance and funding sufficient to effect change. The committee believes the longstanding cultural and organizational tendencies toward stovepiping are no longer acceptable: Disease surveillance needs to be integrated, developed, and implemented across sectors and disciplines. It would be useful to have a system that requires reporting and action; however, other incentives will surely be necessary to ensure full participation.

Locations where the drivers of emerging zoonotic infectious diseases are most active currently coincide with the developing regions of the world, precisely where the resources for disease surveillance and response are the most limiting. This is a global concern because the impact of zoonotic disease emergence is global, not just local. Because of this interconnectedness, this requires commitment among all nations to share in the cost of developing effective disease surveillance and to sustain and continually improve the technical capabilities of this system. This also requires countries to ensure that responses intended to prevent spread and limit the impact on human and animal health, including the financial and economic consequences of a local or global outbreak, are evidence-based and prompt. The recommendations in this report are broad in their reach and involve technical, financial, and organizational inputs, and they include significant changes in the way global governance of zoonotic disease surveillance and response should be handled. They are based on the full implementation of IHR 2005, and will necessitate significant changes in the way animal disease surveillance, reporting, and response is conducted. None of this will be simple to accomplish, but efforts need to begin now.

Future Research Needs and Considerations

The disease surveillance and response system is never static. As a component of continuous quality improvement and because it is a dynamic system, additional research and other considerations will be needed to evaluate the impact of integrated disease surveillance. Zoonotic disease surveillance and response would benefit from research in the following areas:

- Developing global standards and evaluation criteria for integrated zoonotic disease surveillance systems;
- Adapting evolving methodologies from other basic science disciplines that could be incorporated into integrated surveillance systems;
- Identifying future pathogens (microbiome-type projects) to guide the development of diagnostics, vaccines, and drugs;

- Determining efficacy of and resistance to antibiotics and antivirals;
- Evaluating the effectiveness of laws and regulations on compliance with reporting requirements;
- Evaluating the effectiveness of integrated zoonotic disease prevention programs;
- Identifying and evaluating social and economic incentives to comply with and disincentives to ignore reporting requirements;
- Identifying incentives for communicating, cooperating, and building trust across sectors and disciplines;
- Evaluating the timeliness and level of response that results from early warning systems and the separation of accurate reports from false-positive ones;
- Developing community-based participatory research in the epidemiology of zoonotic diseases;
- Developing social incentives at the local level to promote early disease reporting, avoid outbreak concealment, and engage in effective responses;
- Evaluating how communities understand zoonoses epidemiology, prevention, and treatment in order to foster local participation in disease reporting and surveillance activities; and
- Tracking media representations of zoonotic diseases and assessing how information is produced, circulated, and responded to by the community and policymakers.

To evaluate its progress and impact, it is essential to periodically conduct an in-depth review of how the zoonotic disease surveillance and response system is implemented and functions. Such an evaluation needs to be conducted by an independent, multisectoral, scientific body encompassing human, animal, and environmental expertise to monitor and evaluate the progress of this report's recommendations. As part of that evaluation, an interim report card should be issued by 2012, which coincides with the target date for full-implementation of IHR 2005, and a full report should be issued by 2016 to allow time for responding to the interim evaluation.

Closing Thoughts

The 12 recommendations in this report represent the committee's consensus view on how to systematically address the multiple requirements needed for an effective and sustainable system. In its deliberations, the committee attempted to ensure that its recommendations are pragmatic, focused, informed, and supported by the review of evidence, even when they may challenge convention in some areas. This report reflects the broad disciplinary experience necessary to accomplish the goals it addresses, informed by the broad disciplinary expertise of the committee members. In many ways, the committee is a microcosm of the expertise needed to achieve the sustainable integrated disease surveillance and response called for by the report, and also demonstrates that reaching consensus is no simple task yet is possible. It is essential to begin the process now toward achieving this system. U.S. agencies, in particular USAID, can and should take a lead role—working together with international, intergovernmental, and multinational partners from the public and private domains—in moving from discussion to action.

Global sustainability of zoonotic disease surveillance is predicated on a system that assures international exchange and collaboration to contain the spread of zoonotic diseases through the creation of an atmosphere of transparency, trust, and accountability. The system needs to assist developing countries through relevant capacity building, enabling countries to appropriately contribute in improving global disease surveillance and using information to promptly implement the necessary evidence-based responses. For countries to assume responsibility for zoonotic disease surveillance, the system needs to survive within available national resources and be sustainable. It also needs to be adaptable and flexible enough to meet the needs of each country's changing national demands and priorities and be acceptable to its stakeholders.

Reaching the goal of a sustainable and better integrated global human and animal surveillance and response system for zoonotic emerging diseases depends on a number of preconditions: sufficient political and social will to accomplish it; allocation of necessary financial and technical resources in a sustainable and continuous way; and ensuring that human and animal health officials have the authority and resources to regulate the drivers associated with zoonotic disease emergence, to report emerging events as they occur, and to determine the proper interventions based on the specific nature of the agent and the circumstances of the emergence itself. This is certainly a tall order, but given that political will and financial resources have been individually marshaled for one emerging zoonotic disease after the other, the committee believes that it is possible to create a reliable and sustainable zoonotic disease surveillance system that is flexible, has assured funding, is efficiently implemented, and is acceptable to all stakeholders.

REFERENCES

AVMA (American Veterinary Medical Association). 2008. *One health: A new professional imperative—One Health Initiative Task Force final report*. Washington, DC: AVMA.

Fidler, D. P. 2005. From international sanitary conventions to global health security: The new International Health Regulations. *Chinese J of International Law* 4(2):325–392.

GAO (U.S. Government Accountability Office). 2001. *Challenges in improving infectious disease surveillance systems*. GAO-01-722. Washington, DC: GAO.

Hadorn, D. C., and K. D. Stark. 2008. Evaluation and optimization of surveillance systems for rare and emerging infectious diseases. *Vet Res* 39(6):57.

IPCC (Intergovernmental Panel on Climate Change). 2009. *Organization*. http://www.ipcc.ch/organization/organization.htm (accessed July 20, 2009).

Maugh, T. H., II. 2009. WHO will urge most countries to stop H1N1 testing. *The Los Angeles Times*, July 8.

Murphy, F. A. 1998. Emerging zoonoses. *Emerg Infect Dis* 4(3):429–435.

NRC (National Research Council). 2005. Critical needs for research in veterinary science. Washington, DC: The National Academies Press.

OIE (World Organization for Animal Health). 2008. *OIE tool for the evaluation of performance of veterinary services*, 3rd ed. Paris, France: OIE.

Pappaioanou, M., M. Malison, K. Wilkins, B. Otto, R. A. Goodman, R. E. Churchill, M. White, and S. B. Thacker. 2003. Strengthening capacity in developing countries for evidence-based public health: The data for decision-making project. *Soc Sci Med* 57(10):1925–1937.

Schwabe, C. W. 1996. Ancient and modern veterinary beliefs, practices and practitioners among Nile Valley peoples. In *Ethnoveterinary research and development*, edited by C. M. McCorkle, E. Mathias, T. W. Schillhorn van Veen. London, UK: Intermediate Technology Publications.

Appendix A

Glossary of Terms

Anthropogenic—Caused or produced by humans.

Arboviruses—Viruses transmitted mainly by arthropods.

Biosafety levels—Recommended containment or biosafety levels (BSL) that describe safe methods for managing infectious materials in the laboratory environment where they are being handled or maintained.

Biosecurity—A strategic and integrated approach that encompasses the policy and regulatory frameworks (including instruments and activities) that analyze and manage risks in the sectors of food safety, animal life and health, and plant life and health, including associated environmental risk. Biosecurity covers the introduction of plant pests, animal pests and diseases, and zoonoses, the introduction and release of genetically modified organisms and their products, and the introduction and management of invasive alien species and genotypes.

BSL laboratory designations (BSL 1–4)—There are four BSLs, with BSL-1 representing a basic level of containment relying on standard microbiological practices and BSL-4 representing the most advanced containment when working with dangerous and exotic agents that pose a high individual risk of life-threatening disease (which may be transmitted via the aerosol route and for which no vaccine or therapy is available). The increasing numbers correspond to the increasing levels of protection for personnel and the environment. The purpose is to reduce or eliminate exposure of laboratory workers, other persons, and the outside environment to potentially hazardous agents. Each combination is specifically appropriate for the operations performed, the documented or suspected routes of transmission of the infectious agents, and the laboratory function or activity.

Bush animals—Species include apes, other primates, ungulates, rodents, and birds. The species hunted depends on the geographical area and the hunters' preferences, cultural practices, and prohibitions.

Bushmeat—Term commonly used for meat of terrestrial wild animals, killed for subsistence or commercial purposes throughout the humid tropics of the Americas, Asia, and Africa.

CITES—The Convention on International Trade in Endangered Species of Wild Fauna and Flora is an international agreement between governments. Its aim is to ensure that international trade in specimens of wild animals and plants does not threaten their survival. CITES was drafted as a result of a resolution adopted in 1963 at a meeting of members of IUCN (The World Conservation Union). The text of the Convention was agreed upon at a meeting of representatives of 80 countries in Washington, DC, on March 3, 1973, and on July 1, 1975, CITES entered in force. It is an international agreement to which countries adhere voluntarily, and is now made up of 175 parties.

Codex Alimentarius—A collection of internationally recognized standards, codes of practice, guidelines, and other recommendations relating to food, food production, and food safety. Its texts are developed and maintained by the Codex Alimentarius Commission, a body that was established in 1963 by the Food and Agriculture Organization of the United Nations (FAO) and the World Health Organization (WHO). The Commission's main aims are stated as being to protect the health of consumers and ensure fair practices in the international food trade. The Codex Alimentarius is recognized by the World Trade Organization as an international reference point for the resolution of disputes concerning food safety and consumer protection.

Domestic animal—Animals that have been bred selectively in captivity and thereby modified from their ancestors for use by humans who control the animals' breeding and food supply.

Driver—A factor that causes a zoonotic disease to emerge or reemerge.

Emerging infection—Either a newly recognized, clinically distinct infectious disease, or a known infectious disease whose reported incidence is increasing in a given place or among a specific population.

Endemic—Restricted or peculiar to a locality or region. Endemic infection refers to a sustained, relatively stable pattern of infection in a specified population.

Epidemic—The occurrence of an illness (or other health-related event) in a community or region clearly in excess of normal expectancy.

Food security—Comprises access, availability, and utilization issues. Nutrition security is achieved when reliable access to appropriately nutritious food is coupled with a sanitary environment, adequate health services, and care to ensure a healthy and active life for all household members.

GDP—Gross domestic product is the market value of all final goods and services made within the borders of a nation in a year.

Globalization—A widely used term to describe the process by which people of the world are unified into a single society and function together. This process is usually recognized as being driven by a combination of economic, technological, sociocultural, political, and biological factors.

Host—Person or other living animal that affords subsistence or lodgment to an infectious agent under natural conditions.

Hotspot—Region where factor(s) are most densely aggregated, most highly prevalent, and where risk of a (disease) event is most intense.

Human–animal interface—Ways in which humans and animals interact, which may include, but are not limited to, cohabitation (domestic and exotic animals as pets or harvesting parts such as wool to make products for human use) or coexistence (with juxtaposed or integrated habitats), the production of food animals or hunting, scientific research, wildlife conservation, and public education (in zoos or sanctuaries).

Integrated disease surveillance system for emerging zoonotic diseases—A system of shared and/or integrated, linked, clinical, epidemiological, laboratory, and risk behaviour components of human and animal disease surveillance systems, such that the processes of information collection, management, collation, analysis, presentation/reporting, and dissemination of data from human and animal systems are brought together to be used in decisionmaking for response by human and animal health authorities from local through international levels.

International Health Regulations—Originally adopted in 1969, World Health Organization (WHO) member states approved a revised set of these regulations (known as IHR 2005) that went into effect in 2007. IHR 2005 establishes WHO's central role in coordinating the control of disease and facilitating disease surveillance and response efforts against the spread of disease at the global level. Under the regulations, WHO requires member states to report all events that may constitute a "public health emergency of international concern," which includes a (1) human and animal health risk to other states through the international spread of disease, and (2) an event that potentially could require a coordinated international response.

Necropsy—An examination and dissection of a dead body to determine cause of death or the changes produced by disease.

Notifiable disease—A disease for which regular, frequent, timely information on individual cases is considered necessary to prevent and control that disease. Each year a list of nationally notifiable diseases is agreed on and maintained by the Council of State and Territorial Epidemiologists and the Centers for Disease Control and Prevention (CDC). Diseases that are considered nationally notifiable may or may not be designated by a given state as notifiable (reportable) in the state. States may use the national notifiable diseases list as well as other information, such as state-specific health priorities, to guide their determination of which conditions/diseases to make notifiable in their state. Thus, the list of state-specific notifiable diseases may vary across states and in a given state; the list may vary over time as well. Disease reporting is currently mandated by legislation or regulation only at the local or state level.

One health—The American Veterinary Medical Association defines "one health" as the collaborative efforts of multiple disciplines working locally, nationally, and globally to attain optimal health for people, animals, and our environment. The concept was first proposed by veterinary epidemiologist and parasitologist Dr. Calvin W. Schwabe, who used the term "one medicine" in the 1960s to capture the vital importance of considering medical and veterinary issues jointly in the study of zoonotic diseases. This multidisciplinary approach has been captured by recent "one health" initiatives. An example of such an initiative, "One world-one health," which is a trademark of the Wildlife Conservation Society, has developed a series of symposiums under this concept.

Pandemic—A global disease outbreak.

Pathogen—Biological agent capable of causing disease.

Pathogenesis—The entry, primary replication, spread to target organs, and establishment of infection.

Prion—A microscopic protein particle similar to a virus but lacking nucleic acid, thought to be the infectious agent responsible for scrapie and certain other degenerative diseases of the nervous system.

Production system—A production system clusters production units (herds, farms, ranches), which, because of the similar environment in which they operate, can be expected to produce according to similar production functions. This similar environment can be characterized by the physical (climate, soils, and infrastructure) and biological environments (plant biomass production, food animal species composition), economic and social conditions (prices, population pressure and markets, human skills, and access to technology and other services), and policies (land tenure, trade, and subsidy policies).

Reemerging—Known diseases that have reappeared after a significant decline in incidence.

Regime—Principles, norms, rules, and decisionmaking procedures around which actor expectations converge in a given issue or area.

Reservoir—Any person, animal, arthropod, plant, soil, or substance (or combination of these) in which an infectious agent may reside.

Response—Interventions that involve human and animal health systems and practitioners using disease surveillance information to plan and execute activities that prevent infectious diseases from affecting human and animal populations, protect such populations against exposure to pathogenic microbes that evade prevention measures, and control morbidity and mortality among populations infected by pathogenic agents.

Risk assessment—The process of quantifying the probability of a harmful effect to individuals or populations from certain human activities.

Sanction(s)—General trade restrictions between nations. Economic sanctions include trade bans, tariffs, import duties, and import or export quotas.

Smallholder poultry keeper—Describes the practice of individuals and families keeping small flocks of poultry or other fowl in their backyards for their consumption or as a means of economic livelihood.

Spillover—Spillover occurs when epidemics in a host population are driven not by transmission within that population but by transmission from a reservoir population. A pathogen typically reaches high prevalence in a reservoir and then spills over into the other host.

SPS Agreement—The Agreement on the Application of Sanitary and Phytosanitary Measures is an international treaty of the World Trade Organization (WTO). It was negotiated during the Uruguay Round of the General Agreement on Tariffs and Trade (GATT) and entered into force with the establishment of WTO at the beginning of 1995. Under the SPS Agreement, WTO sets constraints on member states' policies relating to food safety as well as animal and plant health about imported pests and diseases. It contains specific rules for countries that want to restrict trade to ensure food safety and the protection of human life from zoonoses, although it is a fundamental requirement that member states have a scientific basis to justify trade measure aimed at mitigating a health risk.

Surveillance system—A system for public health surveillance is a group of integrated and quality-assured, cost-effective, and legally and professionally acceptable processes designed for the purpose of identifying in an ongoing, flexible, standardized, timely, simple, sensitive, and predictive manner the emergence of meaningful epidemiologic phenomena and their specific associations. These processes include human, animals, laboratory, and informatics activities to skillfully manage information derived from an entire defined community (or a subgroup thereof that is sufficiently representative and large) and to disseminate that information in a timely and useful manner to those able to implement appropriate public health interventions.

Sustainability—In a broad sense, sustainability refers to the capacity for systems to remain diverse and productive over time. It requires the integration of social, economic, and environmental spheres such that the needs of the present are met without compromising the needs of future generations. A sustainable surveillance system would include long-term financial investment and infrastructure development and maintenance.

TBT Agreement—The Agreement on Technical Barriers to Trade is an international treaty of the World Trade Organization (WTO). It was negotiated during the Uruguay Round of the General Agreement on Tariffs and Trade (GATT) and entered into force with the establishment of WTO at the beginning of 1995. It is in place to ensure that regulations, standards, testing, and certification procedures do not create unnecessary obstacles to trade. The agreement prohibits technical requirements created in order to limit trade, as opposed to technical requirements created for legitimate purposes such as consumer or environmental protection.

Transboundary—Diseases that are of significant economic, trade, and/or food security importance for a considerable number of countries, which can easily spread to other countries and reach epidemic proportions and where control or management, including exclusion, requires cooperation between several countries.

TRIPS Agreement—The Agreement on Trade Related Aspects of Intellectual Property Rights is an international agreement administered by the World Trade Organization (WTO) that sets down minimum standards for many forms of intellectual property regulation. It was negotiated at the end of the Uruguay Round of the General Agreement on Tariffs and Trade (GATT) in 1994. It contains several provisions that enable governments to implement their intellectual property regimes in a manner which takes account of immediate and longer-term public health considerations.

Vector—An organism, such as an insect, that transmits a pathogen from one host to another.

Vector-borne disease—A disease in which the pathogenic microorganism is transmitted from an infected individual to another individual by an arthropod or other agent, sometimes with other animals serving as intermediary hosts.

Virulence—Degree of pathogenicity of an infectious agent, indicated by the case fatality rates and/or its ability to invade and damage tissue of the host.

Xenotransplantation—Any procedure that involves the transplantation, implantation, or infusion into a human recipient of either (1) live cells, tissues, or organs from a nonhuman animal source, or (2) human body fluids, cells, tissues, or organs that have had ex vivo contact with live nonhuman animal cells, tissues, or organs.

Wild animal—Terrestrial animals that are untamed or undomesticated. They are killed for subsistence or commercial purposes throughout the humid tropics of the Americas, Asia, and Africa.

Zoonoses—Any infection or infectious disease transmissible under natural conditions from animals to humans or those shared between humans and animals. Zoonoses may be bacterial, viral, or parasitic, or may involve unconventional agents.

Zoonotic disease surveillance—The ongoing systematic and timely collection, analysis, interpretation, and dissemination of information about the occurrence, distribution, and determinants of diseases transmitted between humans and animals. Zoonotic disease surveillance reaches its full potential when it is used to plan, implement, and evaluate responses to reduce infectious disease morbidity and mortality through a functionally integrated human and animal health system.

Appendix B

Surveillance and Response of Select Zoonotic Disease Outbreaks

WILDLIFE TRADE AND THE HUMAN MONKEYPOX OUTBREAK IN THE UNITED STATES

Multistate Epidemiological Investigation

The first U.S. outbreak of human monkeypox was reported in May 2003 and initially included cases in Wisconsin, Indiana, and Illinois. By the end of the outbreak in June 2003, there were reports of cases in Missouri, Kansas, and Ohio (CDC, 2003a). As of July 31, 2003, there were 72 reported cases, of which 37 were laboratory confirmed (CDC, 2003b). Epidemiological and trace-back investigation by local, state, and federal public health authorities found that patients acquired the disease from prairie dogs in contact with human monkeypox-infected African rodents (CDC, 2003b). These prairie dogs were housed together with infected African rodents in an Illinois wholesale pet store. Approximately 200 prairie dogs were in this facility and possibly exposed to human monkeypox in the period between when the Illinois animal distributor purchased the African rodents and the first reported human case of human monkeypox. A Texan animal distributor legally imported the infected rodents (762 rodents that included rope squirrels, tree squirrels, Gambian giant rats, brush-tailed porcupines, dormice, and striped mice) from Accra, Ghana (CDC, 2003a). These rodents were not screened for disease before or after they entered the United States. Of this shipment, 23 percent of the imported rodents could not be traced beyond the port of entry because records were not available. Before laboratory confirmation, trace-forward investigations suspected these rodents were the source of human monkeypox. This investigation determined that no other U.S. animals besides prairie dogs were infected with human monkeypox (CDC, 2003b). Finally, clinical studies concluded that respiratory and direct mucocutaneous exposures were important routes of transmissions between infected prairie dogs and humans (Guarner et al., 2004).

CDC and FDA Restrictions of Rodents from Africa

On June 11, 2003, the Centers for Disease Control and Prevention (CDC) and Food and Drug Administration (FDA) jointly issued an order pursuant to 42 C.F.R. 70.2 and 21 C.F.R. 120.30, respectively, to restrict the "transportation or offering for transportation in interstate commerce, or the sale, offering for sale, or offering for any other type of commercial or public distribution, including the release into the environment of" prairie dogs, tree squirrels, rope squirrels, Gambian giant rats, dormice, brush-tailed porcupines, and striped mice (CDC, 2003a; FDA, 2008). In addition, pursuant to 42 C.F.R. 71.32(b), CDC implemented an immediate embargo on the importation of all rodents from Africa. Because the actions taken by state health authorities were insufficient to prevent the spread of human monkeypox, CDC and FDA issued an interim final rule (42 C.F.R. 71.56 and 21 C.F.R. 1240.63 respectively) under section 361 of the Public Health Service (PHS) Act that was intended to prevent future introduction, establishment, and spread of the human monkeypox virus in the United States. Based on risk-assessment of the further transmission of the human monkeypox virus, FDA removed its regulation in 21 C.F.R. 1240.63 in 2008 and concluded that CDC's interim final rule and routine state disease surveillance and preventive measures were sufficient to prevent new human and animal cases of human monkeypox. Under section 368(a) of the PHS Act, any person who violates a regulation prescribed under the Act may be punished by imprisonment for up to 1 year or fined up to $100,000 per violation if death has not resulted from the violation or up to $250,000 per violation if death has resulted. Organizations may be fined up to $200,000 per violation not resulting in death and $500,000 per violation resulting in death (FDA, 2008).

Reemergence of Human Monkeypox in Africa

The virus that causes human monkeypox was first isolated in 1958 from monkeys and recognized as a new virus of the genus *Orthopoxvirus* (same genus as the smallpox virus although different epidemiologically and biologically) (Guarner et al., 2004). Human monkeypox, however, was first identified in humans in 1970 in the tropical areas of the Democratic Republic of the Congo (DRC) (Breman, 2000; CDC, 2003a). The first outbreaks of human monkeypox occurred in the period of 1970–1980 in the DRC, Côte d'Ivoire, Liberia, Nigeria, and Sierra Leone. Active disease surveillance was implemented with the assistance of the World Health Organization (WHO) in 1981–1986 in the DRC, where most of the human cases during this period occurred. Reporting of human monkeypox decreased, and after 1992 no new cases were reported to WHO. Failure to maintain disease surveillance of human monkeypox contributed to the reemergence of the disease in the DRC in 1996. Epidemiological and laboratory investigation of the DRC outbreak concluded that the disease was mild but highly transmissible (Heymann et al., 1998).

UK GOVERNMENT REGULATORY RESPONSE TO CONTROL BOVINE SPONGIFORM ENCEPHALOPATHY (BSE)[1]

In 1986, when BSE was identified as a new disease, the Ministry of Agriculture, Fisheries and Food (MAFF) was the government agency responsible for overseeing state veterinary services under the State Veterinary Service for Great Britain, composed of the Veterinary Investigation Service (VI Service), the Veterinary Field Service, and the Central Veterinary Laboratory (CVL). The VI Service implemented surveillance and provided expert advice for veterinary surgeons in private practice about unknown animal diseases. Employees of the VI Service reported to the assistant chief veterinarian and the chief veterinary officer at MAFF.

MAFF and its agencies, prior to the identification of BSE, relied on a passive surveillance system for the identification of new diseases in animals. The surveillance of nonnotifiable diseases was based on the observations of an astute farmer and veterinarian, who would voluntarily notify one of the many Veterinary Investigation Centers (VICs) of the VI Service.

December 22, 1984—David Bee, a local private veterinarian, was called to examine Cow 133, owned by Peter Stent of Pitsham Farm in Sussex. Cow 133 developed a head tremor and a lack of coordination before dying on February 11, 1985. Bee sought assistance from J. M. Watkin-Jones, a veterinarian at the Winchester VIC, one of the branches of the VI Service.

September 13, 1985—Carol Richardson, the pathologist on duty at the CVL, received samples of brain, spinal cord, and kidney of Cow 142 and examined them. Cow 142 was also owned by Stent and was showing nervous clinical signs similar to Cow 133. Richardson shared the sample with her colleagues at the CVL Pathology Department. Initially, the pathological examination suggested that the cause of the disease was not acute, but chronic bacteremia or endotoxemia.

[1]The BSE Inquiry (2000).

> April 1985—Colin Whitaker, a private veterinarian, was called to Plurenden Manor Farm in Kent to look at a cow. He sent the cow to the University of Bristol Veterinary School at Langford, and postmortem examination showed "progressive nervous signs, hyperesthesia, tremors, mania and hind leg ataxia."
>
> November 1986—Whitaker consulted Dr. Carl Johnson, veterinary officer of the Wye VIC, who referred the brains of the three animals to the CVL in November and December 1986.
>
> December 11, 1986—The CVL also received brain samples from a cow that was referred by Langford VIC, in Bristol. In December 1986, after the identification of BSE as a novel disease by CVL scientists, Dr. Watson, CVL director, informed William Rees, chief veterinary officer at MAFF, about BSE. In June 1987, new knowledge regarding the pathology and epidemiology of the disease led to the notification of this new disease to other ministers and government officials. After the preponderance of the evidence at the time regarding the risks posed by this novel disease to human health, the government implemented a series of regulations aimed at the animal feed and rendering industry as well as slaughterhouses.

Because it was initially understood that BSE was spread through animal feed, in June 1988, the Bovine Spongiform Encephalopathy Order 1988 introduced a ban of ruminant feed in addition to the compulsory notification of BSE. The Order, which came into effect on July 18 and only applied to Great Britain, required farmers or their veterinarians to notify the local Divisional Veterinary Officer if they suspected an animal was affected by BSE. At this point, MAFF would send one of its own veterinarians to investigate. The ruminant feed ban included the following provisions:

> (1) No person shall knowingly sell or supply for feeding to animals any feedstuff in which he knows or has reason to suspect any animal protein has been incorporated.
>
> (2) No person shall feed to an animal any feedstuff in which he knows or has reason to suspect that any animal protein has been incorporated.

On August 8, 1988, two further Orders came into effect: The Bovine Spongiform Encephalopathy (Amendment) Order 1988 and The Bovine Spongiform Encephalopathy Compensation Order 1988. They introduced a policy of compulsory slaughter of BSE-infected animals and payment of compensation to the owner of the slaughtered animal.

On November 13, 1989, the Bovine Offal (Prohibition) Regulations 1989 came into effect in England and Wales. This regulation prohibited the use of specified bovine offal (SBO) in the preparation of food for human consumption after findings showed that particular cattle organs were most likely to carry the infective agent. SBO included the brain, spinal cord, thymus, spleen, tonsils, and intestines from a bovine animal more than 6 months of age. One of the unintended consequences of this ban was that renderers (rendering is the process of converting animal byproducts into more useful materials, e.g., purifying fatty tissue into lard or tallow) demanded that mechanically recovered bovine meat (MRM) should not contain SBO, and thus no longer welcomed cow heads containing the brain. As a result, a practice rapidly developed at many slaughterhouses of splitting the skull and removing the brain. This practice gave rise to problems of contamination. Later on, review of the SBO ban revealed a concern about the risk that slaughterhouse practices would result in the contamination of MRM. MAFF officials assumed that the regulations up to 1989 would have reduced the scale of infection to a fraction of that at the height of the epidemic.

However, many more animals born after the ban (BAB) were diagnosed with BSE, which showed proof of the limitation and problems in the implementation of the BSE legislation up to 1989. The first case of BSE in an animal born after the introduction of the ruminant feed ban was not confirmed until March 1991. By September 1992, the number of BABs had risen to 220. By September 1994, the total number of BABs had reached 12,860. It was concluded that BABs had been fed contaminated feed. Based on the finding, more aggressive regulatory measures followed in the period 1994–1996 to prevent the spread of the disease and to protect human health.

The UK BSE epidemic forced changes in different sectors of the animal and food production industry. First, regulations of the rendering industry changed the rendering processes, and, as a result of BSE, meat and bone meal is no longer used in the United Kingdom in animal feed or as fertilizer. Second, the introduction of regulations of the animal feed industry affected the industry but was an essential part of control of the disease. Third, it was considered essential that slaughterhouses separated SBO from those parts of the carcass that were going to enter the human food chain. All these interventions underscore that the risk of disease or contamination was in the processing of animal materials, which put humans at risk at many different points of this process.

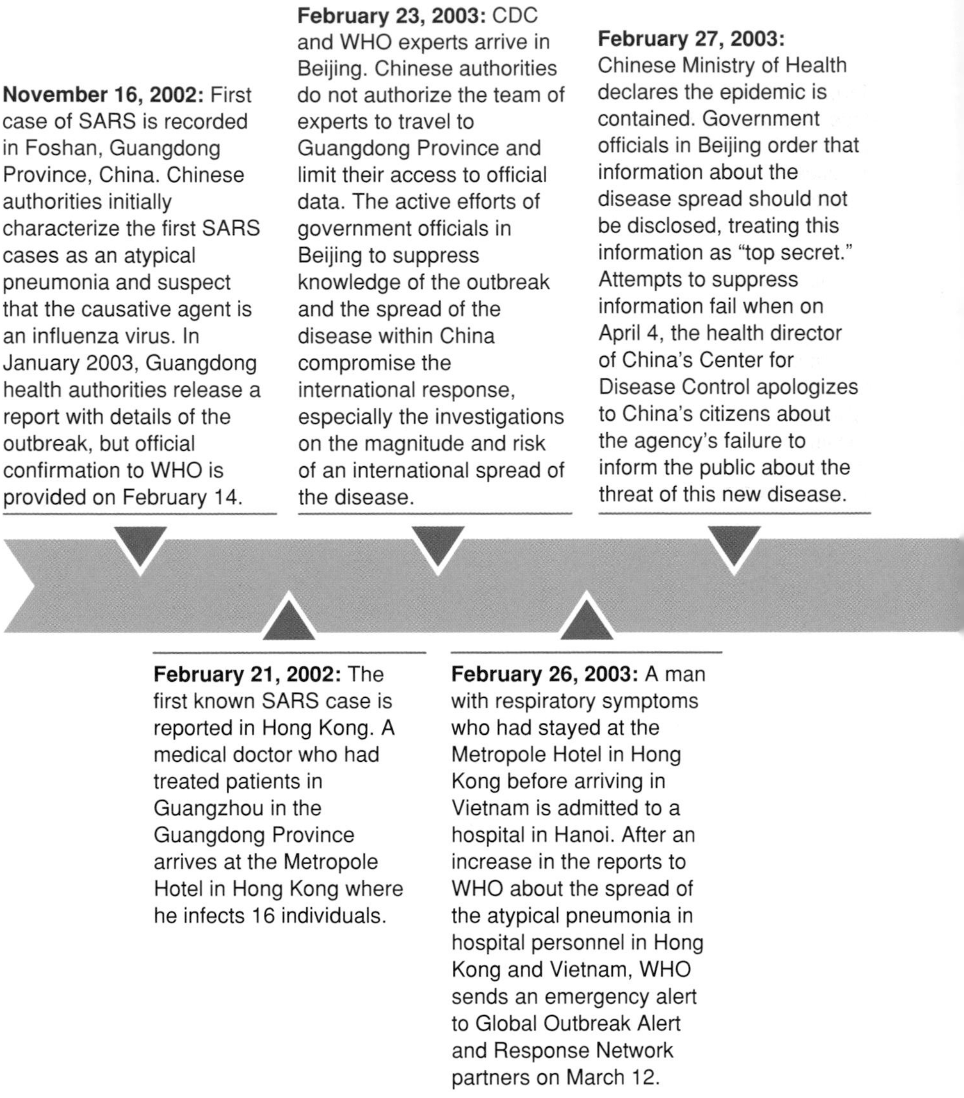

November 16, 2002: First case of SARS is recorded in Foshan, Guangdong Province, China. Chinese authorities initially characterize the first SARS cases as an atypical pneumonia and suspect that the causative agent is an influenza virus. In January 2003, Guangdong health authorities release a report with details of the outbreak, but official confirmation to WHO is provided on February 14.

February 21, 2002: The first known SARS case is reported in Hong Kong. A medical doctor who had treated patients in Guangzhou in the Guangdong Province arrives at the Metropole Hotel in Hong Kong where he infects 16 individuals.

February 23, 2003: CDC and WHO experts arrive in Beijing. Chinese authorities do not authorize the team of experts to travel to Guangdong Province and limit their access to official data. The active efforts of government officials in Beijing to suppress knowledge of the outbreak and the spread of the disease within China compromise the international response, especially the investigations on the magnitude and risk of an international spread of the disease.

February 26, 2003: A man with respiratory symptoms who had stayed at the Metropole Hotel in Hong Kong before arriving in Vietnam is admitted to a hospital in Hanoi. After an increase in the reports to WHO about the spread of the atypical pneumonia in hospital personnel in Hong Kong and Vietnam, WHO sends an emergency alert to Global Outbreak Alert and Response Network partners on March 12.

February 27, 2003: Chinese Ministry of Health declares the epidemic is contained. Government officials in Beijing order that information about the disease spread should not be disclosed, treating this information as "top secret." Attempts to suppress information fail when on April 4, the health director of China's Center for Disease Control apologizes to China's citizens about the agency's failure to inform the public about the threat of this new disease.

FIGURE B-1 National and international response to the SARS outbreak.

March 15, 2003: WHO issues Global Travel Advisory. Before the identification of the causative agent, the virus spreads within 6 months to 30 countries and administrative regions. The virus transmission along five major airline routes by the symptomatic individuals traveling from Hong Kong to Beijing, Hanoi, Singapore, Taiwan, and Toronto accelerates the global spread of SARS.

March 27, 2003: WHO issues recommendation of exit screening of passengers at airports.

April 3, 2003: WHO expert team finally arrives in Guangdong Province. The next day, U.S. President George W. Bush signs executive order adding SARS to the list of quarantinable communicable diseases, which provides CDC with legal authority to implement isolation and quarantine measures.

April 16, 2003: WHO laboratory network announces conclusive identification of new coronavirus as the causative agent for SARS. The Chinese government increases transparency through the release of number of cases in each Province, in addition to daily updates. Moreover, based on media reports, more than 120 officials were dismissed, including the health minister and Beijing's mayor, or penalized for ineffective response to the outbreak.

July 5, 2003: WHO announces the global containment of the SARS outbreak.

HIGHLY PATHOGENIC AVIAN INFLUENZA H5N1 SURVEILLANCE IN HONG KONG AND VIETNAM

H5N1 Virus Evolution

Highly pathogenic avian influenza (HPAI) H5N1 virus is the causative agent for millions of bird deaths in Southeast Asia, the Middle East, Europe, and Africa. The natural reservoir of the influenza type A virus is wild waterfowl, but the virus can also infect domestic poultry and humans and cause illness and death, thus the high pathogenicity of the virus (Nguyen et al., 2005). Virus typing and serological identification tests have indicated that different strains of type A influenza H5N1 are responsible for illness and death in humans in Southeast Asia since the Hong Kong outbreak in 1997 (Wan et al., 2008). This outbreak was characterized by a 10 percent incidence of HPAI H5N1 infection in live-bird market poultry workers exposed to infected domestic birds housed in close contact with wild waterfowl (Nguyen et al., 2005). After the Hong Kong outbreak, the virus spread to other countries in the region more likely through the poultry trade. Although the H5N1 virus has reassorted many times, all these viruses carry the same H5 hemagglutinin (HA) gene, which has a central role in antigenic drift. In Vietnam, isolation of the H5N1 virus shows six different HA clades, thus suggesting that the virus has been introduced at least six times since the first isolation in poultry in 2001 (Wan et al., 2008).

Response to HPAI H5N1 Outbreak in Hong Kong

In 1997, Hong Kong health authorities quickly instituted strong control measures in poultry to minimize or stop human exposures (Webster, 2004). These measures included slaughter of 1.6 million chickens present in wholesale facilities or vendors within Hong Kong; banning importation of chickens from neighboring areas; instituting serological monitoring of chickens in Hong Kong; marketing chickens separately from other avian species; separating chickens and ducks for transport to market; slaughtering chickens and ducks separately; changing the operation and management of the live market system such that aquatic birds were no longer housed and sold in Hong Kong live bird markets, rather they were made available for sale only as killed, chilled poultry; serologically screening all poultry imported for sale in Hong Kong for avian influenza virus H5 subtype antibody prior to release for sale; and instituting measures to improve hygiene in the markets. Further interventions were instituted that included establishing surveillance in live poultry markets and in poultry at the Chinese border at which each arriving flock was quarantined, tested, and held for 2 days, flocks with one or more sick birds were rejected, and clean flocks were moved to a central wholesale warehouse and held for 2 or more days. Culling was carried out on an ongoing basis as necessary; monthly rest days in live bird markets were instituted where unsold poultry in retail markets were killed, markets were left empty a whole day, cleaned, and restocked with fresh poultry the next day; and birds were vaccinated in outbreak situations. Transmission of the virus and further outbreaks in poultry were controlled and stopped.

During 1998–2003, isolated outbreaks of HPAI H5N1 in poultry occurred in Southeast Asia; however, it was not until mid-2003, when more widespread outbreaks in poultry occurred in South Korea. There were significant delays in international reporting, and weaker response measures were instituted and the virus began to spread across Southeast Asia. Additional outbreaks in poultry and human cases of HPAI were next identified in Vietnam in 2003.

HPAI H5N1 Outbreaks in Humans and Animals in Vietnam

HPAI infection in humans was first officially reported in Vietnam in January 2004; subsequently the country has endured six waves of epizootics of HPAI H5N1 in poultry (Vu, 2009). In 2003–2004, two waves of outbreaks in poultry affected many provinces (56 provinces reported outbreaks during the first wave and 17 provinces during the second wave), which resulted in the death by infection or culling of more than 44 million birds (Sims, 2007; Vu, 2009). During the third wave from December 2004 to April 2005, outbreaks were reported in 36 provinces, with about 2 million birds killed. At this time, the government implemented a pilot vaccination campaign and was recommending a nationwide vaccination (Vu, 2009). During the fourth wave from October to December 2005, 21 provinces reported outbreaks in poultry, which resulted in a loss of 4 million birds. In 2006, the virus activity was low mainly due to mass vaccination of poultry in the previous year; however, new reassorted viruses were still circulating at low levels (Wan et al., 2008). From December 2006 to November 2007, the reemergence of virus was reported in poultry in more than 20 provinces and resulted in a loss of 270,000 birds. Recent reports indicate that a sixth wave of outbreaks occurred from December 2007 to March 2008 (Vu, 2009). In March 2009, the government reported eight outbreaks in six more provinces.

Vaccination of Poultry in 2005

In October 6, 2005, the government of Vietnam launched the vaccination campaign nationwide (Vu, 2009). The goal of vaccination was to reduce the number of susceptible poultry, raise the immunological resistance to virus, and reduce the amount of the virus that immune-infected poultry can excrete (Sims, 2007). In mid-2005, the government also introduced a number of control measures such as banning of duck breeding, public awareness campaigns, and the closure of urban markets in addition to restricting culling to known infected flocks in order to reduce the risk of infection to HPAI H5N1 viruses (Van Nam, 2007). The next year, in 2006, scientists suggested that the lower activity of virus was due to vaccination. Postvaccination disease surveillance provided evidence that the mass vaccination program and other control measures were successful in controlling transmission to humans, as there were no reported cases of disease in humans in 2006. On the other hand, vaccination activities resulted in a shift to passive disease surveillance of the virus, which assumed the eradication of the virus in Vietnam and bordering countries. However, the emergence of the virus in 2007 and later years demonstrates the systematic failure to detect the new variants circulating in Vietnam, although at lower levels, previous to the 2007 outbreak. Although vaccination was important in reducing the virus genetic reservoir, the experience in Vietnam demonstrates that strengthening disease surveillance in poultry is an essential component of the strategy to be able to prevent and control the introduction of new HPAI H5N1 strains into the country (Wan et al., 2008).

Response Measures and Impacts on Human and Animal Morbidity and Mortality

Although Vietnam is one of the countries most affected by HPAI H5N1, the response measures have been successful in controling new infections in poultry and preventing transmission to humans. Additionally, there was a cost benefit of implementing vaccination in poultry versus mass culling of poultry. However, the World Organization for Animal Health has recently emphasized the need for an exit strategy in places where vaccination is being used as a control measure that have been able to improve veterinary services and biosecurity measures (OIE, 2009). Some experts believe eradication of HPAI H5N1 would be difficult to achieve and thus many countries will continue to use vaccination for many years (Sims, 2007). This means that a mass vaccination program may be unsustainable, especially due to the high costs and the limited number of field staff as in the case of Vietnam. However, Vietnam has taken steps to review their current vaccination policies and explore the option of more targeted vaccination of poultry. It has also recognized the importance of postvaccination disease surveillance, especially in monitoring the effects of vaccination on emergence of virus variants. As of July 1, 2009, 436 human cases of HPAI H5N1 and 262 deaths have been reported from 15 countries (WHO, 2009a). Despite further surveillance and response efforts instituted in poultry by human and health authorities of affected countries, poultry outbreaks and human cases continue to occur.

WEST NILE VIRUS OUTBREAK IN NEW YORK CITY[2]

West Nile virus (WNV) first appeared in birds around mid-June of 1999 when veterinarians at Bayside veterinary clinic in the Flushing neighborhood of Queens identified neurological disorders in crows. By mid-August, dead crows were sent to the state Department of Environmental Conservation (DEC), which had jurisdiction over wildlife, for necropsy examinations. Parallel to the bird deaths in Queens, numerous crows and other birds were dying in and around the Bronx Zoo, prompting veterinarians at the zoo to send dead birds to the DEC for examination. The chief pathologist at the Bronx Zoo, however, believed that the DEC wildlife pathologist's determination of the cause of deaths of bird specimens from Queens was not correct since it was not based on histopathology, and therefore decided to initiate her own necropsy on zoo birds, which showed possible encephalitis. Only days later, a separate epidemiological investigation of suspected human cases of viral encephalitis was initiated by the New York City Department of Health (DOH) Bureau of Communicable Disease. An initial investigation by city public health officials revealed a cluster of human cases with the same symptoms; subsequently the city DOH notified the state health department and CDC for additional assistance. After conversations with CDC and the state health department, city health officials sent patient specimens to the state virology laboratory for examination. Field investigations revealed the presence of *Culex pipiens* mosquito breeding sites and larvae in many of the patients' homes and in the Queens neighborhood, reinforcing the theory of viral encephalitis.

In early September 1999, public health and veterinary authorities continued to conduct two separate investigations. The human outbreak investigations involved multiple laboratory facilities (state and federal), public health officials at the local, state, and federal level, and city government officials. On the animal side, mainly state wildlife scientists and Bronx Zoo veterinarians conducted investigations of the deaths in birds. On September 2, 1999, state laboratory tests were positive for a flavivirus; specifically the test showed a strong serological reaction to St. Louis Encephalitis (SLE) virus, results that were confirmed the next day by the CDC Division of Vector-Borne Infectious Disease laboratory in Fort Collins, Colorado. The same day, city officials announced CDC's confirmation of an SLE outbreak in New York City and the decision to initiate mosquito control activities.

At this point of the human outbreak investigation, communications

[2]GAO (2000); Fine and Layton (2001); Scott (2002).

between federal, state, and local public health officials were consistent. However, key public health officials were not aware of the early events in birds, especially the neurological disorders identified by local veterinarians and the increased deaths of birds in the Bronx Zoo. City public health officials first became aware of the bird deaths after news report on the SLE outbreak resulted in calls to the bureau hotline. On the other hand, after listening to news reports of an SLE outbreak in Queens, the Bronx Zoo chief pathologist began to suspect a possible link between the bird deaths and human cases of SLE and decided to send specimens directly to the National Veterinary Services Laboratory (NVSL), a U.S. Department of Agriculture (USDA) reference laboratory located in Ames, Iowa.

After multiple efforts by the Bronx Zoo chief pathologist to send zoo specimens to the CDC laboratory in Fort Collins, the laboratory scientists accepted to examine the bird specimens from the Bronx Zoo. Still the priority of the CDC laboratory in Fort Collins was to not only process thousands of samples from hospitals but also confirm the initial SLE diagnosis through lengthy viral neutralization tests, which required isolation of the virus and polymerase chain reaction (PCR). Although some of these tests reveal questions on the accuracy of the diagnostic tools, the CDC laboratory did not reconsider the SLE diagnosis until the NVSL notified them that they had successfully isolated a flavivirus from one of the Bronx specimens and other specimens received from the state DEC. At the same time, independent analyses of human specimens by the New York State (NYS) DOH virology laboratory resulted in a negative PCR reaction for SLE. In addition, in a meeting between state and city health officials and CDC the participants raised questions about the accuracy of the results from serologic tests performed on specimens of suspected and confirmed cases. The issue of test accuracy was again raised in a meeting of an independent working group studying encephalitis from unknown origin in which NYS health officials and CDC participated. By the end of the meeting, it was agreed that the NYS DOH would share specimens of human brain tissue with Dr. Ian Lipkin, an academic researcher from University of California at Irvine attending the meeting. Around the same time, independent efforts by the Bronx Zoo chief pathologist resulted in the involvement of the U.S. Army Medical Research Institute of Infectious Diseases (USAMRIID).

Parallel to the independent investigations by Dr. Lipkin and USAMRIID, a Fort Collins scientist began PCR testing on human specimens after PCR tests on bird specimens received from NVSL resulted in high reactivity to West Nile virus. Almost simultaneously to the CDC laboratory's confirmation of WNV in birds, Dr. Lipkin informed the NYS DOH that the identity of the flavivirus could be either WNV or Kunjin virus. Finally, on September 27, 1999, CDC announced that the human outbreaks in New York City were due to West Nile virus.

Although the delay in diagnosis of WNV did not have an effect on the response to the human outbreak, especially since mosquito control activities had been implemented, some experts suggest that the failure of public health and veterinary authorities to recognize the unexpected increase of neurological disorders and deaths in birds as potential index cases of a new outbreak in that animal population lead to the establishment of WNV in the area and ultimately its spread. Moreover, the need for laboratory facilities able to test for animal diseases and the insistence of the Bronx Zoo pathologist in the linkage between the deaths in birds and the SLE outbreak resulted in the convergence of these parallel investigations. The lack of communication linkages between the animal and human health sectors at the time was an additional barrier. After the WNV outbreak, many steps have been taken to close this gap and to integrate animal and human health surveillance in New York City. In December 2001, the New York City Health Code added to the Communicable Disease Control Section new animal disease reporting requirements, which established new procedures for reporting and controling of animal diseases that are transmittable to humans or any animal disease of public health importance. In addition, an invitation to join the Health Alert Network (HAN), an e-mail-based alert system, was extended to veterinarians and other animal or wildlife specialists who wish to receive veterinary alerts from the New York City DOH. Furthermore, the NYS DOH has sponsored several meetings, jointly with the Veterinary Medical Association, on animal disease surveillance as part of the efforts to enhance relations with the animal health community (practicing veterinarians, wildlife specialists, zoo veterinarians, agriculture agencies, etc.).

INFLUENZA A(H1N1) PANDEMIC, 2009

In the United States, seasonal influenza infections result in high morbidity and mortality in humans, resulting in approximately 36,000 excess deaths annually. For the most part, these deaths occur in older and younger people having less developed or compromised immune systems. Pandemic influenza events have occurred every 40–50 years over the past several hundred years. During the 20th century, pandemic influenza has occurred in 1918, 1957, and 1968. The 1918 influenza pandemic caused extremely high rates of morbidity and mortality, especially in healthy adults between 20–40 years of age. Since 2003, continuing human influenza infections from HPAI H5N1 have caused great concern over the potential of this virus to result in the next pandemic. With the high mortality rate observed among infected patients, human health officials have been worried that should this virus become easily transmissible, a pandemic with this virus would also be accompanied by severe illness, morbidity, and mortality.

Detection and Identification of Risk Factors

In March and April 2009, influenza caused by a novel influenza A(H1N1) virus was detected in human populations in Mexico and the United States (Brownstein et al., 2009). Early results from phylogenetic studies aimed at determining the genetic composition of the virus (Smith et al., 2009) found that the virus is derived from combinations of swine viruses that have been circulating over the past 20 years. Because of this, the virus was quickly referred to as the "swine flu" by human health authorities and the media, even though to date, there has been only one known occurrence of an outbreak caused by this virus in pigs. The specifics of when and how this virus emerged, in what populations, how long its circulation has gone undetected, as well as the source of exposure for the outbreak in pigs remain the focus of ongoing investigations. This outbreak highlights a need for more strategic and systematic surveillance of influenza in pigs.

Response

As the new virus was detected and began to spread, health officials in Mexico quickly decided to close schools and take other actions to limit opportunities for large groups of people to come together where person-to-person spread could easily occur. Health officials in Mexico and the United States quickly launched outbreak investigations to learn more about the virus, routes of and risk factors for transmission, and the potential for severe morbidity and mortality. In the United States, local health authorities based decisions for school closures on information that CDC was providing in daily updates. As increased numbers of cases were detected, it was determined that this new virus was spread the same way that seasonal influenza is transmitted, and although morbidity rates among the exposed were high, mortality was relatively low. Unfortunately, the name used to refer to the disease as "swine flu" was not based on actual detection of the virus in pigs or from human cases resulting from contact with sick pigs. And despite a joint press release from OIE/FAO and WHO, many people quickly associated wrongly that they could become infected by eating pork. The negative impact of the inappropriate naming of the virus on the pork industry was significant and is described below. However, once this impact became known to human health authorities, they quickly responding by changing the name of the virus and infection to novel influenza A(H1N1) 2009, which was greatly appreciated by animal health authorities and the swine industry. The media, however, continued to inappropriately refer to the virus as the "swine flu," causing public confusion about the actual risk factors for exposure and therefore leading policymakers to base their responses on factors other than evidence. For example, despite the lack of evidence of any swine infections, Egypt responded to the outbreak by culling more than 250,000 pigs in the country.

Outcome and Impact

Human infections are being monitored globally. On June 11, 2009, WHO declared an influenza pandemic caused by this virus. As of July 6, 2009, pandemic A(H1N1) 2009 virus had been officially reported in 94,512 human cases, caused 429 deaths, and had been found in 99 countries, territories, and areas (WHO, 2009b). Those under the age of 50 years appeared to be at increased risk of infection. Work has begun to develop a vaccine that would be available by the 2009–2010 winter season in North America. By incorrectly naming a virus transmitted between humans as swine flu, this has resulted in trade bans and reductions in pork consumption, ultimately causing losses of approximately $28 million per week to the swine industry (Snelson, 2009; TVMDL, 2009). Given the importance of encouraging disease surveillance, reporting, and response by the livestock industry in an effective emerging zoonotic disease surveillance system, these losses based on misinformation are unfortunate and serve to discourage future cooperation in an integrated surveillance and response effort. The accurate naming of influenza viruses is significant in reporting and response and critical in effectively conveying information to protect public health.

REFERENCES

Breman, J. G. 2000. Monkeypox: An emerging infection for humans. In *Emerging infections 4*, edited by M. Scheld, W. A. Craig, and J. M. Hughes. Washington, DC: ASM Press.

Brownstein, J. S., C. C. Freifeld, and L. C. Madoff. 2009. Influenza A (H1N1) virus, 2009—Online monitoring. *N Engl J Med* 360(21):2156.

CDC (Centers for Disease Control and Prevention). 2003a. Multistate outbreak of monkeypox—Illinois, Indiana, and Wisconsin, 2003. *MMWR* 52(23):537–540.

CDC. 2003b. Update: Multistate outbreak of monkeypox—Illinois, Indiana, Kansas, Missouri, Ohio, and Wisconsin, 2003. *MMWR* 52(27):642–646.

FDA (Food and Drug Administration). 2008. Control of communicable diseases; restrictions on African rodents, prairie dogs, and certain other animals, final rule. *Federal Register* 73(174):51912–51919.

Fine, A., and M. Layton. 2001. Lessons from the West Nile viral encephalitis outbreak in New York City, 1999: Implications for bioterrorism preparedness. *Clin Infect Dis* 32(2):277–282.

GAO (U.S. General Accountability Office). 2000. *West Nile virus outbreak: Lessons for public health preparedness*. Washington, DC: U.S. Government Printing Office.

GAO. 2004. *Emerging infectious diseases: International Asian SARS outbreak challenged national and international responses*. Washington, DC: U.S. Government Printing Office.

Guarner, J., B. J. Johnson, C. D. Paddock, W. J. Shieh, C. S. Goldsmith, M. G. Reynolds, I. K. Damon, R. L. Regnery, and S. R. Zaki. 2004. Monkeypox transmission and pathogenesis in prairie dogs. *Emerg Infect Dis* 10(3):426–431.

Heymann, D. L., M. Szczeniowski, and K. Esteves. 1998. Re-emergence of monkeypox in Africa: A review of the past six years. *Br Med Bull* 54(3):693–702.

Nguyen, D. C., T. M. Uyeki, S. Jadhao, T. Maines, M. Shaw, Y. Matsuoka, C. Smith, T. Rowe, X. Lu, H. Hall, X. Xu, A. Balish, A. Klimov, T. M. Tumpey, D. E. Swayne, L. P. Huynh, H. K. Nghiem, H. H. Nguyen, L. T. Hoang, N. J. Cox, and J. M. Katz. 2005. Isolation and characterization of avian influenza viruses, including highly pathogenic H5N1, from poultry in live bird markets in Hanoi, Vietnam, in 2001. *J Virol* 79(7):4201–4212.

OIE (World Organization for Animal Health). 2009. Avian influenza and vaccination: What is the scientific recommendation?—The OIE repeats critical requisites for best use of vaccination in birds and for a systematic exit strategy, OIE Press Release, March, 4.

Price-Smith, A. T. 2009. *Contagion and chaos: Disease ecology and national security in the era of globalization.* Cambridge, Massachusetts: MIT Press.

Scott, E. 2002. *The West Nile virus outbreak in New York City (a): On the trail of a killer virus.* Document C16-02-1645.0, Kennedy School of Government Case Program. Boston, MA: Harvard University.

Sims, L. D. 2007. Lessons learned from Asian H5N1 outbreak control. *Avian Diseases* 50:174–181.

Smith, G. J., D. Vijaykrishna, J. Bahl, S. J. Lycett, M. Worobey, O. G. Pybus, S. K. Ma, C. L. Cheung, J. Raghwani, S. Bhatt, J. S. Peiris, Y. Guan, and A. Rambaut. 2009. Origins and evolutionary genomics of the 2009 swine-origin H1N1 influenza A epidemic. *Nature* 459(7250):1122–1125.

Snelson, H. 2009. *H1N1 update and economic impact.* http://www.aasv.org/news/story.php?id=3562 (accessed July 13, 2009).

The BSE Inquiry. 2000. *The BSE Inquiry: The inquiry into BSE and variant CJD in the United Kingdom.* London, UK: Stationery Office. http://www.bseinquiry.gov.uk (accessed January 14, 2009).

TVMDL (Texas Veterinary Medical Diagnostics Laboratory). 2009. *H1N1 economic impact May 4, 2009—Price and value impact since April 24.* http://tvmdl.tamu.edu/content/main/articles/h1n1/h1n1_impact.php (accessed July 13, 2009).

Van Nam, H. 2007. *Vaccination for control of H5N1 avian influenza in Viet Nam.* Presentation, International Ministerial Conference on Avian and Pandemic Influenza, New Delhi, December 4–6. http://www.fao.org/docs/eims/upload//237239/ah706e.pdf (accessed August 11, 2009).

Vu, T. 2009. *The political economy of avian influenza response and control in Vietnam.* Working Paper 19. Brighton, UK: STEPS Centre.

Wan, X. F., T. Nguyen, C. T. Davis, C. B. Smith, Z. M. Zhao, M. Carrel, K. Inui, H. T. Do, D. T. Mai, S. Jadhao, A. Balish, B. Shu, F. Luo, M. Emch, Y. Matsuoka, S. E. Lindstrom, N. J. Cox, C. V. Nguyen, A. Klimov, and R. O. Donis. 2008. Evolution of highly pathogenic H5N1 avian influenza viruses in Vietnam between 2001 and 2007. *PLoS ONE* 3(10):e3462.

Webster, R. G. 2004. Wet markets—A continuing source of severe acute respiratory syndrome and influenza? *Lancet* 363(9404):234–236.

WHO (World Health Organization). 2003a. *Update 95—SARS: Chronology of a serial killer.* http://www.who.int/csr/don/2003_07_04/en/ (accessed on October 17, 2008).

WHO. 2003b. *Consensus document on the epidemiology of severe acute respiratory syndrome (SARS).* WHO/CDS/CSR/GAR/2003.11. Geneva, Switzerland: WHO.

WHO. 2009a. *Cumulative number of confirmed human cases of avian influenza A/(H5N1) reported to WHO.* http://www.who.int/csr/disease/avian_influenza/country/cases_table_2009_07_01/en/index.html (accessed August 11, 2009).

WHO. 2009b. *Pandemic (H1N1) 2009—Update 58.* http://www.who.int/csr/don/2009_07_06/en/index.html (accessed July 22, 2009).

Appendix C

Novel Human Pathogen Species

TABLE C-1 List of 87 Novel Human Pathogen Species Discovered Since 1980

Year	Pathogen Species	Year	Pathogen Species
1980	Puumala virus	1987	Suid herpevirus 1
	Human T-lymphotropic virus 1		Sealpox virus
1981	*Microsporidian africanum*		Dhori virus
1982	Seoul virus	1988	Picobirnavirus
	Human T-lymphotropic virus 2		Barmah Forest virus
	Borrelia burgdorferi	1989	Hepatitis C virus
1983	Human immunodeficiency virus 1		European bat lyssavirus 1
	Human adenovirus F		*Corynebacterium amycolatum*
	Hepatitis E virus	1990	*Vittaforma corneae*
	Helicobacter pylori		Trubanaman virus
	Capnocytophaga canimorsus		Semliki Forest virus
	Candiru virus		Reston Ebola virus
1984	*Scedosporium prolificans*		Gan gan virus
	Rotavirus B		Banna virus
	Human torovirus	1991	*Nosema ocularum*
1985	*Pleistophora ronneafiei*		Guanarito virus
	Enterocytozoon bieneusi		*Encephalitozoon hellem*
	Borna disease virus		*Ehrlichia chaffeensis*
1986	Rotavirus C	1992	Dobrava-Belgrade virus
	Kokobera virus		*Bartonella henselae*
	Kasokero virus	1993	Sin Nombre virus
	Human immunodeficiency virus 2		*Gymnophalloides seoi*
	Human herpesvirus 6		*Encephalitozoon intestinalis*
	European bat lyssavirus 2		*Bartonella elizabethae*
	Cyclospora cayetanensis		

continued

TABLE C-1 Continued

Year	Pathogen Species	Year	Pathogen Species
1994	Sabia virus	1998	*Trachipleistophora anthropophthera*
	Human herpesvirus 8		Menangle virus
	Human herpesvirus 7		*Brachiola vesicularum*
	Hendra virus	1999	TT virus
	Anaplasma phagocytophila		Nipah virus
1995	New York virus		*Ehrlichia ewingii*
	Hepatitis G virus		*Brachiola algerae*
	Côte d'Ivoire Ebola virus	2000	Whitewater Arroyo virus
	Black creek canal virus	2001	*Cryptosporidium felis*
	Bayou virus		Human metapneumovirus
1996	Usutu virus		Baboon cytomegalovirus
	Trachipleistophora hominis	2002	*Cryptosporidium hominis*
	Metorchis conjunctus	2003	SARS coronavirus
	Juquitiba virus	2004	Human coronavirus NL63
	Ehrlichia canis	2005	Human T-lymphotropic virus 4
	BSE agent		Human T-lymphotropic virus 3
	Australian bat lyssavirus		Human coronavirus HKU1
	Andes virus		Human bocavirus
1997	Laguna Negra virus	2007	Ki virus
	Bartonella clarridgeiae		Wu virus
			Melaka virus

NOTE: Approximately 80 percent of these are associated with nonhuman reservoirs or origins.
SOURCE: Adapted from Woolhouse, M., and E. Gaunt (2007). Ecological Origins of Novel Human Pathogens. *Crit Rev Microbiol* 33(4):231–242.

Appendix D

Public Committee Meeting Agendas

MEETING ONE

June 25–26, 2008
Washington, DC

June 25, 2008

9:30–10:15 am	**Registration and Check-In** All participants must check in at the security desk

SESSION I: CHARGE AND STATEMENT OF THE PROBLEM

10:15–10:25 am	**Welcome and Opening Remarks** *Marguerite Pappaioanou and Gerald Keusch, committee co-chairs*
10:25–10:45 am	**Charge to the Committee from the Sponsor** *Dennis Carroll and Murray Trostle, USAID*
10:45–11:15 am	**Keynote Presentation:** **Convergence of Forces Behind Emerging and Reemerging Zoonoses, and Future Trends in Zoonoses** *Tracee Treadwell, CDC*

11:15 am–12:00 pm **Panel Discussion:**
The Need for a Global and Sustainable Disease Surveillance System for Zoonoses, and Roles of Various International Organizations
Moderator: Gerald Keusch, committee co-chair
Panelists
Nancy Cox, CDC
Stephane de La Rocque, FAO
Marlo Libel, Pan American Health Organization, on behalf of David Heymann, WHO
Alejandro Thiermann, OIE
Tracee Treadwell, CDC

12:00–1:00 pm **Lunch**

SESSION II: ACTIVE SURVEILLANCE SYSTEMS FOR DETECTING ZOONOSES

Moderator: Mark Woolhouse, committee member

1:00–1:15 pm **Global Early Warning System (GLEWS) and Transboundary Disease Surveillance Program**
Stephane de La Rocque, FAO

1:15–1:30 pm **OIE Standards for Identifying/Diagnosing Diseases, Diagnostic Confirmation, Data Collection and Reporting from Countries, Network of Reference Laboratories, Relationships with Chief Veterinary Officers—Committee Work, Food Safety (*Codex Alimentarius*)**
Alejandro Thiermann, OIE

1:30–2:00 pm **Surveillance and Outbreak Investigation of Wildlife—Terrestrial and Marine Animals, Birds, Wildlife Disease Information Node**
- Wildlife Disease Information Node
  Joshua Dein (on NBII), USGS National Wildlife Health Center
- Outbreak Investigation
  Scott Wright, USGS National Wildlife Health Center

2:00–2:15 pm	**Ebola Surveillance in Nonhuman Primates** *Pierre Rollin, CDC*
2:15–2:30 pm	**Surveillance of Bats** *Peter Daszak, Consortium for Conservation Medicine*
2:30–2:45 pm	**Surveillance of Bushmeat and Exotic Animal Consumption and GAINS** *William Karesh, Wildlife Conservation Society*
2:45–3:00 pm	**Surveillance of Infectious Diseases in Companion Animals** *Larry Glickman, Purdue University*
3:00–3:45 pm	**Panel Discussion:** **Active Surveillance Systems with Presenters from Session II**
3:45–4:00 pm	**Break**

SESSION III: EARLY WARNING SYSTEMS FOR ZOONOTIC DISEASES IN HUMANS

Moderator: Mo Salman, committee member

4:00–5:00 pm **Overview of Early Warning Systems**

- Global Public Health Intelligence Network (GPHIN)
  Marlo Libel, PAHO
- Global Outbreak Awareness and Response Network (GOARN)
  Marlo Libel, PAHO
- ProMED-Mail
  Peter Cowen, North Carolina State University
- U.S. Department of Defense, Global Emerging Infections Surveillance and Response System (DoD-GEIS)
  Tracy DuVernoy, U.S. Department of Defense
- ArboNET
  Marc Fischer, CDC (via teleconference)
- Emerging Infections Network (IDSA)
  Philip Polgreen, University of Iowa

5:00–6:00 pm	**Panel Discussion:** **Early Warning Systems with Presenters from Session III**
6:00 pm	**Adjourn for the Day**
6:30–8:30 pm	**Committee Working Dinner**

June 26, 2008

8:00–8:30 am	**Registration and Check-In**
8:30–8:45 am	**Recap of Day 1 and Overview of Day 2 of the Workshop** *Gerald Keusch and Marguerite Pappaioanou, committee co-chairs*

SESSION IV: LABORATORY AND EPIDEMIOLOGICAL CAPACITY

Moderator: Terry McElwain, committee member

8:45–9:00 am	**Broad View of Veterinary/Agricultural Laboratory Capacity in Resource-Constrained Countries (Clinical and Field Training, BSL-3 Labs, Biosecurity Issues)** *James Pearson, former director of National Veterinary Services Lab (retired)*
9:00–9:15 am	**Reference Lab Perspective—Experience Serving as an OIE Reference Laboratory and Providing Technical Assistance and Training to Countries in Africa on Avian Influenza; International Policies for Sharing Specimens and Resources and Lab Data** *Ilaria Capua, OIE*
9:15–9:30 am	**Training and Deployment of Assays in Other Countries and Standardization of Assays Worldwide** *Barbara Martin, coordinator for the U.S. National Animal Health Laboratory Network*

9:30–9:45 am	**Experience and Challenges in Establishing and Sustaining Operation of Laboratories in Tanzania with High-Quality Assurance** *Mmeta Grasford Yongolo, virology department of the Animal Diseases Research Institute*
9:45–10:00 am	**Integrated Emerging Infectious Disease Surveillance in Nairobi, Kenya** *Robert Breiman, CDC International Emerging Infectious Diseases Program*
10:00–10:15 am	**Clinical Laboratory and Epidemiological Field Training in Southeast Asia** *Jeremy Farrar, Oxford University Clinical Research Unit*
10:15–11:00 am	**Break**
11:00 am–12:00 pm	**Panel Discussion:** **Laboratory and Epidemiological Capacity with Presenters from Session IV**
12:00–1:15 pm	**Lunch**

SESSION V: FACILITATING INFORMATION EXCHANGE, IMPROVING COMMUNICATION, AND IMPROVING POLICIES

1:15–2:30 pm	**Facilitating Information Exchange, Improving Communication, and Improving Policies** **Moderated Panel Discussion (20 minutes):** *Moderator: Gerald Keusch, committee co-chair* Panelists *Ilaria Capua, OIE* *Stephane de La Rocque, FAO* *Marlo Libel, WHO/PAHO* *Alejandro Thiermann, OIE*
2:30–3:00 pm	**Break**

SESSION VI: DEVELOPING A GLOBAL AND SUSTAINABLE SURVEILLANCE SYSTEM

3:00–5:15 pm	**Moderated General Discussion: Developing Global Sustainable Surveillance and Response to Emerging Zoonoses** *Moderators: Gerald Keusch and Marguerite Pappaioanou, committee co-chairs*
5:15–5:30 pm	**Closing Remarks** *Gerald Keusch and Marguerite Pappaioanou, committee co-chairs*
5:30 pm	**Adjourn**

MEETING TWO

September 11, 2008
Washington, DC

9:30–9:45 am	**Welcome and Introductions** *Marguerite Pappaioanou, committee co-chair*
9:45–10:15 am	**Presentation** *Dr. David Nabarro, assistant secretary-general of the United Nations, senior United Nations system coordinator for avian and human influenza*
10:15–11:30 am	**Question-and-Answer Session with Committee Members**
11:30–11:45 am	**Wrap-Up and Adjourn**

MEETING THREE

September 30, 2008
Woods Hole, MA

1:00–2:00 pm	**Economic Impact of Disease and the Case for Surveillance** *Bruce Y. Lee, University of Pittsburgh*

2:00–3:00 pm	**Economic Consequence Modeling on Foreign Animal and Zoonotic Diseases** *Yanhong Jin, Rutgers University* *Bruce McCarl, Texas A&M University (via videoconference)*
3:00–3:15 pm	**Break**
3:15–4:15 pm	**Impact of Zoonotic Disease on Trade and Small-Scale Producers** *Karl Rich, American University in Cairo (via videoconference)*
4:15–5:15 pm	**The Role of Public–Private Partnerships in Strengthening Food Systems Globally** *Gary Ades, Safe Supply of Affordable Food Everywhere, Inc.*
5:15–5:30 pm	**Additional Questions and Answers and Closing Remarks**
5:30 pm	**Adjourn**

MEETING FOUR

November 18, 2008
Washington, DC

1:30–1:45 pm	**Welcome and Introductions** *Gerald Keusch, committee co-chair*
1:45–2:00 pm	**Presentation** *Dr. Nirmal Ganguly, former director general, Indian Council of Medical Research*
2:00–2:15 pm	**Committee Question-and-Answer Session with Dr. Ganguly**
2:15–2:30 pm	**Presentation** *Dr. Mark Smolinski, director, Predict and Prevent Initiative, Google.org*

2:30–2:45 pm	**Committee Question-and-Answer Session with Dr. Smolinski**
2:45–3:00 pm	**Wrap-Up and Adjourn**

MEETING FIVE

December 1, 2008
Washington, DC

10:00–10:15 am	**Welcome and Meeting Objectives** *Gerald Keusch and Marguerite Pappaioanou, committee co-chairs*
10:15–11:30 am	**Implementation of IHRs** *David Heymann, Max Hardiman, Michael Ryan, and Paul Gully, WHO (World Health Organization)*
11:30 am–12:40 pm	**Reaction to IHR implementation** *Alejandro Thiermann (for Bernard Vallat), OIE (World Organization for Animal Health)*
12:40–1:15 pm	**Lunch**
1:15–2:30 pm	**Regulating Companion and Lab Animal Imports** *Nina Marano, CDC*
2:30–3:20 pm	**Monitoring Wildlife Trade** *Sheila Einsweiler, U.S. Fish and Wildlife Service*
3:20 pm	**Adjourn Open Session of Meeting**

Appendix E

Committee Biosketches

Gerald T. Keusch, M.D. (*Co-Chair*), is Associate Director of the National Emerging Infectious Diseases Laboratory at Boston University, and Special Assistant to the University President for Global Health. Prior to joining Boston University, Dr. Keusch served as Director of the Fogarty International Center at the National Institutes of Health (NIH) and Associate Director for International Research in the office of the NIH Director. A graduate of Columbia College and Harvard Medical School, he is board certified in internal medicine and infectious diseases. He has been involved in clinical medicine, teaching, and research for his entire career, most recently as Professor of Medicine at Tufts University School of Medicine and Senior Attending Physician and Chief of the Division of Geographic Medicine and Infectious Diseases at the New England Medical Center in Boston, Massachusetts. His research has ranged from the molecular pathogenesis of tropical infectious diseases to field research in nutrition, immunology, host susceptibility, and the treatment of tropical infectious diseases and HIV/AIDS. He was a Faculty Associate at Harvard Institute for International Development in the Health Office. Dr. Keusch is the author of more than 300 original publications, reviews, and book chapters, and he is the editor of 8 scientific books. He is the recipient of the Squibb, Finland, and Bristol awards for research excellence of the Infectious Diseases Society of America, and has delivered numerous named lectures on topics of science and global health at leading institutions around the world. He is presently involved in international health research and policy with the NIH, the United Nations, and the World Health Organization. He is an elected member of the Institute of Medicine (IOM) and has been a member of several IOM consensus committees. He is currently a member of the IOM Board on Global Health, the IOM Forum on Microbial Threats, and the National Research Council's (NRC's) Roundtable on Science and Technology for Sustainability.

Marguerite Pappaioanou, D.V.M., M.P.V.M., Ph.D. (*Co-Chair*), is the Executive Director of the Association of American Veterinary Medical Colleges (AAVMC) in Washington, DC. Prior to joining the AAVMC, Dr. Pappaioanou held a joint appointment as Professor of Infectious Disease Epidemiology in the School of Public Health and College of Veterinary Medicine at the University of Minnesota. While at the University of Minnesota, she was Principal Investigator at the National Institutes of Health Center of Excellence for Influenza Research and Surveillance, and at the Centers for Disease Control and Prevention (CDC) Avian Influenza Cooperative Research Center. Dr. Pappaioanou also held numerous positions at CDC, most recently Acting Deputy Director in the Office of Global Health in 2004 and Associate Director for Science and Policy from 1999–2004. She co-coordinated CDC's international response to the SARS and avian flu outbreaks in 2003 and served as the point of contact at CDC for Department of Health and Human Services' activities in Afghanistan and Iraq. As Chief of Surveillance and Evaluation—Special Projects, AIDS Program, and as Assistant Chief for Science, she led studies of AIDS and HIV infection and survey design for a national system of HIV surveillance in 39 U.S. cities. She received the Charles C. Shepard Science Award for coauthorship of the scientific paper *Prevalence of HIV Infection in Childbearing Women in the United States*. Dr. Pappaioanou has received numerous awards, including the U.S. Public Health Service Commendation and Outstanding Service Medals; Award of Recognition, Association of Teachers of Public Health and Preventive Medicine; and the Robert Dyar Labrador Memorial Lectureship, University of California (UC) Davis, 2002. She is a Diplomate of the American College of Veterinary Preventive Medicine and an honorary Diplomate of the American Veterinary Epidemiology Society for her contributions to progress in public health. Dr. Pappaioanou received her Ph.D. in comparative pathology and M.P.V.M. from UC, Davis, and her D.V.M. and B.Sc. from Michigan State University. She recently served on the National Research Council's Committee on Methodological Improvements to the Department of Homeland Security's Biological Agent Risk Analysis.

Corrie Brown, D.V.M., Ph.D., is the Josiah Meigs Distinguished Teaching Professor in the College of Veterinary Medicine, University of Georgia. Her research interests include pathogenesis of infectious disease in food-producing animals through the use of immunohistochemistry and in situ hybridization. She is active in the fields of emerging diseases and international veterinary medicine and currently serves as Coordinator of Activities for the College of Veterinary Medicine. Prior to joining University of Georgia in 1996, she worked at the U.S. Department of Agriculture (USDA) Plum Island Foreign Animal Disease Center for 10 years, conducting pathogenesis and control studies on many of the foreign animal diseases. Her bench research interests at University of Georgia have been focused on poultry diseases, and she works closely with the USDA facility in Athens that is dedicated to foreign diseases of poultry. In educational research, she has several grants to help promote awareness of foreign animal diseases and global issues in veterinary curricula and beyond. Dr. Brown is a Diplomate of the American College of Veterinary Pathologists. She has served on several NRC committees, including the Committee on Assessing the Nation's Framework for Addressing Animal Diseases and the Committee on Genomics Databases for Bioterrorism Threat Agents: Striking a Balance for Information Sharing. She also trained veterinarians in Afghanistan to perform animal autopsies to help prevent the spread of bird flu. She has published or presented more than 250 scientific papers and has testified to Congress on issues involving agroterrorism. Dr. Brown has served on many industrial and federal panels and been a technical consultant to numerous foreign governments on issues involving infectious diseases and animal health infrastructure. Dr. Brown received her Ph.D. in veterinary pathology with a specialization in infectious diseases from UC Davis, and her D.V.M. from the University of Guelph.

John S. Brownstein, Ph.D., is Assistant Professor of Pediatrics at Harvard Medical School and Director of the Computational Epidemiology Group at the Children's Hospital Boston Informatics Program of the Harvard-MIT Division of Health Sciences. Dr. Brownstein was trained as an epidemiologist in the Department of Epidemiology and Public Health at Yale University. His research is dedicated to statistical and informatics approaches aimed at improving public health surveillance and practice. This research has focused on a variety of infectious disease systems including malaria, HIV, dengue, West Nile virus, Lyme disease, RSV, salmonella, and influenza. He is also leading the development of several novel disease surveillance systems, including HealthMap.org, an Internet-based global infectious disease intelligence system. The system is currently in use by the CDC, WHO, DHS, DOD, HHS, and EU. Dr. Brownstein has advised the World Health Organization, the Institute of Medicine, the U.S. Departments of Health and Human Services and Homeland Security, and the White House on real-time public health surveillance. He has used this experience in his role as Vice President of the International Society for Disease Surveillance. He has authored more than 40 articles in the area of public health surveillance. This work has been reported on widely including in pieces in the *New*

England Journal of Medicine, *Science*, *Nature*, *The New York Times*, *The Wall Street Journal*, CNN, National Public Radio, and the BBC.

Peter Daszak, Ph.D., is President of Wildlife Trust, an international conservation and health nongovernmental organization (NGO). His research addresses the link between anthropogenic environmental change, wildlife diseases, public health, and conservation. He is especially involved in research on emerging diseases, in trying to understand their ecology and the factors that drive emergence. Dr. Daszak's work includes studying the ecology of West Nile virus, Nipah virus (a disease that emerged from fruit bats to kill more than 100 humans in Malaysia recently), SARS (identifying bats as the reservoir for SARS-like coronaviruses), H5N1 avian influenza, H1N1 influenza and other diseases that cross the wildlife–livestock–human boundary. His group has developed new ways to predict disease emergence and spread and recently produced the first global map of emerging disease "hotspots." Dr. Daszak also works on wildlife emerging diseases that have conservation significance (e.g., amphibian chytridiomycosis, Partula snail microsporidiosis, testing hypothesized examples of extinction by infection). Dr. Daszak has a number of research projects investigating the role of trade in the spread of wildlife and human pathogens and the impact of this on public health and conservation. He recently served on the NRC's Committee on National Needs for Research in Veterinary Science. He is originally from Britain, where he earned a B.Sc. in zoology and a Ph.D. in parasitology.

Cornelis de Haan graduated with a degree in animal science from Wageningen University, the Netherlands, in 1966. From 1966 to 1967, he worked in dairy research and development in Ecuador and in smallholder agriculture in Peru. He then moved to Africa, where until 1983 he occupied the posts of Senior Scientist and later Deputy Director General (research) of the International Livestock Center for Africa in Addis Ababa. He joined the World Bank in Washington, DC, in 1983, initially as Senior Livestock Specialist for West Africa and later for Eastern Europe and the Middle East. From 1992 to 2001 he occupied the post of Senior Advisor for Livestock Development, responsible for the livestock development policies of the World Bank. Mr. de Haan is now retired but still works as a consultant on animal agriculture for the World Bank. His main interests are institutional aspects of livestock development, livestock and the environment, food safety issues, and livestock and poverty reduction. He is currently also part of a World Bank/United Nations System Influenza Coordinator (UNSIC) task force "Beyond HPAI," which will recommend institutional and funding mechanisms for a more permanent control of pandemics and other zoonotic diseases.

Christl A. Donnelly, Sc.D., M.Sc., is Professor of Statistical Epidemiology at Imperial College, London. Prior to her work at the Imperial College she was the Head of the Statistics Unit at the University of Oxford Wellcome Trust Centre for Infectious Disease Epidemiology (1995–2000) and a Lecturer in Statistics at the University of Edinburgh (1992–1995). Her research focuses on the synthesis of methods combining sound statistical principles and insights from biomathematical models of disease transmission. She has considerable experience with severe acute respiratory syndrome (SARS), bovine tuberculosis (TB), and avian influenza. Dr. Donnelly has been a member of the Bill & Melinda Gates Foundation's Schistosomiasis Control Initiative (SCI) Technical Committee since 2002. She was the Deputy Chairman of the Independent Scientific Group on Cattle TB from 1998 to 2007 and contributed to the Office of Science and Innovation's project *Foresight—Infectious Diseases: Preparing for the Future* from 2004 to 2006. She was an advisor to the Spongiform Encephalopathy Advisory Committee (SEAC) from 1996 to 2003, and she was a "BSE and Sheep Subgroup" member from 1998 to 1999. Dr. Donnelly was also a member of the Foot and Mouth Disease Official Science Group and the Joint Royal Society/Academy of Medical Sciences Working Group on the Science of transmissible spongiform encephalopathies (TSEs) in 2001. She was awarded with the Distinguished Alum Award by the Harvard School of Public Health's Department of Biostatistics in 2005 and with the Franco-British prize by the Académie des Sciences in Paris in 2002. Dr. Donnelly received her Sc.D. and M.Sc. in biostatistics from Harvard University and her B.A. in mathematics from Oberlin College.

David P. Fidler, J.D., M.Phil., B.C.L., is James Louis Calamaras Professor of Law at Indiana University School of Law and is Director of the Indiana University Center on American and Global Security. Professor Fidler is one of the world's leading experts on international law and public health, with an emphasis on infectious diseases. His books in this area include *International Law and Infectious Diseases* (Clarendon Press, 1999), *International Law and Public Health* (Transnational Publishers, 2000), *SARS, Governance, and the Globalization of Disease* (Palgrave Macmillan, 2004), and *Biosecurity in the Global Age: Biological Weapons, Public Health, and the Rule of Law* (Stanford University Press, 2008) (with Lawrence O. Gostin). Professor Fidler has acted as an international legal consultant to the World Health Organization, the World Bank, the U.S. Centers for Disease Control and Prevention, the U.S. Department of Defense, and various nongovernmental organizations involved with global health or arms control issues.

Kenneth H. Hill, Ph.D., is Visiting Professor of Population Practice at the Harvard School of Public Health. His research interests have been in the development of demographic measurement methods (particularly for demographic outcomes that are hard to measure, such as child and adult mortality, unmet need for family planning, undocumented migration); the measurement of child mortality (with particular emphasis on tracking national trends and linking them to other changes); the exploration of links between demographic parameters and economic crisis; the impact of policy and programs on demographic change; the role of gender preferences on child health behaviors and fertility; the demography of sub-Saharan Africa; the role of development, particularly child mortality change, on fertility decline; the measurement of demographic parameters for populations undergoing complex emergencies; and measurement of adult mortality in the developing world: Africa, Asia, Middle East, Latin America. Dr. Hill has also served on several National Research Council committees or panels and has chaired both the Panel on the Population Dynamics of Sub-Saharan Africa and the Working Group on Demographic Effects of Economic and Social Reversals.

Ann Marie Kimball, M.D., M.P.H.,[1] is Professor of Epidemiology and Health Services and an Adjunct in Medicine and Biomedical and Health Informatics at the University of Washington. She also serves as the Director of the Asia Pacific Economic Cooperation Emerging Infections Networks and the Director of the Amauata Global Informatics research and training program. Dr. Kimball has devoted her career to studying health issues and has worked in numerous positions in the United States and abroad. Her research interests are primarily in international health, trade, HIV/AIDS, emerging infections, maternal and child health, and health informatics. In 2006, she published *Risky Trade: Infectious Disease in an Era of Global Trade* (Ashgate). She has previously served as a member of the Institute of Medicine's Forum on Emerging Infections, as a member of the Department of Health Emerging and Reemerging Diseases Strategic Planning Task Force, as regional adviser for the Pan American Health Organization in HIV/AIDS, and as Chair of the National Alliance of State and Territorial AIDS Directors in the United States. She has served as a U.S. delegate to the American Pacific Economic Council Health Working Group. Dr. Kimball received her M.D. and M.P.H. from the University of Washington and her B.S. in biology and humanities from Stanford University. She is a Fellow of the American College of Physicians.

Ramanan Laxminarayan, Ph.D., M.P.H.,[1] is a Senior Fellow at Resources for the Future, where he directs the Center for Disease Dynamics, Economics, and Policy, and a visiting scholar and lecturer at Princeton University.

[1]Appointed in September 2008.

His research deals with the integration of epidemiological models of infectious diseases and drug resistance into the economic analysis of public health problems. He has worked to improve the understanding of drug resistance as a problem of managing a shared global resource. Dr. Laxminarayan has worked with WHO and the World Bank on evaluating malaria treatment policy, vaccination strategies, the economic burden of tuberculosis, and control of noncommunicable diseases. He has served on a number of advisory committees at WHO, CDC, and IOM. In 2003–2004, he served on the NRC/IOM Committee on the Economics of Antimalarial Drugs and subsequently helped create the Affordable Medicines Facility for malaria, a novel financing mechanism for antimalarials. His work has been covered in major media outlets including Associated Press, BBC, CNN, *Economist*, *Los Angeles Times*, NBC, NPR, Reuters, *Science*, *The Wall Street Journal*, and *National Journal*. Dr. Laxminarayan received his undergraduate degree in engineering from the Birla Institute of Technology and Science in Pilani, India, and his M.P.H. and Ph.D. in economics from the University of Washington, Seattle.

Terry F. McElwain, D.V.M., Ph.D., is a professor in the School for Global Animal Health and holds administrative appointments as Executive Director of the Washington Animal Disease Diagnostic Laboratory and Director of the Animal Health Research Center in the College of Veterinary Medicine at Washington State University. He is Past President of the American Association of Veterinary Laboratory Diagnosticians and serves on the Board of Directors of the World Association of Veterinary Laboratory Diagnosticians. Dr. McElwain has been a key architect in the creation and development of the National Animal Health Laboratory Network and has been closely involved in the development of the new School for Global Animal Health at Washington State University. He interacts with CDC and is also a member of the governor's emergency preparedness task force in the state of Washington. Dr. McElwain is an elected member of the IOM. He recently served on the NRC's Committee on Assessing the Nation's Framework for Addressing Animal Diseases. Dr. McElwain has a long and established research record in the field of veterinary infectious diseases, especially those of agricultural animals. He received his D.V.M. from the College of Veterinary Medicine, Kansas State University, in 1980, and his Ph.D. from Washington State University in 1986.

Mark Nichter, M.P.H., Ph.D., is Regents Professor of Anthropology at the University of Arizona, holding joint appointments in the Departments of Family Medicine and Public Health. He has pioneered the use of ethnographic methods in the fields of medicine, ethnomedicine, and public health. Professor Nichter has conducted extensive research in developing countries as well as in the United States, and his research and writing has shaped the field of medical anthropology and addressed such issues as child survival, infectious and vector-borne disease, women's health, pharmaceutical use and drug resistance, tobacco use and nicotine dependency, and emerging diseases. At the University of Arizona, he has built a doctoral program in medical anthropology and has helped train health social scientists and medical and public health researchers in India, Sri Lanka, Thailand, and the Philippines. He also played a pivotal role in developing an international clinical epidemiology network that operates in over 41 different countries. Professor Nichter has received some of the most prestigious awards in his discipline, including the Radcliffe-Brown Award from the Royal Anthropological Society and the Margaret Mead Award from the American Anthropological Association. The Society for Medical Anthropology awarded him the Virchow Award and most recently its Career Achievement Award. Professor Nichter served as the President for the Society of Medical Anthropology and has served as a member of two IOM committees: one focusing on tobacco use among youth in the United States, and the other on Americans' use of complementary and alternative medicine.

Mo Salman, B.V.M.S., M.P.V.M., Ph.D., is Professor of Veterinary Epidemiology in the Animal Population Health Institute of College of Veterinary Medicine and Biomedical Sciences at Colorado State University. He holds appointments in the Department of Clinical Science and Department of Environmental Health and Radiological Sciences. He established the Animal Population Health Institute at Colorado State University in 2002 and served as its Director until 2006. Prior to the establishment of the Animal Population Health Institute, he served as Director of the Center of Veterinary Epidemiology and Animal Disease Surveillance Systems and Director of the Center of Economically Important Infectious Animal Diseases. His educational background is in veterinary medicine, preventive veterinary medicine, and comparative pathology. He received his veterinary medical degree from the University of Baghdad, Iraq, and a master's degree in preventive veterinary medicine and a Ph.D. from UC Davis. He is a Diplomate on the American College of Veterinary Preventive Medicine and Fellow of the American College of Epidemiology. Dr. Salman is the author of more than 230 peer-reviewed papers in scientific journals and has participated in numerous conferences and national and international meetings during his more than 25 years as a faculty member. He is Editor-in-Chief of *Preventive Veterinary Medicine* and has served on the boards of scientific journals such as the *American Journal of Veterinary Research*. Dr. Salman is engaged in research and outreach projects in more than 15 countries around the world. Dr. Salman's research interests are on the methodology of surveillance and survey for animal diseases with emphasis on infectious diseases. He is the recipient of the 2007 AVMA XII International Veterinary Congress Prize for his contributions to international understanding of veterinary medicine.

Oyewale Tomori, D.V.M., Ph.D., is Vice-Chancellor of Redeemer's University in Nigeria. Professor Tomori received his D.V.M. from the Ahmadu Bello University, Zaria, and his Ph.D. in virology from the University of Ibadan. He is a Fellow of the Royal College of Pathologists of the United Kingdom and a Fellow of the Academy of Science in Nigeria. Dr. Tomori is also a Fellow of the College of Veterinary Surgeons of Nigeria. Dr. Tomori worked as a Regional Virologist in Africa with WHO for many years and is a virologist of international repute in the Africa Region. Within the past 30 years, he has carried out meaningful research studies on a wide range of human viruses and zoonotic and veterinary viruses, which are of immense public health importance in Nigeria and Africa as a whole. The studies involve epidemiological and serological surveys for viral infections, the control of viral epidemics, the development of diagnostic tests for viral infections, the immunology of viruses, the pathology and pathogenesis of viruses, the development of viral vaccines, and the characterization and ecology of viruses. Prominent among the viruses he has studied are the Yellow fever virus, the Lassa fever virus, the poliomyelitis virus, the measles virus, the Ebola virus, and a hitherto unknown virus, the Orungo virus, of which he elucidated the properties and registered with the International Committee of Virus Taxonomy. This discovery is considered an outstanding contribution to the discipline of virology. Professor Tomori is recognized as one of Africa's frontline Lassa fever researchers. He has developed a unique diagnostic virus neutralization test for the Lassa fever. His major contribution on Yellow fever was the development of a technique for forecasting impending outbreaks of the disease, which has helped to put Nigeria in a state of preparedness to combat the epidemic.

Kevin D. Walker, M.S., Ph.D., is a professor with the National Food Safety and Toxicology Center at Michigan State University, College of Veterinary Medicine at Michigan State University. Current initiatives include food and its sustainability in a highly connected, complex world, and design and implementation of strategic initiatives where animal health, public health and the environment intersect at the national and global level. Dr. Walker's areas of expertise include animal diseases, economics, food safety, international trade standards and agreements, leadership, and policy. He previously spent 8 years as Director of Agricultural Health and Food Safety within the Inter-American Institute for Cooperation in Agriculture, based in Costa Rica, where he worked with national governments in the 34 countries in the Americas to enhance public infrastructure, leadership development, emerging issues assessments, and implementation of international trade standards and agreements. Prior to working overseas, he was the Director of the Centers for Emerging Issues within the USDA Animal and Plant Health Inspection Service (APHIS) Veterinary Services. During this time the Center carried out a variety of national risk analyses for emerging issues including BSE, *E. coli* O157:H7, avian influenza, and tuberculosis. Dr. Walker has collaborated and worked with a large number of organizations including the World Trade Organization, the World Organization for Animal Health, the United Nations Food and Agricultural Organization, the International Plant Protection Convention, and the Codex Alimentarius, and he recently served on the NRC's Committee on Assessing the Nation's Framework for Addressing Animal Diseases. He is also a Fellow with the Kellogg Foundation.

Mark Woolhouse, Ph.D., is Professor of Infectious Disease Epidemiology at the University of Edinburgh in Scotland. He trained as a population biologist with a B.A. from Oxford University, an M.Sc. from the University of York, and a Ph.D. from Queen's University before turning to epidemiology. He held research posts at the University of Zimbabwe, Imperial College London (MRC Training Fellowship), the University of Oxford (Beit Memorial Fellowship and Royal Society University Research Fellowship), and now Edinburgh (initially in the School of Veterinary Studies). He has worked on a variety of infectious disease systems: human schistosomiasis, involving extensive field work in rural Zimbabwe; verocytotoxigenic *E. coli* in cattle in rural Scotland; the epidemiology and transmission biology of foot-and-mouth disease in livestock; trypanosomiasis in humans, cattle, and tsetse in east and southern Africa; and transmissible spongiform encephalopathies in cattle (BSE) and in sheep (scrapie). He has published more than 150 scientific papers on these and other topics. He advises the UK government on both animal and human health, and his work during the UK 2001 foot-and-mouth disease epidemic led to an Officer of the British Empire (OBE) award in 2002. Dr. Woolhouse is a Fellow of the Royal Society of Edinburgh.